Foundations of Engineering Mechanics

A.G. Gorshkov, D.V. Tarlakovsky
Transient Aerohydroelasticity of Spherical Bodies

Springer-Verlag Berlin Heidelberg GmbH

http://www.springer.de/engine/

A.G. Gorshkov, D.V. Tarlakovsky

Transient Aerohydroelasticity of Spherical Bodies

Translated by E. Evseev and V. Balmont

With 65 Figures

 Springer

Series Editors:
V. I. Babitsky
Department of Mechnical Engineering
Loughborough University
LE11 3TU Loughborough, Leicestershire
United Kingdom

J. Wittenburg
Institut für Technische Mechanik
Universität Karlsruhe (TH)
Kaiserstraße 12
76128 Karlsruhe / Germany

Authors:
Anatoly G. Gorshkov
Moscow Aviation Institute
Dept. of Applied Mechanics
Volokolamskoe Shosse 4
125871 Moscow, Russia
e-mail: bugaev@mai.ru

Dmitry V. Tarlakovsky
Moscow Aviation Insitute
Dept. of Applied Mechanics
Volokolamskoe Shosse 4
125871 Moscow, Russia
e-mail: bugaev@mai.ru

Translators:
Evgeny G. Evseev
Shodnenskaya Street, 50-34
123363 Moscow, Russia

Vladimir V. Balmont
Kosmodemiyanskikh Street, 6-125
125171 Moscow, Russia

Cataloging-in-Publication data applied for
Die Deutsche Bibliothek - CIP-Einheitsaufnahme
Gorškov, Anatolij G.:
Transient aerohydroelasticity of spherical bodies / A. G. Gorshkov; D.V. Tarlakovsky. Transl. by E. Evseev and V. Balmont.
Berlin; Heidelberg; New York; Barcelona; Hong Kong; London, Milan; Paris; Tokyo: Springer, 2001
  (Foundations of engineering mechanics)
  (Engineering online library)

ISBN 978-3-642-53626-7   ISBN 978-3-540-45159-4 (eBook)
DOI 10.1007/978-3-540-45159-4

Springer-Verlag Berlin Heidelberg New York
a member of BertelsmannSpringer Science + Business Media GmbH

http://www.springer.de

© Springer-Verlag Berlin Heidelberg 2001
Ursprünglich erschienen bei Springer-Verlag Berlin Heidelberg 2001
Softcover reprint of the hardcover 1st edition 2001

Typesetting: Camera ready copies from translators
Cover-Design: de'blik, Berlin
Printed on acid-free paper   SPIN 10796996   62/3020 Rw 5 4 3 2 1 0

Preface

The problems of transient interaction of deformable bodies with surrounding media are of great practical and theoretical importance. When solving the problems of this kind, the main difficulty is in the necessity to integrate jointly the system of equations which describe motion of the body and the system of equations which describe motion of the medium under the boundary conditions predetermined at the unknown (movable) curvilinear interfaces. At that, the position of these interfaces should be determined as part of the solution process. That is why, the known exact solutions in this area of mechanics of continuum have been derived mainly for the cases of idealized rigid bodies.

Different aspects of the problems of transient interaction of bodies and structures with continuum (derivation of the efficient mathematical models for the phenomenon, development of the theoretical and experimental methods to be used for study of the transient problems of mechanics, etc.) were considered in the books by S.U. Galiev, A.N. Guz, V.D. Kubenko, V.B. Poruchikov, L.I. Slepyan, A.S. Volmir, and Yu.S. Yakovlev. The results presented by these authors make interest when solving a great variety of problems and show a necessity of joint usage of the results obtained in different areas: aerohydrodynamics, theory of elasticity and plasticity, mechanics of soils, theory of shells and plates, applied and computational mathematics, etc.

We start this book with a presentation of the main dependences and relationships of the theory of transient interaction of elastic bodies with continuum (Chap. 1).

Throughout, we consider only the bodies and compound structures having spherical surfaces. For many cases of their interaction with elastic and acoustic media, we derive exact analytical solutions (Chapters 2–4 and 6). These solutions, both in the cases of external and internal problems, are derived in the form of series expansion in terms of the Legendre polynomials. We propose an algorithm for determination of the coefficients of series expansion of the displacements and stresses into Legendre polynomials in a form of superposition of *generalized spherical waves*, and this is a generalization of the known method of summation of elementary spherical and plane waves used for solving one-dimensional problems (Chapters 3 and 5). At that, the governing equations for the problems under consideration are the recursive relationships in the space of the Laplace transform with respect to time for arbitrary functions corresponding to converging and diverging waves. Contrary to the traditional approximate computation of the integrals of convolution type, we present here an exact algorithm of inversion of the Laplace transform for the class of the problems under study (Sects. 2.2 and 3.5).

We derive the new representations for the transition functions of diffraction of acoustic waves by spherical obstacles in a form of the solution of an ordinary differential equation derived using the integral for generalized spherical waves, and in a form of the inproper integral derived using the spectrum theory of self-ajoint differential operators (Sect. 4.1).

Based on the solution of the general problem, we obtain the solutions for various limiting cases of great practical importance (Chapters 2, 5, and 6).

Because of the complexity of derivation of analytical solutions of the problems on penetration of deformable structures into fluid (the complexity is mainly caused by significant change in shape of contacting and free surfaces, appearance and propagation of the zones of cavitation in the fluid, and elastic–plastic deformations in the structure's material), we regard the numerical methods, particularly, the methods of finite differences and finite elements, to be the most promising.

In Chap. 7, we consider one of the possible variants of numerical solution of the problem on penetration of elastic spherical shell into a half-space occupied by an ideal compressible fluid. At that, we use the finite difference scheme of the second order of approximation with respect to time and the spatial coordinates.

The authors are very thankful to N.I. Drobyshevsky, who performed numerical calculations, the results of which are presented in Chap. 7.

The monograph is addressed to scientists, institutional and industrial researchers, lecturers, and graduate students.

Moscow, October 2000

Anatoly Gorshkov
Dmitry Tarlakovsky

Table of Contents

1. Basic Theory of Transient Aerohydroelasticity of Spherical Bodies

This chapter should be regarded as an auxiliary one. We present the equations of motion of some models of continuum, which are then used for analysis of corresponding dynamic problems. We also present the statements of problems of interaction of deformable media, as well as the body of mathematics applied when investigating the propagation of waves.

At that, in accounting for the class of the models to be studied, we analyze only the models of the linear theory of elasticity of homogeneous isotropic bodies, acoustic media, and thin-walled elastic homogeneous shells, selecting them out of the multiple models of continuum currently known. Taking into consideration a vast number of publications devoted to these problems, we present practically all the relationships without derivations making references to corresponding sources.

Wherever it is possible, we present the relationships in vector form. Since we study spherical bodies mainly, a transition to the coordinate form of representation, where it is acceptable, is provided only for the rectangular Cartesian coordinate system (x_1, x_2, x_3) and the spherical coordinate system (R, θ, ϑ) with the center at the point O. The meaning of the designations used is illustrated in Fig. 1.1.

1.1 Equations of Motion of Elastic Media

An elastic medium is a continuous deformable medium for which the process of deformation is reversible. In addition, there exists a one-to-one correspondence between the stress state of the medium and its strain state, between the stress tensor and the strain tensor. We shall consider the process of deformation to be either isothermal or adiabatic (Ilyushin (1978), Sedov (1984)). Moreover, we shall confine ourselves to the assumptions that the elastic medium is homogeneous and isotropic and the deformations of this medium are small.

Then, the equations of motion of elastic medium in displacements (the Lamé equations; these are also known as the Navier equations) in vector form are as follows (Love (1959), Slepyan (1972), Amenadze (1976), Ilyushin (1978), Sedov (1984)):

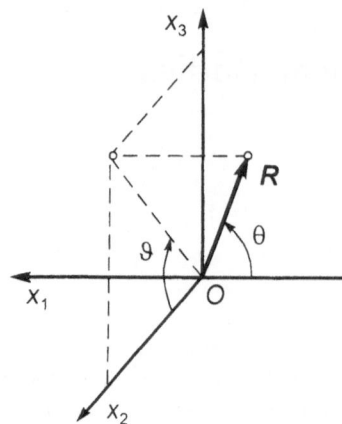

Fig. 1.1. The spherical coordinate system

$$(\lambda + \mu)\,\mathrm{grad\,div}\,\mathbf{u} + \mu\Delta\mathbf{u} + \rho\mathbf{F} = \rho\,\frac{\partial^2\mathbf{u}}{\partial t^2}\,; \qquad (1.1)$$

here, λ and μ are the Lamé elastic constants, $\mathbf{u}$ is the displacement vector of the elastic medium, $\rho\mathbf{F}$ is the vector of volume force (from here on, we shall assume $\mathbf{F} \equiv 0$), ρ is the medium density, and t is time. We will employ the following conventional designation for the gradient, the divergence, and the Laplacian, respectively:

$$\mathrm{grad}\,f = \nabla f\,, \qquad \mathrm{div}\,\mathbf{a} = (\nabla,\,\mathbf{a})\,, \qquad \Delta f = (\nabla,\,\nabla)\,f = \nabla^2 f\,;$$

the operator $\nabla = \nabla_i \mathbf{e}^i$ involves covariant derivatives ∇_i and contravariant components of the basis $\mathbf{e}^i$ (from here on, summation from 1 to 3 will be made over repeated index).

The components of the small-strain tensor ϵ_{ij} represented in an arbitrary curvilinear coordinate system $Ox^1x^2x^3$ are related to the covariant components of the displacement vector u_1, u_2, u_3 as follows:

$$\epsilon_{ij} = \frac{1}{2}\left(\nabla_i u_j + \nabla_j u_i\right), \qquad i,\,j = 1,\,2,\,3\,. \qquad (1.2)$$

For an isotropic medium, a relationship between the stress tensor and the strain tensor obeys Hooke's law

$$\sigma_{ij} = \lambda I_1 g_{ij} + 2\mu\epsilon_{ij}\,, \qquad i,\,j = 1,\,2,\,3\,, \qquad (1.3)$$

where $I_1 = \epsilon_{ij}g_{ij}$; here, g_{ij} and g^{ij} are the covariant and contravariant components of the metric tensor, respectively.

It is known that elastic waves described by the Lamé equation (1.1) can be represented in the form of a superposition of waves of two types (Slepyan (1972), Sedov (1984)): the *longitudinal waves* (known also as the *compression, dilatational,* or *irrotational* waves) and the *transverse waves* (known also as the *shear* or *rotational* waves). Let us notice that the term *waves of shear* is relatively conventional since shear strains cannot propagate when no compression strains take place (Slepyan (1972)). We can make such a separation of elastic waves owing to the following representation of the displacement vector:

$$\mathbf{u} = \operatorname{grad}\varphi + \operatorname{rot}\boldsymbol{\psi};\qquad(1.4)$$

here, $\operatorname{rot}\mathbf{a} = \operatorname{curl}\mathbf{a} = [\nabla, \mathbf{a}]$; the function φ is termed the *scalar elastic potential* of displacements and the function ψ is termed the *vector elastic potential* of displacements. At that, in order to determine the potential ψ uniquely, it is possible, without the loss of generality, to set $\operatorname{div}\boldsymbol{\psi} = 0$.

It can be shown that in the absence of mass forces, the following system of homogeneous wave equations with respect to the potentials φ and ψ is the analogy to the Lamé equation:

$$c_1^2\,\Delta\varphi = \frac{\partial^2\varphi}{\partial t^2}, \qquad c_1^2 = \frac{\lambda + 2\mu}{\rho},$$
$$c_2^2\,\Delta\psi = \frac{\partial^2\psi}{\partial t^2}, \qquad c_2^2 = \frac{\mu}{\rho}.\qquad(1.5)$$

Hence, the compression waves propagate in the medium at the speed c_1 and the shear waves propagate at the speed c_2.

From (1.5) it follows that the waves of different types propagate in a boundless space independently of each other. However, when the bounds are present, each type of wave generates, as a rule, the reflected waves of both types. From a mathematical point of view, when solving a particular problem, it is necessary to solve the two equations (1.5) simultaneously since the functions φ and ψ appear to be interrelated via the boundary conditions.

Let us now derive the expressions for the operators grad, div, rot, and Δ and the formulas which describe interrelation of the physical components of the displacement tensor and the stress tensor with the physical components of the displacement vector $\mathbf{u}$ and the potentials φ and ψ.

Let the vectors $\mathbf{e}_1$, $\mathbf{e}_2$, and $\mathbf{e}_3$ be the unit vectors of the rectangular Cartesian coordinate system $Ox_1x_2x_3$. Then, we have

$$\mathbf{u} = u_1\mathbf{e}_1 + u_2\mathbf{e}_2 + u_3\mathbf{e}_3 = (u_1, u_2, u_3),$$

$$\boldsymbol{\psi} = \psi_1\mathbf{e}_1 + \psi_2\mathbf{e}_2 + \psi_3\mathbf{e}_3 = (\psi_1, \psi_2, \psi_3),$$

$$\epsilon_{ij} = \frac{1}{2}\left(\frac{\partial u_i}{\partial x_j} + \frac{\partial u_j}{\partial x_i}\right),$$

$$\sigma_{ij} = \lambda I_1\delta_{ij} + 2\mu\epsilon_{ij}, \qquad i, j = 1, 2, 3,$$

$$\operatorname{grad}\varphi = \frac{\partial\varphi}{\partial x_1}\mathbf{e}_1 + \frac{\partial\varphi}{\partial x_2}\mathbf{e}_2 + \frac{\partial\varphi}{\partial x_3}\mathbf{e}_3,$$

$$\operatorname{div} \boldsymbol{\psi} = \frac{\partial \psi_1}{\partial x_1} + \frac{\partial \psi_2}{\partial x_2} + \frac{\partial \psi_3}{\partial x_3},$$

$$\operatorname{rot} \boldsymbol{\psi} = \begin{vmatrix} \mathbf{e}_1 & \mathbf{e}_2 & \mathbf{e}_3 \\ \dfrac{\partial}{\partial x_1} & \dfrac{\partial}{\partial x_2} & \dfrac{\partial}{\partial x_3} \\ \psi_1 & \psi_2 & \psi_3 \end{vmatrix},$$

$$\Delta \varphi = \frac{\partial^2 \varphi}{\partial x_1^2} + \frac{\partial^2 \varphi}{\partial x_2^2} + \frac{\partial^2 \varphi}{\partial x_3^2}. \tag{1.6}$$

The Kronecker delta $\delta_{ij} \equiv \delta_i^j \equiv \delta^{ij}$ is employed here, which denotes the value 1 when i is equal to j and zero when i is unequal to j.

Let the vectors $\mathbf{e}_r$, $\mathbf{e}_\theta$, and $\mathbf{e}_\vartheta$ be the unit vectors of the spherical coordinate system (Fig.1.1). Then, we have

$$\mathbf{u} = u\mathbf{e}_r + v\mathbf{e}_\theta + w\mathbf{e}_\vartheta = (u, v, w),$$

$$\boldsymbol{\psi} = \psi_r \mathbf{e}_r + \psi_\theta \mathbf{e}_\theta + \psi_\vartheta \mathbf{e}_\vartheta = (\psi_r, \psi_\theta, \psi_\vartheta),$$

$$\epsilon_{rr} = \frac{\partial u}{\partial R},$$

$$\epsilon_{\vartheta\vartheta} = \frac{1}{R\sin\theta} \frac{\partial w}{\partial \vartheta} + \frac{u}{R} + \frac{\cot\theta}{R} v,$$

$$\epsilon_{\theta\theta} = \frac{1}{R} \frac{\partial v}{\partial \theta} + \frac{u}{R},$$

$$\epsilon_{\vartheta r} = \epsilon_{r\vartheta} = \frac{1}{2}\left(\frac{1}{R\sin\theta} \frac{\partial u}{\partial \vartheta} + \frac{\partial w}{\partial R} - \frac{w}{R}\right),$$

$$\epsilon_{\vartheta\theta} = \epsilon_{\theta\vartheta} = \frac{1}{2}\left(\frac{1}{R} \frac{\partial w}{\partial \vartheta} - \frac{\cot\theta}{R} w + \frac{1}{R\sin\theta} \frac{\partial v}{\partial \vartheta}\right),$$

$$\epsilon_{\theta r} = \epsilon_{r\theta} = \frac{1}{2}\left(\frac{\partial v}{\partial R} - \frac{v}{R} + \frac{1}{R} \frac{\partial u}{\partial \theta}\right), \tag{1.7}$$

$$\sigma_{rr} = (\lambda + 2\mu)\frac{\partial u}{\partial R} + \lambda\left(\frac{2u}{R} + \frac{\cot\theta}{R} v + \frac{1}{R} \frac{\partial w}{\partial \theta} + \frac{1}{R\sin\theta} \frac{\partial w}{\partial \vartheta}\right),$$

$$\sigma_{\vartheta\vartheta} = (\lambda + 2\mu)\left(\frac{1}{R\sin\theta} \frac{\partial w}{\partial \vartheta} + \frac{\cot\theta}{R} v\right) + 2(\lambda + \mu)\frac{u}{R} + \lambda\left(\frac{\partial u}{\partial R} + \frac{1}{R} \frac{\partial v}{\partial \theta}\right),$$

$$\sigma_{\theta\theta} = (\lambda + 2\mu)\frac{1}{R} \frac{\partial v}{\partial \theta} + 2(\lambda + \mu)\frac{u}{R} + \lambda\left(\frac{1}{R\sin\theta} \frac{\partial w}{\partial \vartheta} + \frac{\cot\theta}{R} v\right),$$

$$\sigma_{\vartheta r} = \sigma_{r\vartheta} = \mu\left(\frac{1}{R\sin\theta} \frac{\partial u}{\partial \vartheta} + \frac{\partial w}{\partial R} - \frac{w}{R}\right),$$

$$\sigma_{\theta r} = \sigma_{r\theta} = \mu\left(\frac{\partial v}{\partial R} - \frac{v}{R} + \frac{1}{R} \frac{\partial u}{\partial \theta}\right),$$

$$\sigma_{\vartheta\theta} = \sigma_{\theta\vartheta} = \mu\left(\frac{1}{R}\frac{\partial w}{\partial\vartheta} - \frac{\cot\theta}{R}w + \frac{1}{R\sin\theta}\frac{\partial v}{\partial\vartheta}\right), \tag{1.8}$$

$$\operatorname{grad}\varphi = \frac{\partial\varphi}{\partial R}\mathbf{e}_r + \frac{1}{R}\frac{\partial\varphi}{\partial\theta}\mathbf{e}_\theta + \frac{1}{R\sin\theta}\frac{\partial\varphi}{\partial\vartheta}\mathbf{e}_\vartheta,$$

$$\operatorname{div}\boldsymbol{\psi} = \frac{1}{R^2}\frac{\partial}{\partial R}\left(R^2\psi_r\right) + \frac{1}{R\sin\theta}\frac{\partial}{\partial\theta}\left(\psi_\theta\sin\theta\right) + \frac{1}{R\sin\theta}\frac{\partial\psi_\vartheta}{\partial\vartheta},$$

$$\operatorname{rot}\boldsymbol{\psi} = \frac{1}{R\sin\theta}\left[\frac{\partial}{\partial\theta}\left(\psi_\vartheta\sin\theta\right) - \frac{\partial\psi_\theta}{\partial\vartheta}\right]\mathbf{e}_r$$

$$+ \frac{1}{R}\left[\frac{1}{\sin\theta}\frac{\partial\psi_r}{\partial\vartheta} - \frac{\partial(R\psi_\vartheta)}{\partial R}\right]\mathbf{e}_\theta + \frac{1}{R}\left[\frac{\partial(R\psi_\theta)}{\partial R} - \frac{\partial\psi_r}{\partial\theta}\right]\mathbf{e}_\vartheta,$$

$$\Delta\varphi = \frac{1}{R^2}\frac{\partial}{\partial R}\left(R^2\frac{\partial\varphi}{\partial R}\right) + \frac{1}{R^2\sin\theta}\frac{\partial}{\partial\theta}\left(\sin\theta\frac{\partial\varphi}{\partial\theta}\right) + \frac{1}{R^2\sin^2\theta}\frac{\partial^2\varphi}{\partial\vartheta^2},$$

$$\Delta\boldsymbol{\psi} = \left\{\Delta\psi_r - \frac{2\psi_r}{R^2} - \frac{2}{R^2\sin\theta}\left[\frac{\partial}{\partial\theta}\left(\psi_\theta\sin\theta\right) + \frac{\partial\psi_\vartheta}{\partial\vartheta}\right]\right\}\mathbf{e}_r$$

$$+ \left[\Delta\psi_\theta + \frac{2}{R^2}\frac{\partial\psi_r}{\partial\theta} - \frac{1}{R^2\sin^2\theta}\left(\psi_\theta + 2\cos\theta\frac{\partial\psi_\vartheta}{\partial\vartheta}\right)\right]\mathbf{e}_\theta$$

$$+ \left(\Delta\psi_\vartheta + \frac{2}{R^2\sin\theta}\frac{\partial\psi_r}{\partial\vartheta} + 2\frac{\cot\theta}{R^2\sin\theta}\frac{\partial\psi_\theta}{\partial\vartheta} - \frac{\psi_\vartheta}{R^2\sin^2\theta}\right)\mathbf{e}_\vartheta, \tag{1.9}$$

$$u = \frac{\partial\varphi}{\partial R} + \frac{1}{R\sin\theta}\left[\frac{\partial}{\partial\theta}\left(\psi_\vartheta\sin\theta\right) - \frac{\partial\psi_\theta}{\partial\vartheta}\right],$$

$$v = \frac{1}{R}\frac{\partial\varphi}{\partial\theta} + \frac{1}{R}\left[\frac{1}{\sin\theta}\frac{\partial\psi_r}{\partial\vartheta} - \frac{\partial(R\psi_\vartheta)}{\partial R}\right],$$

$$w = \frac{1}{R\sin\theta}\frac{\partial\varphi}{\partial\vartheta} + \frac{1}{R}\left[\frac{\partial(R\psi_\theta)}{\partial R} - \frac{\partial\psi_r}{\partial\theta}\right]. \tag{1.10}$$

Let us introduce the following dimensionless values:

$$r = \frac{R}{b}, \qquad \tau = \frac{c_* t}{b}, \qquad \gamma = \frac{c_*}{c_1}, \qquad \eta = \frac{c_*}{c_2},$$

$$\kappa = \frac{\lambda}{\lambda + 2\mu}, \qquad \beta = \frac{\lambda + 2\mu}{\lambda_* + 2\mu_*} = \frac{\rho c_1^2}{\rho_* c_*^2}, \qquad \tilde{\sigma}_{\alpha\beta} = \frac{\sigma_{\alpha\beta}}{\lambda + 2\mu},$$

$$\tilde{u} = \frac{u}{b}, \qquad \tilde{v} = \frac{v}{b}, \qquad \tilde{w} = \frac{w}{b}; \qquad \tilde{\varphi} = \frac{\varphi}{b^2}, \qquad \tilde{\psi}_\alpha = \frac{\psi_\alpha}{b^2},$$

$$\alpha, \beta = R, \theta, \vartheta. \tag{1.11}$$

Here, b is some characteristic linear dimension, λ_* and μ_* are the Lamé parameters, c_* is the speed of propagation of compression waves, and ρ_* is the density of the elastic medium. From here on, when it will make no misunderstanding, we shall omit tildes in dimensionless values without any supplementing references.

If the motion of an elastic medium is axially symmetric with respect to the axis Ox_1, then the elastic potentials and the components of stress and strain tensors do not depend on the angle ϑ. Moreover, it is necessary to set $w \equiv 0$ and $\psi_r = \psi_\theta \equiv 0$. At that, we designate $\psi_\vartheta = \psi$. Then, taking into account (1.5) and (1.9), the equations of motion represented in the dimensionless variables with respect to the scalar potential φ and the nonzero component ψ of the vector potential $\boldsymbol{\psi}$ are as follows:

$$\gamma^2 \frac{\partial^2 \varphi}{\partial \tau^2} = \Delta \varphi,$$

$$\eta^2 \frac{\partial^2 \psi}{\partial \tau^2} = \Delta \psi - \frac{\psi}{r^2 \sin \theta};$$

(1.12)

here,

$$\Delta = \frac{1}{r^2} \frac{\partial}{\partial r} \left(r^2 \frac{\partial}{\partial r} \right) + \frac{1}{r^2 \sin \theta} \frac{\partial}{\partial \theta} \left(\sin \theta \frac{\partial}{\partial \theta} \right).$$

At that, it follows from (1.8) and (1.10) that the dimensionless stresses and displacements are related to the potentials by the following formulas:

$$u = \frac{\partial \varphi}{\partial r} + \frac{1}{r \sin \theta} \frac{\partial}{\partial \theta} (\psi \sin \theta),$$

$$v = \frac{1}{r} \left[\frac{\partial \psi}{\partial r} - \frac{\partial}{\partial \theta} (r \psi) \right],$$

$$\sigma_{r\theta} = \beta \frac{1-\kappa}{2} \left(\frac{1}{r} \frac{\partial u}{\partial \theta} + \frac{\partial v}{\partial r} - \frac{v}{r} \right),$$

$$\sigma_{r\vartheta} = \sigma_{\vartheta\theta} \equiv 0,$$

$$\sigma_{\theta\theta} = \beta \left[\kappa \frac{\partial u}{\partial r} + (1+\kappa) \frac{u}{r} + \frac{\kappa}{r} \frac{\partial v}{\partial \theta} + \frac{v}{r} \cot \theta \right],$$

$$\sigma_{rr} = \beta \left[\frac{\partial u}{\partial r} + 2\kappa \frac{u}{r} + \frac{\kappa}{r} \left(\frac{\partial v}{\partial \theta} + v \cot \theta \right) \right],$$

$$\sigma_{\vartheta\vartheta} = \beta \left[\kappa \frac{\partial u}{\partial r} + (1+\kappa) \frac{u}{r} + \frac{1}{r} \frac{\partial v}{\partial r} + \kappa \frac{v}{r} \cot \theta \right].$$

(1.13)

In order to determine the potentials φ and ψ uniquely, as well as the components of the displacement vector and the stress tensor, it is necessary to complete the system of equations (1.12) by boundary and initial conditions.

In the theory of elasticity, three types of boundary conditions are distinguished. We shall use the third one, which is the most general one and from which the first and second cases follow being its particular cases.

Let us assume that an elastic body occupies the region G bounded by the surface S which consists of two surfaces S_1 and S_2 such that $S = S_1 \cup S_2$.

We designate the vector of unit external normal to the surface S by $\mathbf{n}$. On the surface S_1, the displacement vector $\mathbf{U}$ is predetermined, that is,

$$u^k \mathbf{e}_k \big|_{S_1} = \mathbf{U}, \tag{1.14}$$

and on the surface S_2, the stress vector $\mathbf{T}$ is predetermined, that is,

$$\sigma^{ik} n_i \mathbf{e}_k \big|_{S_2} = \mathbf{T}, \qquad \mathbf{n} = (n_1, n_2, n_3). \tag{1.15}$$

In the formulas (1.14) and (1.15), σ^{mk} and u^k are the contravariant components of the stress tensor and the displacement vector, respectively, in the curvilinear coordinate system with the covariant basis $(\mathbf{e}_1, \mathbf{e}_2, \mathbf{e}_3)$; n_1, n_2, and n_3 are the covariant components of the vector $\mathbf{n}$.

If the surface S coincides with the surface S_1, we get a boundary value problem with the first type of boundary conditions. However, for $S = S_2$, we get a boundary value problem with the second type of boundary conditions.

In the case of an axially symmetric deformation of the elastic body, when the surface S coincides with one of the coordinate spheres, the conditions (1.14) and (1.15) become significantly simpler:

$$
\begin{aligned}
u\big|_{S_2} &= U_r, & v\big|_{S_2} &= U_\theta, & \mathbf{U} &= U_r \mathbf{e}_r + U_\theta \mathbf{e}_\theta, \\
\sigma_{rr}\big|_{S_1} &= T_r, & \sigma_{r\theta}\big|_{S_1} &= T_\theta, & \mathbf{T} &= T_r \mathbf{e}_r + T_\theta \mathbf{e}_\theta.
\end{aligned} \tag{1.16}
$$

If the elastic medium under consideration occupies a boundless region, then, because of a hyperbolic character of the system of equations (1.5), we assume no disturbances at infinity.

In order to state the initial conditions to be imposed on the components of the stress and strain tensors, we shall prescribe the displacement vector and the speed of displacements at the initial time $t = t_0$ (it is usually assumed that $t_0 = 0$) in the whole region G:

$$
\begin{aligned}
\mathbf{u}(x^1, x^2, x^3, t)\big|_{t=t_0} &= \mathbf{u}_0(x^1, x^2, x^3), \\
\frac{\partial}{\partial t} \mathbf{u}(x^1, x^2, x^3, t)\big|_{t=t_0} &= \mathbf{v}_0(x^1, x^2, x^3), \qquad (x^1, x^2, x^3) \in G.
\end{aligned} \tag{1.17}
$$

If the initial conditions are zeros ($\mathbf{u}_0 = 0$ and $\mathbf{v}_0 = 0$), that is, if the process of the medium movement starts from rest, then, for example, in the case of an axially symmetric deformation, it is sufficient to set

$$\varphi\big|_{t=t_0} = \psi\big|_{t=t_0} = 0,$$

$$\frac{\partial \varphi}{\partial t}\bigg|_{t=t_0} = \frac{\partial \psi}{\partial t}\bigg|_{t=t_0} = 0. \tag{1.18}$$

As it has been proven in the courses of the theory of elasticity (Love (1959), Amenadze (1976), Sedov (1984)), the mixed initial and boundary value problem has a unique solution at appropriate conditions imposed on the functions which enter into (1.14)–(1.16).

1.2 An Acoustic Medium

Observing the mathematical models of continuous media successfully applied in studies of diffraction, we can mark out the model of ideal fluid (ideal liquid or ideal gas). Ideal fluid can be determined as a continuous medium in which a stress vector at any small plane element is orthogonal to this element (Sedov (1984)). A stress tensor for such a medium is a spherical tensor, and in an arbitrary curvilinear coordinate system its mixed components have the following form:

$$\sigma_k^i = -p\,\delta_k^i\,. \tag{1.19}$$

Here, p is the pressure in the fluid. The sign in (1.19) has been taken because of the assumption that the stress produced by compression is positive, since the media for which the model of ideal fluid is valid are typically in a compressed state at $p > 0$ and cannot withstand tension without generation of surfaces of discontinuity. It can be proven that for a given point of the ideal fluid, the pressure p at an arbitrarily oriented plane element is constant.

Let us analyze the equation of motion of an ideal fluid in Euler variables, that is, let us study the motion of media at individual points of the immovable space (Mises (1966), Kochin et al. (1963), Sedov (1984)).

The equation of motion of an ideal fluid has the following form:

$$\frac{\mathrm{d}\mathbf{V}}{\mathrm{d}t} = \mathbf{F} - \frac{1}{\rho}\,\mathrm{grad}\,p\,; \tag{1.20}$$

here, ρ is the fluid density, $\mathbf{F}$ is the density of mass force, $\mathbf{V}$ is the fluid velocity at the point of observation, and t is time.

The vector equation (1.20) should be supplemented by the equation of continuity

$$\frac{\mathrm{d}\rho}{\mathrm{d}t} + \rho\,\mathrm{div}\,\mathbf{V} = 0\,. \tag{1.21}$$

The system of equations (1.20) and (1.21) is not complete. It consists of four scalar equations with respect to five unknown scalar functions: the pressure p, the density ρ, and three projections of the velocity vector $\mathbf{V}$ onto the axes of the coordinate system. In order to analyze the class of problems to be studied in this book, it is sufficient to assume that the fluid is barotropic, that is, the pressure depends on the density only,

$$p = \Psi(\rho)\,. \tag{1.22}$$

At that, the thermodynamic properties of the medium need not be specified. In the general case, however, the condition (1.22) is not valid. An example of the barotropic flow is an isothermal flow, in which the temperature is constant at all points of the medium. Other examples are the isentropic and, generally, polytropic flows of a perfect gas (Mises (1966)).

For a wide class of problems of fluid dynamics, we can consider the fluid flow to be potential. In that case, we can assume that in the region occupied by the medium there exists the velocity potential Φ such that

$$\mathbf{V} = \operatorname{grad} \Phi. \tag{1.23}$$

Let us note that at an appropriate smoothness of the vector function $\mathbf{V}(x^1, x^2, x^3)$, the necessary and sufficient condition of the existence of potential flow is

$$\operatorname{rot} \mathbf{V} = 0. \tag{1.24}$$

For the potential flows, we can derive a first integral of the Euler equations (1.20) (Sedov (1984). Let us reduce (1.20) to Gromeka–Lamb form. We shall use then the Cartesian coordinate system $Ox_1x_2x_3$ and write the total derivative of the velocity $\frac{d\mathbf{V}}{dt}$ in the form

$$\frac{d\mathbf{V}}{dt} = \frac{\partial \mathbf{V}}{\partial t} + V_1 \frac{\partial \mathbf{V}}{\partial x_1} + V_2 \frac{\partial \mathbf{V}}{\partial x_2} + V_3 \frac{\partial \mathbf{V}}{\partial x_3} = \frac{\partial \mathbf{V}}{\partial t} + (\mathbf{V}, \nabla) \mathbf{V},$$

$$\mathbf{V} = V_1 \mathbf{e}_1 + V_2 \mathbf{e}_2 + V_3 \mathbf{e}_3, \qquad V_i = \frac{dx_i}{dt}, \tag{1.25}$$

where $(\mathbf{a}, \mathbf{b})$ is the scalar product of the vectors $\mathbf{a}$ and $\mathbf{b}$.

The derivative

$$\frac{\partial \mathbf{V}}{\partial t}$$

is termed the *local acceleration*, and the term

$$(\mathbf{V}, \nabla) \mathbf{V}$$

is termed the *convective* (or *translational*) *acceleration*.

From vector analysis it is known that

$$(\mathbf{V}, \operatorname{grad}) \mathbf{V} = [\operatorname{rot} \mathbf{V}, \mathbf{V}] + \operatorname{grad} \left(\frac{V^2}{2} \right). \tag{1.26}$$

Substituting (1.25) into (1.20) and taking into account (1.26), we get the equation of motion of an ideal fluid in Gromeka–Lamb form

$$\frac{\partial \mathbf{V}}{\partial t} = \operatorname{grad} \left(\frac{V^2}{2} \right) + [\operatorname{rot} \mathbf{V}, \mathbf{V}] = \mathbf{F} - \frac{1}{\rho} \operatorname{grad} p. \tag{1.27}$$

Assuming the potentiality of the fluid motion ($\mathbf{V} = \operatorname{grad} \Phi$, $\operatorname{rot} \mathbf{V} = 0$) and its barotropic character ($p = \Psi(\rho)$), as well as the existence of the potential of the external mass forces U ($\mathbf{F} = \operatorname{grad} U$), we can introduce the *pressure function*

$$P(p) = \int \frac{dp}{\rho(p)}, \qquad \frac{1}{\rho} \operatorname{grad} p = \operatorname{grad} P. \tag{1.28}$$

At that, we transform (1.27) into the following equation:

$$\mathrm{grad}\left(\frac{\partial \Phi}{\partial t} + \frac{V^2}{2} + P - U\right) = 0.$$ (1.29)

Then, we get the Cauchy–Lagrange integral

$$\frac{\partial \Phi}{\partial t} + \frac{V^2}{2} + P - U = f(t),$$ (1.30)

where $f(t)$ is some arbitrary function. Since the potential Φ is determined up to an arbitrary additive constant (at that, the velocity field does not vary), we can set $f(t) \equiv 0$.

Differentiating the pressure function with respect to t in accounting for the equality (1.28), we get

$$\frac{\mathrm{d}P}{\mathrm{d}t} = \frac{1}{\rho}\frac{\mathrm{d}p}{\mathrm{d}t} = \frac{1}{\rho}\frac{\mathrm{d}p}{\mathrm{d}\rho}\frac{\mathrm{d}\rho}{\mathrm{d}t} = \frac{c^2}{\rho}\frac{\mathrm{d}\rho}{\mathrm{d}t}, \qquad c = \sqrt{\frac{\mathrm{d}p}{\mathrm{d}\rho}},$$ (1.31)

where c is the local velocity of the speed of sound in the fluid. When deriving (1.31), we have assumed that the function $\Psi(\rho)$ is a monotonically increasing one.

Then, using (1.21) and (1.30), assuming no mass forces ($U \equiv 0$), and taking into account the equality

$$\mathrm{div}(\mathrm{grad}\,\Phi) = \Delta\Phi,$$ (1.32)

we get the system of equations for the barotropic ideal fluid

$$\frac{1}{c^2}\frac{\mathrm{d}P}{\mathrm{d}t} + \Delta\Phi = 0,$$
$$\frac{\partial \Phi}{\partial t} + \frac{1}{2}(\mathrm{grad}\,\varphi)^2 + P - U = 0.$$ (1.33)

This system is complete. The unknowns are the pressure function P and the velocity potential Φ. However, in the general case, integration of this nonlinear system of differential equations presents difficulties.

It is necessary to point out that there exists an important class of potential motions of fluids for which the methods of study of the system of equations (1.33) are well developed. This is the motion of an acoustic medium. We shall define an *acoustic medium* as an ideal barotropic fluid, in which the variations of pressure, density, speed, and their derivatives with respect to time and space coordinates relatively to some undisturbed state are small. Neglecting small values of the order higher than one in (1.33), we get the system of linear differential equations

$$\frac{1}{c_0^2}\frac{\mathrm{d}P}{\mathrm{d}t} + \Delta\Phi = 0, \qquad c_0^2 = \left.\frac{\mathrm{d}p}{\mathrm{d}\rho}\right|_{\rho=\rho_0},$$
$$\frac{\partial \Phi}{\partial t} + P - U = 0.$$ (1.34)

Here, $c_0 = \text{const}$ is the speed of sound, ρ_0 and $p_0 = \Psi(\rho_0)$ are the density and the pressure in the fluid at undisturbed state.

If the mass forces are time-independent, we can reduce the system of equations (1.34) to the wave equation with respect to the velocity potential Φ

$$\frac{1}{c_0^2} \frac{\partial^2 \Phi}{\partial t^2} = \Delta \Phi. \tag{1.35}$$

Linearizing the Cauchy–Lagrange integral, we obtain the relationship between the parameter p and the potential Φ for a disturbed motion:

$$p = -\rho_0 \frac{\partial \Phi}{\partial t}. \tag{1.36}$$

Despite the strong limitations, which we have imposed on the motions described by (1.35), the acoustic approximation allows us to solve a wide class of problems of aerohydroelasticity (including the diffraction problem). The limitations of application of the model of acoustic medium for simulation of the real phenomena are discussed in Zamyshlyaev, Yakovlev (1967), Slepyan (1972), and Grigolyuk, Gorshkov (1974).

In addition to the dimensionless values introduced by (1.11), let us introduce the following dimensionless values:

$$\tilde{\Phi} = \frac{\Phi}{c_* b}, \qquad \gamma_0 = \frac{c_*}{c_0}, \qquad \beta_0 = \frac{\rho_0 c_0^2}{\rho_* c_*^2},$$

$$\tilde{p} = \frac{p}{\rho_* c_*^2}, \qquad \tilde{V} = \frac{V}{c_*}. \tag{1.37}$$

Here, a tilde designates dimensionless value. From here on, we omit tildes in dimensionless values.

Then, the wave equation and the relationships which interrelate the velocity potential, the speed, and the pressure, are to be as follows:

$$\gamma_0^2 \frac{\partial^2 \Phi}{\partial \tau^2} = \Delta \Phi, \qquad p = -\beta_0 \gamma_0^2 \frac{\partial \Phi}{\partial \tau}, \qquad V = \text{grad}\, \Phi. \tag{1.38}$$

The relationships (1.38) are valid in any coordinate system. In order to make a transition to the Cartesian or spherical coordinate systems (particularly, in the case of axial symmetry), it is necessary to use the corresponding formulas for the Laplacian Δ and the gradient presented in Sect. 1.1.

The barotropic ideal fluids can be regarded as a particular case of elastic media. This is the reason why R. Mises termed them the *elastic liquids* by analogy with the elastic body, in which the stress field determines the strain field and vice versa (Mises (1966)). For this reason, the acoustic fluids can be regarded to be a particular case of the linearly-elastic media. Consequently, the relationships (1.38) can be derived from the equations of motion of the linearly-elastic medium in the case of no shear strength ($\mu = 0$). In the dimensionless variables (1.11), this brings us to the following limiting case:

$$c_2 = 0, \qquad \eta \to \infty, \qquad \kappa = 1, \qquad \psi \equiv 0. \tag{1.39}$$

From Hooke's law (1.3) and the representation (1.4) it follows that

$$\sigma_j^i = \lambda I_1 \delta_j^i, \qquad i, j = 1, 2, 3,$$
$$\mathbf{u} = \operatorname{grad} \varphi. \tag{1.40}$$

Using the Cartesian coordinate system $Ox_1x_2x_3$, we obtain

$$I_1 = \epsilon_{11} + \epsilon_{22} + \epsilon_{33} = \frac{\partial u_1}{\partial x_1} + \frac{\partial u_2}{\partial x_2} + \frac{\partial u_3}{\partial x_3} = \operatorname{div} \mathbf{u}; \tag{1.41}$$

it means that the invariant of the deformation tensor I_1 will be a relative volume deformation of the media at the point.

From (1.40), (1.41), and (1.5) it follows that

$$\sigma_j^i = \lambda \operatorname{div}(\operatorname{grad} \varphi) \, \delta_j^i = \lambda \, \Delta\varphi \, \delta_j^i = \frac{\lambda}{c_1^2} \frac{\partial^2 \varphi}{\partial t^2} \delta_j^i = \rho \frac{\partial^2 \varphi}{\partial t^2} \delta_j^i, \qquad c_1^2 = \frac{\lambda}{\rho}. \tag{1.42}$$

Taking into account the relationships between the velocity vector $\mathbf{V}$ and the displacement vector $\mathbf{u}$, between the displacement potential φ and the velocity potential Φ, and between the pressure p and the stresses σ_{ij},

$$\mathbf{V} = \frac{\partial \mathbf{u}}{\partial t}, \qquad \Phi = \frac{\partial \varphi}{\partial t}, \qquad \sigma_j^i = -p\,\delta_j^i, \tag{1.43}$$

we obtain

$$p = -\rho \frac{\partial^2 \varphi}{\partial t^2} = -\rho \frac{\partial \Phi}{\partial t}; \tag{1.44}$$

this expression corresponds to the formula (1.36) for the acoustic medium.

Taking into account (1.43) and the formula

$$c_0^2 = c_1^2 = \frac{\lambda}{\mu}$$

and differentiating the first equation of (1.5), we obtain the wave equation (1.35) for the velocity potential Φ.

In order to determine the velocity potential and the pressure in the acoustic fluids, it is necessary to set the initial and boundary conditions.

Let us consider the boundary conditions of the third type. We shall assume that the acoustic medium occupies the region G bounded by the surface S ($S = S_1 \cup S_2$) with the unit normal $\mathbf{n}$. On the surface S_2, the pressure p_n is set (pressure is assumed to be positive if it produces compression), that is,

$$p|_{S_2} = p_{\mathbf{n}}. \tag{1.45}$$

On the surface S_1, the normal component of velocity is set, that is,

$$\left. \frac{\partial \Phi}{\partial \mathbf{n}} \right|_{S_1} = V_{\mathbf{n}} \,, \tag{1.46}$$

where

$$\frac{\partial \Phi}{\partial \mathbf{n}}$$

is the derivative of the velocity potential Φ along the vector $\mathbf{n}$.

If the surface S coincides with S_2, we obtain the boundary conditions of the second type. Such boundary conditions appear, for example, when the surface is free. In the case when $S = S_1$, we get the boundary conditions of the first type. Such conditions appear when the boundary surface S moves at some predetermined velocity.

In the cases when the region G is boundless, we assume no disturbances at infinity.

The initial conditions presume setting the velocity field and the pressure at each point of the region G at the initial instant $t = t_0$:

$$\begin{aligned} \mathbf{V} \left(x^1, x^2, x^3, t\right)\big|_{t=t_0} &= \mathbf{V}_0 \left(x^1, x^2, x^3\right) , \\ p \left(x^1, x^2, x^3, t\right)\big|_{t=t_0} &= p_0 \left(x^1, x^2, x^3\right) , \qquad \left(x^1, x^2, x^3\right) \in G . \end{aligned} \tag{1.47}$$

If the disturbed motion starts from the quiescence ($p_0 = 0$, $\mathbf{V}_0 = 0$), then it is sufficient to impose the following conditions on the velocity potential:

$$\Phi|_{t=t_0} = \left. \frac{\partial \Phi}{\partial t} \right|_{t=t_0} = 0 \,. \tag{1.48}$$

In the classical courses of fluid mechanics, it is proven that the solution of the mixed initial and boundary value problem formulated above exists and this solution is unique (Mises (1966), Kochin et al. (1963), Sedov (1984)).

1.3 Equations of Motion of Thin-Walled Shells

Currently, there exists a vast number of publications devoted to the general theory of shells. The systems of equations presented in these publications differ from each other. These differences are caused by the assumptions made in the derivation of the kinematic and static relationships; they depend on the peculiarities of the problems to be solved. Here, assuming the simplest model, we shall consider single-layer shells of constant thickness h made from elastic homogeneous isotropic material.

Since any shell is an elastic body bounded by two curvilinear surfaces separated by a small distance h compared to the other characteristic dimensions, then, describing its movements, it is natural to reduce the three-dimensional equations of the theory of elasticity (1.1) to two-dimensional equations.

One of the ways of reduction from a three-dimensional problem to a two-dimensional one is in application of the Kirchhoff–Love hypothesis. In this case, we obtain the classical theory of shells, which was successfully used for the solution of many problems of statics and dynamics (Volmir (1976), Volmir (1979)). One way to modify the classical theory of shells is to take into account the shear and the inertia of rotation of cross section of shell. Such models are usually associated with the name of S.P. Timoshenko, who proposed to apply it for the analysis of beams. In that case, the improved system of equations is a system of hyperbolic type and it allows us to describe the phenomenon of wave propagation in shells.

Besides, the equations derived in accounting for the inertia of rotation and the shear strains of cross section include the derivatives of a lower order. That simplifies the statement of boundary conditions and, when using computers, significantly facilitates numerical algorithms of integration of the equations and improves the quality of calculation procedures and their stability.

Another way to derive basic relationships of the theory of thin-walled shells (both classical and improved) is to use a series expansion of displacements or stresses in terms of powers of the normal coordinate (series in thickness) and saving a particular number of terms of this series; the number of terms depends on the accuracy required and on the problem peculiarities.

Below, we present one of the mostly used variants of the geometrically nonlinear equations of the theory of thin-walled shells of revolution, which takes into account the shear and the inertia of rotation of cross section. We shall determine position of a shell's point by the Gauss coordinates λ_1 and λ_2 of the middle surface and by the coordinate γ directed as the external normal.

Since in this book we consider mainly the axially symmetric problems, let us present the governing equations for this case. During axially symmetric deformations of a shell of revolution, its stress–strain state depends not only on the coordinate λ_1. We shall assume that when a deformation takes place, the longitudinal displacement u_1 (in more exact terms, the displacement along the generatrix of the shell) varies linearly with the shell thickness, and the normal displacement u_3 does not vary (i.e., $u_2 \equiv 0$). Thus, we have

$$u_1(\lambda_1, \gamma, t) = u(\lambda_1, t) - \gamma\,\chi(\lambda_1, t)\,,$$
$$u_3(\lambda_1, \gamma, t) = w(\lambda_1, t)\,, \tag{1.49}$$

where u and w are the displacements of the points of the middle surface along λ_1 and γ, respectively, and χ is the angle of rotation of the normal.

Following Ogibalov (1963), we shall assume that the components of deformation vary linearly with the thickness of shell. We have

$$\epsilon_1^{\gamma}(\lambda_1, \gamma, t) = \epsilon_1(\lambda_1, t) - \gamma\,\kappa_1(\lambda_1, t)\,,$$
$$\epsilon_2^{\gamma}(\lambda_1, \gamma, t) = \epsilon_2(\lambda_1, t) - \gamma\,\kappa_2(\lambda_1, t)\,,$$
$$\omega^{\gamma}(\lambda_1, \gamma, t) = \omega(\lambda_1, t) - \gamma\,\kappa_{12}(\lambda_1, t)\,. \tag{1.50}$$

We shall also assume the following expressions for ϵ_1, ϵ_2, and ω (which are the relative deformations of the middle surface):

$$\epsilon_1 = \frac{1}{A_1} \frac{\partial u}{\partial \lambda_1} + \frac{w}{R_1} + \frac{1}{2} \left(\frac{1}{A_1} \frac{\partial w}{\partial \lambda_1} - \frac{u}{R_1} \right)^2 ,$$

$$\epsilon_2 = \frac{1}{A_1 A_2} \frac{\partial A_2}{\partial \lambda_1} u + \frac{w}{R_2} ,$$

$$\omega = \frac{1}{A_1} \frac{\partial w}{\partial \lambda_1} - \frac{u}{R_1} - \chi ;$$

here, R_1 and R_2 are the radii of curvature of the normal cross sections of the middle surface along the lines λ_1 and λ_2; A_1 and A_2 are the Lamé parameters.

The curvature changes κ_1 and κ_2 and the twist χ of the middle surface can be determined by the expressions

$$\kappa_1 = \frac{1}{A_1} \frac{\partial \chi}{\partial \lambda_1} , \qquad \kappa_2 = \frac{1}{A_1 A_2} \frac{\partial A_2}{\partial \lambda_1} \chi , \qquad \kappa_{12} = 0 .$$

The strains of the middle surface are completely determined by six parameters ϵ_1, ϵ_2, ω, κ_1, κ_2, and κ_{12}, the first three of which characterize the variation of the dimensions of a small element of the middle surface and three others characterize its distortion.

Let us consider a small element enclosed by two pairs of planes, which are normal to the middle surface of the shell and contain its principal curvatures. Let σ_1, σ_2, and $\sigma_{1\gamma}$ be the stresses in the middle surface and σ_1^γ, σ_2^γ, and $\sigma_{1\gamma}^\gamma$ be the stresses in some layer, which is offset by γ from the middle surface.

Let us introduce forces T_1, T_2, and Q and moments M_1 and M_2 using the conventional formulas

$$T_i = \int_{-h/2}^{h/2} \sigma_i^\gamma \, d\gamma ,$$

$$M_i = \int_{-h/2}^{h/2} \sigma_i^\gamma \gamma \, d\gamma , \qquad i = 1, 2 ,$$

$$Q = \int_{-h/2}^{h/2} \sigma_{1\gamma}^\gamma \, d\gamma . \tag{1.51}$$

In the case of a generalized plane stress state, the relationship between stresses and strains are as follows:

$$\sigma_1^\gamma = \frac{E}{1 - \nu^2} (\epsilon_1^\gamma + \nu \epsilon_2^\gamma) , \qquad \sigma_2^\gamma = \frac{E}{1 - \nu^2} (\epsilon_2^\gamma + \nu \epsilon_1^\gamma) ,$$

$$\sigma_{1\gamma}^\gamma = \frac{E}{2(1 + \nu)} \omega^\gamma ; \tag{1.52}$$

here, E is the modulus of elasticity (Young's modulus) and ν is Poisson's ratio of the material of the shell.

Using (1.50)–(1.52) and assuming a parabolic law for the distribution of tangential stresses through the thickness of shell, we obtain

$$T_1 = B\left(\epsilon_1 + \nu\epsilon_2\right), \qquad M_1 = -D\left(\kappa_1 + \nu\kappa_2\right),$$

$$T_2 = B\left(\epsilon_2 + \nu\epsilon_1\right), \qquad M_2 = -D\left(\kappa_2 + \nu\kappa_1\right),$$

$$Q = \frac{B\left(1 - \nu\right)}{2}\, k^2\omega,$$

$$B = \frac{Eh}{1 - \nu^2}, \qquad D = \frac{Eh^3}{12\left(1 - \nu^2\right)}, \qquad k^2 = 0.86.$$

Let us introduce the following dimensionless values:

$$\tilde{u} = \frac{u}{b}, \qquad \tilde{w} = \frac{w}{b}, \qquad \delta = \frac{h}{b}, \qquad \beta = \frac{s}{b}, \qquad \gamma_0^2 = \frac{c_*^2}{c_0^2},$$

$$f = \frac{1}{A_2}\frac{\partial A_2}{\partial \beta}, \qquad \tau = \frac{c_* t}{b}, \qquad c_0^2 = \frac{E}{\rho_0\left(1 - \nu^2\right)},$$

$$\tilde{p} = \frac{p}{E}, \qquad S = \frac{Q}{B}, \qquad \eta = \frac{\partial w}{\partial \beta} - k_1 u, \qquad \tilde{q} = \frac{q}{E},$$

$$N_i = \frac{T_i}{B}, \qquad H_i = \frac{bM_i}{D}, \qquad k_i = \frac{b}{R_i}, \qquad \tilde{\kappa}_i = \kappa_i b, \qquad i = 1, 2;$$

here, a tilde designates dimensional values, ρ_0 is the density of the material of the shell, s is the arc length of the meridian of the middle surface, p and q are the external pressure and the tangential load applied to the shell.

When analyzing a shell of revolution, we can assume the coordinate λ_1 to be s ($A_1 = 1$), and the coordinate λ_2 to be the azimuth. Then, the nonlinear equations of axially symmetric movement of the shell can be represented as follows:

$$\gamma_0^2 \frac{\partial^2 u}{\partial \tau^2} = \frac{\partial}{\partial \beta}\left(N_1 - \chi S\right) + f\left(N_1 - N_2 - \chi S\right)$$

$$+ k_1\left(\eta N_1 + S\right) + \frac{\chi}{\delta}\left(1 - \nu^2\right)p + \frac{1 - \nu^2}{\delta}\, q,$$

$$\gamma_0^2 \frac{\partial^2 w}{\partial \tau^2} = \frac{\partial}{\partial \beta}\left(\eta N_1 + S\right) + f\left(\eta N_1 + S\right)$$

$$- k_1\left(N_1 - \chi S\right) - k_2 N_2 - \frac{1 - \nu^2}{\delta}\, p,$$

$$\gamma_0^2 \frac{\partial^2 \chi}{\partial \tau^2} = -\frac{\partial H_1}{\partial \beta} - f\left(H_1 - H_2\right) - k_1\eta H_1 + \frac{12}{\delta}\, S. \qquad (1.53)$$

When solving particular boundary value problems, it is necessary to complete the governing equations of motion of shell by boundary and initial conditions. Let us obtain the boundary conditions for several variants of constraints of the edges of shell.

If the edge $\beta = \text{const}$ is joined to a weightless diaphragm, which is absolutely stiff in tension and bending, then

$$\epsilon_2 = \chi = 0.$$

Analyzing the character of movement of the edges in the axial direction, we can state the third boundary condition. If the edge $\beta = \text{const}$ is immovable, then

$$u \sin \xi - w \cos \xi = 0,$$

where ξ is the angle between the plane of the diaphragm and the generatrix of the shell.

If one of the diaphragms is a flexible membrane, then the boundary condition with respect to χ is as follows:

$$\frac{\partial \chi}{\partial \beta} + \nu f \chi = 0.$$

If an edge of the shell has no constraints in the axial direction, we get

$$K = (N_1 - \chi S) \sin \xi - (S + \eta N_1) \cos \xi = 0, \qquad K = \frac{K^0}{B},$$

where K^0 is the axial specific force applied to the edge $\beta = \text{const}$.

Thus, for the edge $\beta = \text{const}$, it is possible to write the following combinations of boundary conditions:

(1) the diaphragm is stiff in tension and bending, the edge is immovable, that is,

$$u = w = \chi = 0;$$

(2) the diaphragm is stiff in tension and bending, the edge is loaded by the axial force L, that is,

$$\epsilon_2 = \chi = 0, \qquad K = \frac{L}{2\pi A_2 B};$$

when $L = 0$, the edge is free;

(3) the diaphragm is an elastic membrane, stiff in tension, the edge is immovable, that is,

$$u = w = \frac{\partial \chi}{\partial \beta} + \nu f \chi = 0.$$

There also exist combinations of boundary conditions stated for the edge loaded by bending moments.

If the shell is immovable until an external dynamic pressure p is applied to it, then we can write the initial conditions as follows:

$$u = w = \chi = \frac{\partial u}{\partial \tau} = \frac{\partial w}{\partial \tau} = \frac{\partial \chi}{\partial \tau} = 0 \quad \text{at} \quad \tau = \tau_0. \tag{1.54}$$

Analyzing a spherical shell of radius R in the spherical coordinate system ($\beta = \theta$), we should assume in the equations of motion and other relationships that

$$b = R, \qquad A_2 = R\sin\theta, \qquad k_1 = k_2 = 1, \qquad f = \cot\theta. \tag{1.55}$$

Discarding the nonlinear terms in the system of equations (1.53), we can easily obtain a linear variant of the equations of motion for the shell. In the case of a spherical shell, we should represent the equations in operator form (positive normal pressure is directed as the external normal)

$$\gamma_0^2 \frac{\partial^2 u}{\partial \tau^2} = L_{11}(u) + L_{12}(w) + L_{13}(\chi) + \alpha_0 q,$$

$$\gamma_0^2 \frac{\partial^2 w}{\partial \tau^2} = L_{21}(u) + L_{22}(w) + L_{23}(\chi) + \alpha_0 p,$$

$$\gamma_0^2 \frac{\partial^2 \chi}{\partial \tau^2} = L_{31}(u) + L_{32}(w) + L_{33}(\chi),$$

$$L_{11} = \frac{\partial^2}{\partial \theta^2} + \cot\theta \frac{\partial}{\partial \theta} - (\alpha_1 + \cot^2\theta), \qquad L_{12} = \alpha_2 \frac{\partial}{\partial \theta},$$

$$L_{13} = -\alpha_7,$$

$$L_{21} = -\alpha_2 \left(\frac{\partial}{\partial \theta} + \cot\theta \right), \qquad L_{22} = \beta_2 \left(\frac{\partial^2}{\partial \theta^2} + \cot\theta \frac{\partial}{\partial \theta} \right) - \alpha_3,$$

$$L_{23} = -\alpha_7 \left(\frac{\partial}{\partial \theta} + \cot\theta \right),$$

$$L_{31} = -\alpha_4, \qquad L_{32} = \alpha_4 \frac{\partial}{\partial \theta},$$

$$L_{33} = \frac{\partial^2}{\partial \theta^2} + \cot\theta \frac{\partial}{\partial \theta} - (\alpha_5 + \cot^2\theta),$$

$$\alpha_0 = \frac{1 - \nu^2}{\delta}, \qquad \alpha_1 = \nu + \alpha_7, \qquad \alpha_2 = 1 + \alpha_1,$$

$$\alpha_3 = 2(1 + \nu), \qquad \alpha_4 = \frac{12\alpha_6}{\delta}, \qquad \alpha_5 = \nu + \alpha_4,$$

$$\alpha_6 = \frac{(1 - \nu)k^2}{2\delta}, \qquad \alpha_7 = \delta\alpha_6. \tag{1.56}$$

At that, the dimensionless strains, curvature changes, forces, and displacements are interrelated as follows:

$$\epsilon_1 = \frac{\partial u}{\partial \theta} + w, \qquad \epsilon_2 = u \cot \theta + w, \qquad \omega = -u_0 + \frac{\partial u}{\partial \theta} - \chi,$$

$$\kappa_1 = \frac{\partial \chi}{\partial \theta}, \qquad \kappa_2 = \chi \cot \theta, \qquad N_1 = \epsilon_1 + \nu \epsilon_2, \qquad N_2 = \epsilon_2 + \nu \epsilon_1,$$

$$H_1 = -(\kappa_1 + \nu \kappa_2), \qquad H_2 = -(\kappa_2 + \nu \kappa_1), \qquad S = \alpha_7 \omega. \qquad (1.57)$$

As we have already mentioned, in some cases, when analyzing a stress-strain state of a shell, it is possible to apply a model of the first approximation based on the Kirchhoff–Love hypothesis. Keeping within the framework of this model, we shall represent a brief derivation of the equations of motion for a spherical shallow shell, in which deformations are of an axially symmetric character. These equations will be applied to study the interaction of spherical panels with a medium.

First, let us consider a symmetrical bending of a shallow spherical shell of radius R subjected to the external static pressure p. In that case, we can use the radius r as the only independent variable, which is measured from the axis of rotation of the shell. The positive value of the deflection w coincides with the direction of the internal normal. Then, the differential equations of equilibrium of the shallow spherical shell can be represented in the following form:

$$\frac{d(rT_1)}{dr} - T_2 - \frac{r}{R} Q = 0,$$

$$\frac{d(rQ)}{dr} + \frac{r}{R} (T_1 + T_2) + rp = 0,$$

$$\frac{d(rM_1)}{dr} - M_2 - rQ = 0. \qquad (1.58)$$

The relationships between the resultant stresses, the strain components, and the displacements w and u are

$$\epsilon_1 = \frac{1}{Eh} (T_1 - \nu T_2) = \frac{du}{dr} - \frac{w}{R},$$

$$\epsilon_2 = \frac{1}{Eh} (T_2 - \nu T_1) = \frac{u}{r} - \frac{w}{R},$$

$$T_1 = B \left[\frac{du}{dr} - \frac{w}{R} + \nu \left(\frac{u}{r} - \frac{w}{R} \right) \right],$$

$$T_2 = B \left[\frac{u}{r} - \frac{w}{R} + \nu \left(\frac{du}{dr} - \frac{w}{R} \right) \right],$$

$$M_1 = -D \left(\frac{d^2 w}{dr^2} + \frac{\nu}{r} \frac{dw}{dr} \right),$$

$$M_2 = -D \left(\frac{1}{r} \frac{dw}{dr} + \nu \frac{d^2 w}{dr^2} \right). \qquad (1.59)$$

In the case of a shallow shell, we can neglect the influence of the transverse force Q on the membrane forces in the first equation of (1.59). Then, this equation takes the following form:

$$\frac{d(rN_1)}{dr} - N_2 = 0.$$ (1.60)

Substituting the relationships (1.59) obtained for the forces acting in the middle surface into (1.60), we obtain

$$\frac{d}{dr}\left[\frac{1}{r}\frac{d}{dr}(ur)\right] = \frac{1+\nu}{R}\frac{dw}{dr}.$$

Then, after double integration, we get

$$u = \frac{1+\nu}{R}\frac{1}{r}\int_0^r wr_1\,dr_1 + \frac{Cr}{2}.$$

At the boundary, the displacement u is equal to the displacement of the supporting element u_0 in the tangential direction. If R_0 is the radius of the panel, we get

$$C = \frac{2}{R_0}u_0 - \frac{1+\nu}{R}\frac{2}{R_0^2}\int_0^{R_0} wr_1\,dr_1.$$

Thus, the displacement u is equal to

$$u = u_0\frac{r}{R_0} - \frac{1+\nu}{R}\frac{r}{R_0^2}\int_0^{R_0} wr_1\,dr_1 + \frac{1+\nu}{R}\frac{1}{r}\int_0^r wr_1\,dr_1.$$

Let us introduce the dimensionless coordinate

$$\zeta = \frac{\theta}{\theta_0} = \frac{r}{R_0},$$

where θ_0 is the half-angle of the panel and θ is the angle measured from the vertex of the shell). Then,

$$u = u_0\zeta - \frac{1+\nu}{R}R_0\zeta\int_0^1 w\zeta\,d\zeta + \frac{1+\nu}{R}\frac{R_0}{\zeta}\int_0^\zeta w\zeta\,d\zeta.$$ (1.61)

Eliminating Q from the remaining two equations of the system of equations (1.58), we obtain

$$\frac{d}{dr}\left[\frac{d(rM_1)}{dr} - M_2\right] + \frac{r}{R}(T_1 + T_2) + rp = 0.$$ (1.62)

Substituting the expressions (1.59) obtained for the moments and forces into (1.62) and taking into account (1.61), we obtain one governing equation with respect to the normal displacement w. Supplementing this equation by the inertia forces and assuming the principal state to be momentless, we finally obtain ($R_0 \approx R\theta_0$)

$$\frac{D}{R_0^2}\, \nabla^2 \nabla^2 w - T_0\, \nabla^2 w$$

$$+ \frac{Eh}{1-\nu}\left[(1-\nu)\,\theta_0^2 w + 2\,(1+\nu)\,\theta_0^2 \int_0^1 w\zeta\,\mathrm{d}\zeta - 2\theta_0 u_0\right]$$

$$+ \rho h R_0^2\, \frac{\partial^2 w}{\partial t^2} = R_0^2 p\,, \tag{1.63}$$

where

$$\nabla^2 = \frac{\partial^2}{\partial \zeta^2} + \frac{1}{\zeta}\,\frac{\partial}{\partial \zeta}$$

and T_0 is the initial constant force in the shell.

Equation (1.63) describes the small oscillations of a shallow spherical shell, when closed at its vertex and subjected to an axially symmetric dynamic loading. When deriving (1.63), we have neglected the inertial forces directed along the tangent to the middle surface of the shell. Let us state the boundary conditions at $\zeta = 1$ for the displacements u and w in the following form:

$$w = 0\,,$$

$$\frac{D}{R_0^2}\left(\frac{\partial^2 w}{\partial \zeta^2} + \frac{\nu}{\zeta}\,\frac{\partial w}{\partial \zeta}\right) - M_0 = 0\,,$$

$$\frac{Eh}{1-\nu}\left(u_0 - 2\theta_0 \int_0^1 w\zeta\,\mathrm{d}\zeta\right) + \frac{EF}{R_0}\,u_0 = 0\,,$$

$$M_0 = -\frac{G}{R_0}\,\frac{\partial w}{\partial \zeta}\bigg|_{\zeta=1}\,. \tag{1.64}$$

Here, M_0 is the torque in the supporting ring (rib), F is the cross-sectional area of this ring, and G is the ring stiffness in torsion.

Let us consider a particular case, when the rib stiffness satisfies the following relationships:

$$EF = \frac{EhR_0}{1+\nu}\,, \qquad G = \frac{Eh^3}{12\,(1+\nu)\,R_0}\,.$$

In this case, (1.63) and the boundary conditions (1.64) are simplified.

Let us derive the expressions for the moments and forces in the shell via w. The displacement of the supporting element u_0 is equal to

$$u_0 = (1+\nu)\,\theta_0 \int_0^1 w\zeta\,\mathrm{d}\zeta\,. \tag{1.65}$$

The forces T_1 and T_2 and the moments M_1 and M_2 can be obtained as follows:

$$T_1 = \frac{Eh}{\zeta^2 R} \int_0^\zeta w\zeta \, d\zeta \,,$$

$$T_2 = \frac{Eh}{\zeta^2 R} \int_0^\zeta w\zeta \, d\zeta - \frac{Eh}{R} w \,,$$

$$M_1 = \frac{D}{R_0^2} \left(\frac{\partial^2 w}{\partial \zeta^2} + \frac{\nu}{\zeta} \frac{\partial w}{\partial \zeta} \right) \,,$$

$$M_2 = -\frac{D}{R_0^2} \left(\frac{1}{\zeta} \frac{\partial w}{\partial \zeta} + \nu \frac{\partial^2 w}{\partial \zeta^2} \right) \,. \tag{1.66}$$

Setting $\gamma_0 = 1$ and $W = w/h$ and eliminating u_0, we can represent (1.63) and two first boundary conditions (1.64) as follows:

$$\frac{\partial^2 W}{\partial \tau^2} = -B_1 \nabla^2 \nabla^2 W + B_2 \nabla^2 W - B_3 W + B_4 P \,, \tag{1.67}$$

$$W = \nabla^2 W = 0 \quad \text{at} \quad \zeta = 1 \,. \tag{1.68}$$

Here,

$$\tau = \frac{c_0 t}{R} \,, \qquad B_1 = \frac{k^2}{12\theta_0^4} \,, \qquad B_2 = \frac{T_0 h \, (1 - \nu^2)}{E R_0^2 \delta^2} \,,$$

$$B_3 = 1 - \nu^2 \,, \qquad B_4 = \frac{1 - \nu^2}{\delta^2} \,.$$

In order to integrate (1.67) in the particular cases of loading, we can apply the analytical methods, for example, the Fourier method and the methods of integral transforms (the Hankel transform and the Laplace transform) (Dötsch (1974), Slepyan, Yakovlev (1980)).

1.4 Conditions of Contact for Interacting Media

In the previous sections of this chapter, we have derived the equations of motion for the basic models of continuous media, which are used for solving various dynamic problems. We have also presented the boundary conditions which are valid in the case when we analyze oscillations of the 'homogeneous' medium, that is, when the motion of the medium is described by the equations of only one model. At the same time, in most of practical cases, a medium should be presented as a finite number of regions, each of which requires application of a particular type of the medium model. We can mark out, for example, the problems of dynamics of thin-walled or thick-walled shells containing an elastic medium or an acoustic medium or surrounded by such a medium, the problems of penetration of solid bodies into an elastic space or into an acoustic space with a free surface, etc.

The problems of this type can be considered, on the one hand, similar to the problems of propagation of waves in inhomogeneous media, when both contacting media are described by one and the same model. In that case, it requires integration of equations with discontinuous coefficients. Such an approach should have an advantage when the number of boundary surfaces is relatively large. Another advantage of such an approach is in an opportunity to formulate the boundary conditions in a 'standard', 'conventional' form acceptable for the given model. However, the methods of integration of hyperbolic equations with discontinuous coefficients, particularly in the case of transient problems, are not developed sufficiently, yet.

On the other hand, the problems mentioned above can be interpreted as the problems of dynamic interaction of different media. Just that point of view is used in this book when stating and solving the problems. Motion of each contacting medium can be described by the equations of the models of different type. For example, we can consider interaction of a thin-walled shell and a fluid using the linear theory of thin-walled shells and the equations of acoustic medium, respectively. At that, in addition to the conventional boundary conditions, we should set the so-called conditions of contact of interacting media at the internal boundaries of contact S.

Let us consider different variants of conditions of contact of deformable bodies at the boundary of contact S. We shall confine our studies to a case when the surface S coincides with one of the coordinate surfaces in the orthogonal curvilinear coordinate system x^1, x^2, x^3 (for example, $x^3 = $ const). Then, the coordinate line $[x^3]$ is directed along the normal to the surface S. We shall also assume that the coordinate lines $[x^1]$ and $[x^2]$ coincide with the directions of the principal curvatures of the surface S.

First, let us obtain the conditions of contact of two different elastic media. Let us designate these media and their components of the stress–strain state by the superscripts 1 and 2, respectively. The boundary conditions at the boundary surface S in a generalized form can be represented as follows:

$$u_3^{(1)}\Big|_S = u_3^{(2)}\Big|_S \, ,$$

$$\sigma_{33}^{(1)}\Big|_S = \sigma_{33}^{(2)}\Big|_S \, ,$$

$$\sigma_{13}^{(1)}\Big|_S = \sigma_{13}^{(2)}\Big|_S = k_1 \left(u_1^{(2)} - u_1^{(1)} \right)\Big|_S \, ,$$

$$\sigma_{23}^{(1)}\Big|_S = \sigma_{23}^{(2)}\Big|_S = k_2 \left(u_2^{(2)} - u_2^{(1)} \right)\Big|_S \, . \tag{1.69}$$

Here, the superscripts designate the number of the medium. The relationships (1.69) include the continuity conditions for the normal (σ_{33}) and tangential (σ_{13} and σ_{23}) physical projections of the components of the stress tensor onto the axes of coordinate systems, as well as the continuity conditions for the projections of normal displacements of contacting media onto the surface S.

Two more boundary conditions set the linear relationships between the tangential stresses and the displacements u_1 and u_2. At the finite values of the coupling coefficients k_1 and k_2, we obtain an 'elastic' contact similar to a Winkler foundation. The limiting values of coefficients k_1 and k_2 bring us to two classical cases of boundary conditions: (1) for $k_1 = k_2 = 0$, we have a free sliding of the bounding surfaces; (2) for $k_1, k_2 \to \infty$, we get a rigid adhesion of the elastic media. In the first case, a lamination of the media may take place, which complicates the problem significantly because in this case we get the mixed boundary conditions at the unknown interior boundaries; this is why we do not consider this case of conditions of contact in this book.

Let us consider two other limiting cases of boundary conditions, which follow from the general statement (1.69).

(1) A rigid constraint of the surface S:

$$u_i^{(1)}\Big|_S = 0, \qquad i = 1, 2, 3. \tag{1.70}$$

(2) A free surface S:

$$\sigma_{33}^{(1)}\Big|_S = \sigma_{13}^{(1)}\Big|_S = \sigma_{23}^{(2)}\Big|_S = 0. \tag{1.71}$$

In order to realize the boundary conditions (1.70), we shall set $u_i^{(2)}$, $i = 1, 2, 3$, and pass to the limit as $k_1 \to \infty$ and $k_2 \to \infty$ in (1.69). The relationship for the normal stresses σ_{33} will be eliminated. The same result can be obtained if is assumed that the speeds of propagation of elastic waves in the medium designated by the superscript 2 are as follows: $c_1^{(2)} \to \infty$ and $c_2^{(2)} \to \infty$; this statement, when using the dimensionless values (1.11), is equivalent to the following passages to the limit: $\beta_2 \to \infty$ and $\eta_2 = \gamma_2 = 0$.

In order to switch to the boundary conditions (1.71), it is necessary to set $\sigma_{33}^{(2)} = \sigma_{13}^{(2)} = \sigma_{23}^{(2)} = 0$ and $k_1 = k_2 = 0$ in (1.69). The relationship for the normal displacements u_3 will be eliminated. This result corresponds to equality to zero of the speeds of propagation of elastic waves in the second medium: $c_1^{(2)} = c_2^{(2)} = 0$ (i.e., $\beta_2 = 0$ and $\eta_2, \gamma_2 \to \infty$).

In the case of an axially symmetric motion of the elastic media analyzed in the spherical coordinate system, when the contacting surface S coincides with the sphere $r = 1$, the analogy to the conditions (1.69) in the dimensionless values have the following form:

$$u^{(1)}\Big|_{r=1} = u^{(2)}\Big|_{r=1},$$

$$\sigma_{rr}^{(1)}\Big|_{r=1} = \sigma_{rr}^{(2)}\Big|_{r=1},$$

$$\sigma_{r\theta}^{(1)}\Big|_{r=1} = \sigma_{r\theta}^{(2)}\Big|_{r=1} = k_1 \left(v^{(2)} - v^{(1)} \right)\Big|_{r=1}. \tag{1.72}$$

In the case when the medium 1 is elastic and the medium 2 is acoustic, the conditions of contact can be derived from (1.69) applying the passage to the limit (1.39); in this case, we have

$$\sigma_{33}^{(1)}\Big|_S = -p^{(2)}\Big|_S, \qquad \frac{\partial u^{(1)}}{\partial t}\Big|_S = V_{\mathbf{n}}^{(2)}\Big|_S,$$

$$\sigma_{13}^{(1)}\Big|_S = \sigma_{23}^{(1)}\Big|_S = 0, \tag{1.73}$$

where $p^{(2)}$ is the pressure in the acoustic medium and $V_{\mathbf{n}}^{(2)}$ is the projection of the velocity in the acoustic medium (in more exact terms, the velocity at the surface S) onto the normal $\mathbf{n}$ to this surface.

The boundary conditions for the variant of contact of two acoustic media can be presented similarly. We have

$$p^{(1)}\Big|_S = p^{(2)}\Big|_S, \qquad V_{\mathbf{n}}^{(1)}\Big|_S = V_{\mathbf{n}}^{(2)}\Big|_S. \tag{1.74}$$

The last two conditions express the principle of continuity of continuum for acoustic media.

Solution of many problems requires an analysis of contact of an acoustic medium and the atmosphere. Neglecting the disturbances in the atmosphere and assuming its pressure to be constant and equal to p_0, it is possible to reduce the boundary conditions at the surface of contact S to the condition of continuity of the pressure:

$$p^{(1)}\Big|_S = p_0. \tag{1.75}$$

Let us now switch to the conditions of contact of a thin-walled shell and a continuous medium (indicated by the superscript 1). We shall assume that the middle surface of the shell coincides with the surface S. For the elastic medium, the conditions of contact have the following form:

$$u_i^{(1)}\Big|_S = u_{i0},$$

$$\sigma_{i3}^{(1)}\Big|_S = \pm q_i, \qquad i = 1, 2, 3. \tag{1.76}$$

Here, u_{i0} are the components of the displacement vector and q_i are the components of the vector of density of surface forces, which enter into the right-hand sides of the corresponding equations of motion of the shell (see Sect. 1.3). The *plus* sign in the relationships (1.76) corresponds to the elastic medium, which is placed externally to the surface S. (We define the external surface of a shell as its surface, which corresponds to the positive direction of the normal. At that, we assume the normal displacement to be positive if its direction coincides with the direction of the normal vector.) The *minus* sign should be used in the opposite case.

In order to switch to an acoustic medium, we should retain the normal components of stress σ_{33} in the second group of the equalities (1.76). Then, we get

$$u_i^{(1)}\Big|_S = u_{i0}, \qquad i = 1, 2, 3,$$

$$p_1|_S = \mp q_\mathbf{n}, \tag{1.77}$$

where $q_\mathbf{n} = q_3$. The *minus* sign should be used in the case when the continuous medium is placed from the external side of the surface and the *plus* sign should be used in the case when the medium is placed from the internal side of the surface S.

Let us note the following fact, which is of a great importance when stating the problems of dynamics of continuums. In the general case, the surface of contact of two media subjected to deformation is unknown preliminarily, and should be determined whilst solving the problem. All the boundary conditions considered should be set for this deforming surface S_*. However, as it is assumed in the linear theory of elasticity and for the acoustic media, as well, the surface S_* differs slightly from the original surface S. At that, this difference is estimated by the values of the second order of smallness with respect to the small deformations (Sedov (1984)). Hence, it is reasonable to bring the boundary conditions (including the conditions of contact) to the undistorted (original) surface S.

1.5 General Integral of the Wave Equation in the Spherical Coordinates

One of the methods widely applied when solving the linear problems of propagation of elastic or acoustic waves is the method of incomplete separation of variables. The functions sought are represented in the form of the series or the integrals with respect to one variable and, thus, the dimension of the original partial differential equation is reduced.

In so far as the axially symmetric problems represented in the spherical coordinates are concerned, this method presumes a series expansion of the elastic potentials or the velocity potential of the acoustic medium, as well as the components of the stress–strain state of the elastic medium (or the pressure and the components of the velocity vector for an acoustic medium) into Legendre polynomials $P_n(\cos\theta)$ and functions related to them. Since the equation for the velocity potential of an acoustic medium $\varPhi(r, \theta, \tau)$ (1.38) is similar to the equation which determines the scalar elastic potential $\varphi(r, \theta, \tau)$, we shall confine ourselves to an analysis of the solution of (1.12) only.

In order to derive an analytic solution of these equations, we shall write the functions $\varphi(r, \theta, \tau)$ and $\psi(r, \theta, \tau)$ in the form of the series:

$$\varphi(r, \theta, \tau) = \sum_{n=0}^{\infty} \varphi_n(r, \tau) \, P_n(\cos\theta) \,,$$

$$\psi(r, \theta, \tau) = -\sin\theta \sum_{n=0}^{\infty} \psi_n(r, \tau) \, C_{n-1}^{3/2}(\cos\theta) \,; \tag{1.78}$$

here, $C_n^{\alpha}(x)$ is the Gegenbauer polynomial (the ultraspherical polynomial) (Hobson (1955), Kuznetsov (1965), Abramowitz, Stegun (1965)).

The Legendre polynomials $P_n(x)$ satisfy the following ordinary differential equation (Abramowitz, Stegun (1965)):

$$\left[(1-x^2)\,P_n'(x)\right]' = -n(n+1)\,P_n(x)\,; \tag{1.79}$$

here, a prime designates derivative with respect to x.

It follows that the function $P_n(\cos\theta)$ is a solution of the equation

$$\frac{1}{\sin\theta}\frac{d}{d\theta}\left[\sin\theta\,\frac{dP_n(\cos\theta)}{d\theta}\right] = -n(n+1)\,P_n(\cos\theta)\,. \tag{1.80}$$

Differentiating (1.80), after some rearrangements, we arrive at the equation in the derivative $dP_n(\cos\theta)/d\theta$

$$\left[\frac{1}{\sin\theta}\frac{d}{d\theta}\left(\sin\theta\,\frac{d}{d\theta}\right) - \frac{1}{\sin^2\theta}\right]\frac{dP_n(\cos\theta)}{d\theta} = -n(n+1)\,\frac{dP_n(\cos\theta)}{d\theta}\,. \tag{1.81}$$

Taking into consideration the relationship between the Legendre and Gegenbauer polynomials (Abramowitz, Stegun (1965))

$$P_n'(x) = C_{n-1}^{3/2}(x)\,, \tag{1.82}$$

substituting the series (1.78) into (1.12) and taking into account the completeness of the set of functions $\{P_n(\cos\theta)\}$ and $\{C_{n-1}^{3/2}(\cos\theta)\}$ on the closed interval $[0, \pi]$, as well, we obtain that the coefficients of series φ_n and ψ_n should satisfy the following equations of hyperbolic type:

$$\gamma^2\frac{\partial^2\varphi_n}{\partial\tau^2} = \Delta_n\varphi_n\,, \qquad n \geq 0\,,$$

$$\eta^2\frac{\partial^2\psi_n}{\partial\tau^2} = \Delta_n\psi_n\,, \qquad n \geq 1\,,$$

$$\Delta_n = \frac{1}{r^2}\frac{\partial}{\partial r}\left(r^2\frac{\partial}{\partial r}\right) - \frac{n(n+1)}{r^2}\,. \tag{1.83}$$

Since both equations in (1.83) are similar, we shall study, from here on, only one equation with respect to some function $\Phi(r, \tau)$, which can be represented in the following form:

$$\gamma^2 \frac{\partial^2 \Phi}{\partial \tau^2} = \frac{\partial^2 \Phi}{\partial r^2} + \frac{2}{r} \frac{\partial \Phi}{\partial r} - \frac{n(n+1)}{r^2} \Phi. \tag{1.84}$$

In order to derive an analytic solution that satisfies a physical picture of propagation of elastic waves, whose law of distribution of components with respect to the angle θ corresponds to one term of the series expansion in the Legendre polynomials of order n, we shall derive a general solution of (1.84) (Grigolyuk et al. (1976), Grigolyuk et al. (1979)). This solution can be expressed via some arbitrary functions which correspond to the elastic waves converging and diverging in the spherical coordinate system and which we shall term the *generalized spherical waves*. A spherical wave is termed *converging* if its front travels towards the origin of the spherical coordinate system and the wave is termed *diverging* if its front travels in the opposite direction.

Let us note that for $n = 0$, the solution presented below corresponds to the usual spherical waves. For an arbitrary $n \geq 0$, a representation of the general solution for the diverging waves in a similar form was presented in Love (1905) without derivation. The derivation of these formulas was presented in Lamb (1924). A similar solution for the diverging waves was presented in Maaz (1964) and used for $n = 1$. Particular cases of the general solution corresponding to $n = 0$ and $n = 1$ were used in many works on propagation of elastic waves (Kirchhoff (1876), Lamb (1931), Selberg (1952), Das Gupta (1954), Achenbach, Sun (1966), Chadwick, Trowbridge (1967a), Babichev (1966a), Rakhmatulin et al. (1967a), Rakhmatulin et al. (1967b), Chakraborty (1958), Achenbach (1973), Kasiak, Włodarczyk (1975), Fridman (1976a), Fridman (1976b), Moskalenko (1965), Kasiak et al. (1977), Slepyan (1972)). An interesting mathematical result was obtained by K. Friedrichs (Courant, Hilbert (1966)): spherical waves exist in any odd-dimensional space R_{2k+1}, $k = 0, 1, \ldots$.

Among the hyperbolic equations of the second order there exists a relatively narrow class of equations, which can be integrated in closed form, that is, the solutions of these equations can be represented in the form of a combination of arbitrary functions. Such is, for example, the equation of oscillations of a string. More complicated example is the Euler–Poisson equation (Koshlyakov et al. (1962), Babich et al. (1964), Tricomi (1954)) that can be integrated in closed form only for some particular values of the parameters which enter into the coefficients. The following form of representation of the solution of (1.84) for $n = 1$ is also known:

$$\begin{aligned}
\Phi(r, \tau) = {} & r^{-1} f_1'(\tau - \gamma r) + \gamma^{-1} r^{-2} f_1(\tau - \gamma r) \\
& + r^{-1} f_2'(\tau + \gamma r) - \gamma^{-1} r^{-2} f_2(\tau + \gamma r) ;
\end{aligned} \tag{1.85}$$

here, f_1 and f_2 are arbitrary functions.

Let us generalize the formula (1.85) for the case of an arbitrary number n (Grigolyuk et al. (1976), Grigolyuk et al. (1979)). First, let us present in brief the Laplace method, which can be applied for this purpose. A complete and detailed description of this method, as well as a historical reference on it we can find in the literature (Darboux (1915), Tricomi (1954), Koshlyakov et al. (1962), Petrova (1974)).

We need to integrate a hyperbolic equation of the second order with respect to the function dependent on two arguments. The second canonical form of such an equation is

$$z_{xy} + a(x, y) z_x + b(x, y) z_y + c(x, y) z = 0, \tag{1.86}$$

where $z(x, y)$ is an unknown function of two arguments x and y, and z_{xy}, z_x, and z_y are the partial derivatives of that function.

Let us introduce the new unknowns

$$z_1 = z_y + az, \qquad z_{-1} = z_x + bz. \tag{1.87}$$

Then, (1.87) can be represented in two equivalent forms:

$$\frac{\partial z_1}{\partial x} + bz_1 - hz = 0, \qquad \frac{\partial z_{-1}}{\partial y} + az_{-1} - kz = 0; \tag{1.88}$$

here, h and k are the invariants of the equations (1.88); these invariants are related to the opportunity to make the multiplicative transformation

$$z(x, y) = \lambda(x, y)\, \zeta(x, y), \tag{1.89}$$

where λ is a multiplier. We have

$$h = a_x + ab + c, \qquad k = a_y + ab - c. \tag{1.90}$$

If $h = 0$ or $k = 0$, the corresponding equation of (1.88) can be interpreted as an ordinary differential equation of the first order. Then, substituting the functions z_1 or z_{-1} into the first or the second equation of (1.88), respectively, we get an equation, which can be integrated as a linear differential equation. However, if $h \not\equiv 0$ or $k \not\equiv 0$, then, eliminating z from (1.86) and (1.88), we obtain the following system of equations:

$$z_{1xy} + a_1 z_{1x} + b_1 z_{1y} + c_1 z_1 = 0,$$
$$z_{-1xy} + a_{-1}z_{-1x} + b_{-1}z_{-1y} + c_{-1}z_{-1} = 0; \tag{1.91}$$

here,

$$a_1 = a - (\ln h)_y, \qquad b_1 = b, \qquad c_1 = a_1 b_1 + b_y - h,$$
$$a_{-1} = a, \qquad b_{-1} = b - (\ln k)_x, \qquad c_{-1} = a_{-1}b_{-1} + a_x - k; \tag{1.92}$$

at that, the invariants of these equations are as follows:

$$h_1 = 2h - k - (\ln h)_{xy}, \qquad k_1 = h,$$
$$h_{-1} = k, \qquad k_{-1} = 2k - h - (\ln k)_{xy}. \tag{1.93}$$

Using the procedure described above for the case $h_1 = 0$ or $k_1 = 0$, we can integrate the system of equations (1.91) in closed form and, hence, integrate the initial equation (1.86) in closed form. In the opposite case, the procedure described can be applied to the system of equations (1.91). Repeating these transformations, we shall generate a sequence of pairs of the equations of type (1.91) and the following two corresponding sequences of the invariants, as well:

$$\ldots, h_{-2}, h_{-1}, h = h_0, h_1, h_2, \ldots ,$$
$$\ldots, k_{-2}, k_{-1}, k = k_0, k_1, k_2, \ldots ; \tag{1.94}$$

for these sequences, the following recursive formulas are valid:

$$h_{m+1} = 2h_m - h_{m-1} - (\ln h_m)_{xy} \qquad k_m = h_{m-1}, \tag{1.95}$$

at that, $h_{-1} = k$ and $h_0 = h$.

Let us now turn our attention to the initial equation (1.84). Carrying out the characteristic change of variables

$$\xi = \tau - \gamma r, \qquad \eta = \tau + \gamma r, \tag{1.96}$$

we get the canonical form of (1.84) (the Euler–Darboux equation)

$$\frac{\partial^2 \psi}{\partial \xi \, \partial \eta} - \frac{1}{\xi - \eta} \frac{\partial \psi}{\partial \xi} + \frac{1}{\xi - \eta} \frac{\partial \psi}{\partial \eta} + \frac{n(n+1)}{(\xi - \eta)^2} \psi = 0, \tag{1.97}$$

where

$$\psi(\xi, \eta) = \Phi(r, \tau), \qquad a_0 = -(\xi - \eta)^{-1}, \qquad b_0 = (\xi - \eta)^{-1},$$
$$c_0 = n(n+1)(\xi - \eta)^{-2}, \qquad k_0 = h_0 = -n(n+1)(\xi - \eta)^{-2}. \tag{1.98}$$

Assuming that

$$h_m = A_m(\xi - \eta)^{-2}, \tag{1.99}$$

where A_m is an arbitrary constant to be determined, and using the recursive relationship (1.95), we obtain the finite difference equation for determination of the sequence of A_m

$$A_{m+1} = 2A_m - A_{m-1} + 2 \tag{1.100}$$

with the initial conditions

$$A_0 = A_{-1} = -n(n+1). \tag{1.101}$$

It can be proven that the solution of this equation is

$$A_m = (m+1) A_0 + m A_{-1} + m(m+1)$$

or, in accounting for (1.101),

$$A_m = m^2 + m - n(n + 1).$$ (1.102)

It follows that $A_m = 0$ (i.e., $h_m = 0$) if $m = n$ or if $m = -n - 1$, that is, at the nth step of the transformation, we obtain an equation which can be integrated in closed form. It can be proven that (Babich et al. (1964))

$$\psi_m(\xi, \eta) = \exp\left(-\int a_m \, d\eta\right)$$
$$\times \left[f_1(\eta) + \int f_2(\eta) \exp\left(\int a_m \, d\eta - \int b_m \, d\xi\right) d\eta\right].$$ (1.103)

Let us find the explicit formulas for the coefficients a_m and b_m of the mth equation.

Generalizing the formulas (1.92), we have

$$a_{m+1} = a_m - \frac{\partial}{\partial \eta}(\ln h_m), \qquad b_{m+1} = b_m = b_0 = b.$$ (1.104)

In accounting for (1.98), we get

$$a_{m+1} = a_m - 2(\xi - \eta)^{-1}.$$ (1.105)

However, since $a_0 = a$, we set

$$a_m = B_m(\xi - \eta)^{-1}$$ (1.106)

and, consequently, get the recursive formula for B_m

$$B_{m+1} = B_m - 2$$ (1.107)

with the initial condition $B_0 = -1$.

It is evident that the solution of this equation is

$$B_m = -(2m + 1).$$ (1.108)

Thus,

$$a_m = -(2m + 1)(\xi - \eta)^{-1}.$$ (1.109)

Substituting the values of a_m and b_m into (1.103), we arrive at

$$\psi_n(\xi, \eta) = (\xi - \eta)^{-2n-1}\left[f_1(\xi) + \int f_2(\eta)(\xi - \eta)^{2n} \, d\eta\right].$$ (1.110)

Since $h_m \neq 0$ for $m < n$, then, generalizing the formulas (1.87), we get

$$\psi_{m-1} = h_{m-1}^{-1}\left(\frac{\partial \psi_m}{\partial \xi} + b_{m-1}\psi_m\right);$$

substituting the values of h_{m-1} and b_{m-1} in this relationship, we arrive at

$$\psi_{m-1} = \frac{1}{A_{m-1}(\xi - \eta)^2} \left(\frac{\partial \psi_m}{\partial \xi} + \frac{\psi_m}{\xi - \eta} \right), \qquad m = 0, 1, \ldots, n - 1.$$

$$(1.111)$$

It is evident that (1.111) makes a recursive relationship with the initial condition (1.110), implying that we can obtain the function sought $\psi(\xi, \eta) = \psi_0(\xi, \eta)$ via the arbitrary functions $f_1(\xi)$ and $f_2(\eta)$. Omitting sophisticated mathematical manipulations, we finally get

$$\psi(\xi, \eta) = \sum_{k=0}^{n} (-1)^k \frac{A_{nk} n! \, 2^k}{(\xi - \eta)^{k+1}} f_1^{(n-k)}(\xi) + \sum_{k=0}^{n} \frac{A_{nk} n! \, 2^k}{(\xi - \eta)^{k+1}} f_3^{(n-k)}(\eta),$$

$$f_3^{(2n+1)}(\eta) = f_2(\eta), \qquad\qquad (1.112)$$

where

$$A_{nk} = \frac{(n + k)!}{(n - k)! \, k! \, 2^k}, \qquad 0 \le k \le n,$$

$$A_{nk} = 0, \qquad k < 0, \quad k > n. \qquad\qquad (1.113)$$

Switching to the old variables r and τ and applying the formulas (1.96), we obtain

$$\Phi(r, \tau) = \frac{1}{r^{n+1}} \sum_{k=0}^{n} (\gamma r)^{n-k} A_{nk} g_1^{(n-k)}(\tau - \gamma r)$$

$$+ \frac{(-1)^n}{r^{n+1}} \sum_{k=0}^{n} (-\gamma r)^{n-k} A_{nk} g_2^{(n-k)}(\tau + \gamma r), \qquad (1.114)$$

where

$$g_1(\xi) = -\frac{f_1(\xi) n!}{2\gamma}, \qquad g_2(\eta) = -\frac{f_3(\eta) n!}{2\gamma}.$$

The validity of the formula (1.114) can be also proven by the direct substitution of the function $\Phi(r, \tau)$ into the original equation (1.84). For $n = 1$, except for the notation, the expression (1.114) gives us a formula which is identical to (1.85). For $n = 0$, we obtain the known formula for the spherical waves

$$\Phi(r, \tau) = r^{-1} g_1(\tau - \gamma r) + r^{-1} g_2(\tau + \gamma r). \qquad (1.115)$$

A shorter way to derive the formula (1.114) is as follows. Let us introduce a new function χ such that

$$\psi(\xi, \eta) = (\xi - \eta)^{\alpha} \chi(\xi, \eta) \,. \tag{1.116}$$

Let us choose the parameter α such that (1.97) will transform into the Euler–Darboux equation in $\chi(\xi, \eta)$. Substituting (1.116) into (1.97), we obtain

$$\chi_{\xi\eta} - \frac{1 + \alpha}{\xi - \eta} \chi_{\xi} + \frac{1 + \alpha}{\xi - \eta} \chi_{\eta} - \left[\alpha^2 + \alpha - n(n+1)\right] \frac{\chi}{(\xi - \eta)^2} = 0 \,.$$

If α is an arbitrary solution of the quadratic equation

$$\alpha^2 + \alpha - n(n+1) = 0 \,, \tag{1.117}$$

then the function χ will satisfy the Euler–Darboux equation. Assuming $\alpha = n$, we get (Krylov, Skoblya (1974))

$$\chi_{\xi\eta} - \frac{n+1}{\xi - \eta} \chi_{\xi} + \frac{n+1}{\xi - \eta} \chi_{\eta} = 0 \,. \tag{1.118}$$

We can write a general integral of this equation (Koshlyakov et al. (1962)):

$$\chi(\xi, \eta) = \frac{\partial^{2n}}{\partial \xi^n \partial \eta^n} \left[\frac{f_1(\xi) - f_2(\eta)}{\xi - \eta}\right] \,;$$

here, f_1 and f_2 are arbitrary functions. Then, we obtain the general integral of (1.115):

$$\psi(\xi, \eta) = (\xi - \eta)^n \frac{\partial^{2n}}{\partial \xi^n \partial \eta^n} \left[\frac{f_1(\xi) - f_2(\eta)}{\xi - \eta}\right] \,. \tag{1.119}$$

The last formula can be transformed by the following manner:

$$\psi(\xi, \eta) = (\xi - \eta)^n \left\{ \frac{\partial^n}{\partial \xi^n} \left[\frac{\partial^n}{\partial \eta^n} \left(\frac{f_1(\xi)}{\xi - \eta}\right)\right] - \frac{\partial^n}{\partial \eta^n} \left[\frac{\partial^n}{\partial \xi^n} \left(\frac{f_2(\eta)}{\xi - \eta}\right)\right] \right\}$$

$$= (\xi - \eta)^n n! \left\{ \frac{\partial^n}{\partial \xi^n} \left[\frac{f_1(\xi)}{(\xi - \eta)^{n+1}}\right] - (-1)^n \frac{\partial^n}{\partial \eta^n} \left[\frac{f_2(\eta)}{(\xi - \eta)^{n+1}}\right] \right\} \,. \tag{1.120}$$

Using the Leibniz formula for the derivative of the nth order, we get (1.114); at that, g_1 and g_2 are new arbitrary functions, which are proportional to f_1 and f_2 and are determined by the relationships (1.114).

Thus, the formula (1.114) gives us a general solution of (1.84) expressed via two arbitrary functions $g_1(\tau - \gamma r)$ and $g_2(\tau + \gamma r)$ corresponding to diverging and converging generalized spherical waves, respectively.

In order to compare the solution (1.114) with the formula for diverging waves presented in Love (1905), let us make some manipulations of (1.119). Taking into consideration the formulas

$$\frac{\partial}{\partial \xi} = -\frac{1}{2\gamma} \frac{\partial}{\partial r} + \frac{1}{2} \frac{\partial}{\partial \tau} \,, \qquad \frac{\partial}{\partial \eta} = \frac{1}{2\gamma} \frac{\partial}{\partial r} + \frac{1}{2} \frac{\partial}{\partial \tau} \,, \tag{1.121}$$

we obtain

$$
\begin{aligned}
\frac{\partial^2}{\partial \xi \partial \eta} \left[\frac{f_1(\xi)}{\xi - \eta} \right] &= \frac{\partial}{\partial \xi} \left[\frac{f_1(\tau - \gamma r)}{4\gamma^2 r^2} \right] \\
&= -\frac{1}{2\gamma} \frac{\partial}{\partial r} \left[\frac{f_1(\tau - \gamma r)}{4\gamma^2 r^2} \right] + \frac{1}{8\gamma^2 r^2} \frac{\partial}{\partial r} \left[f_1(\tau - \gamma r) \right] \\
&= -\frac{1}{8\gamma^3} \left\{ \frac{\partial}{\partial r} \left[\frac{f_1(\tau - \gamma r)}{r^2} \right] + \frac{1}{r^2} \frac{\partial}{\partial r} \left[f_1(\tau - \gamma r) \right] \right\} \\
&= -\frac{1}{4\gamma^3 r} \frac{\partial}{\partial r} \left[\frac{f_1(\tau - \gamma r)}{r} \right].
\end{aligned} \tag{1.122}
$$

Similarly, we have

$$
\frac{\partial^2}{\partial \xi \partial \eta} \left[\frac{f_2(\eta)}{\xi - \eta} \right] = \frac{1}{4\gamma^3 r} \frac{\partial}{\partial r} \left[\frac{f_2(\tau + \gamma r)}{r} \right].
$$

Substituting the expressions obtained into (1.119), we finally get

$$
\Phi(r, \tau) = \frac{r^n}{(2\gamma^2)^n} \left(\frac{1}{r} \frac{\partial}{\partial r} \right)^n \left[\frac{f_1(\tau - \gamma r)}{r} - (-1)^n \frac{f_2(\tau + \gamma r)}{r} \right]. \tag{1.123}
$$

Except for the notation, the first part of (1.123), which corresponds to diverging waves, coincides with the general integral presented in Love (1905).

The sums in (1.114) are related to the Bessel functions of imaginary argument. To prove this, we shall use the Laplace integral transform with respect to time τ. Assuming that the function $\Phi(r, \tau)$ satisfies the zero initial conditions, we obtain the following analogy to (1.84) in the space of the Laplace transforms:

$$
\frac{\partial^2 \Phi^L}{\partial r^2} + \frac{2}{r} \frac{\partial \Phi^L}{\partial r} - \left[\frac{n(n+1)}{r^2} + \gamma^2 s^2 \right] \Phi^L = 0; \tag{1.124}
$$

here, the superscript L designates the transform and s is the parameter of transformation.

A general solution of (1.124) has the following form (Watson (1945), Kuznetsov (1965)):

$$
\Phi^L(r, s) = \frac{1}{\sqrt{r}} \left[f_1^L(s) I_{n+1/2}(\gamma r s) + f_2^L(s) K_{n+1/2}(\gamma r s) \right]; \tag{1.125}
$$

here, $I_\nu(x)$ and $K_\nu(x)$ are the modified Bessel functions, $f_1^{(L)}(s)$ and $f_2^{(L)}(s)$ are some arbitrary functions. When $\nu = n+1/2$, we get the following formulas (Watson (1945), Gradshtein, Ryzhik (1971), Abramowitz, Stegun (1965)):

$$I_{n+1/2}(z) = \frac{(-1)^n}{\sqrt{2\pi}\, z^{n+1/2}} \left[e^z R_{n0}(-z) - e^{-z} R_{n0}(z) \right],$$

$$K_{n+1/2}(z) = \sqrt{\frac{\pi}{2}} \frac{e^{-z}}{z^{n+1/2}} R_{n0}(z),$$

$$R_{n0}(z) = \sum_{k=0}^{n} A_{nk} z^{n-k}. \tag{1.126}$$

As a result, we obtain

$$\Phi^{\mathrm{L}}(r,\, s) = \frac{(-1)^n}{r^{n+1}} \sum_{k=0}^{n} (-1)^{n-k} (\gamma r s)^{n-k} A_{nk} g_1^{\mathrm{L}}(s)\, e^{\gamma r s}$$

$$+ \sum_{k=0}^{n} (\gamma r s)^{n-k} A_{nk} g_2^{\mathrm{L}}(s)\, e^{-\gamma r s}, \tag{1.127}$$

where

$$g_1^{\mathrm{L}}(s) = \frac{(-1)^{n+1} f_1^{\mathrm{L}}(s)}{\gamma^n s^n \sqrt{2\pi\gamma s}} + \frac{f_2^{\mathrm{L}}(s)\sqrt{\pi}}{\gamma^n s^n \sqrt{2\gamma s}}, \qquad g_2^{\mathrm{L}}(s) = \frac{f_1^{\mathrm{L}}(s)}{\gamma^n s^n \sqrt{2\pi\gamma s}}.$$

Applying the inverse Laplace transform to (1.127) and using the general properties of integral transforms, we get the relationships (1.114) again.

In Smirnov (1937a) and Smirnov (1937b), the general solution of the wave equation represented in the spherical coordinates

$$\gamma^2 \frac{\partial^2 \varphi}{\partial \tau^2} = \Delta \varphi \tag{1.128}$$

has been obtained for diverging waves; except for the notation, this solution is

$$\varphi(r,\, \theta,\, \vartheta,\, \tau) = \sum_{n=0}^{\infty} Y_n(\theta,\, \vartheta) \int_0^{\tau-\gamma r} \omega_n(\xi) Q_{n+1}\left(\frac{\tau-\xi}{\gamma r}\right) d\xi,$$

$$Q_{n+1}(x) = \int_1^x P_n(x)\, dx, \tag{1.129}$$

where $Y_n(\theta,\, \vartheta)$ is the spherical function of order n, $P_n(x)$ is the Legendre polynomial, and $\omega_n(x)$ is an arbitrary function.

Then, taking into account the axial symmetry of the problem, we can write (1.84) as follows:

$$\Phi(r,\, \tau) = \int_0^{\tau-\gamma r} \omega_n(\xi) Q_{n+1}\left(\frac{\tau-\xi}{\gamma r}\right) d\xi. \tag{1.130}$$

Let us introduce the new function $h^{(n+2)}(\xi) = \omega_n(\xi)$ and assume that

$$h^{(j)}(\xi) \equiv 0, \qquad \xi \le 0, \qquad j = 0,\, 1,\, \dots,\, n+1. \tag{1.131}$$

Integrating (1.130) $n + 2$ times by parts, we obtain

$$\Phi(r, \tau) = h^{(n+1)}(\xi)\, Q_{n+1} \left(\frac{\tau - \xi}{\gamma r} \right) \Big|_0^{\tau - \gamma r}$$

$$+ \frac{1}{\gamma r}\, h^{(n)}(\xi)\, P_n \left(\frac{\tau - \xi}{\gamma r} \right) \Big|_0^{\tau - \gamma r}$$

$$+ \ldots + \frac{1}{(\gamma r)^{n+1}}\, h(\xi)\, P_n^{(n)} \left(\frac{\tau - \xi}{\gamma r} \right) \Big|_0^{\tau - \gamma r}$$

$$= \sum_{k=0}^{n} \frac{h^{(n-k)}(\tau - \gamma r)}{\gamma^{k+1} r^{k+1}}\, P_n^{(k)}(1) . \tag{1.132}$$

Taking into consideration the relationship (Watson (1945))

$$P_n^{(k)}(1) = \frac{\mathrm{d}^k P_n(1)}{\mathrm{d}x^k} = \frac{(2k-1)!!\,(n+k)!}{(2k)!\,(n-k)!} = A_{nk} \tag{1.133}$$

and designating $g_1 = \gamma^{-n-k} h$, we get the formula (1.114) once again.

For the converging and diverging waves in the spherical coordinate system, a general solution of the wave equation for a three-dimensional case has been derived in Mindlin (1940a) and Mindlin (1940b) in the form similar to (1.114); these results have been generalized to the case of n-dimensional space (Mindlin (1947)). A solution similar to (1.114) has been also derived later (Akkas (1977), Dikasov (1979), Dikasov (1982)).

2. Radial Vibrations of Media with Spherical Interfaces

Statement and solution of problems on radial (centrally symmetric) vibrations of the elastic media with spherical interfaces (internal boundaries) can be considered to be a particular case of axially symmetric problems, which we analyze in Chap. 3. Nevertheless, it is useful to study the problems on radial vibrations preliminarily, before making analysis of the more general cases. This can be explained, first, from a historical point of view, since the studies of different problems of dynamics of elastic spherical bodies, as a rule, were always started by the studies of radial vibrations. Secondly, an opportunity to derive the sufficiently simple solutions of the centrally symmetric problems allows us to reveal some useful qualitative effects that take place in the more general cases. Physically, a main simplification consists in the absence of shear waves; mathematically, the number of unknown functions is lower.

The simplest problem of the nonstationary theory of elasticity is a problem on propagation of disturbances from a cavity in an infinite elastic media. The simplicity of this problem solution, as compared to the other dynamic problems, is conditioned by the presence of a system of diverging waves only.

Making analysis of the publications available, it can be noticed that a high number of them is devoted to studies of radial expansion of a spherical cavity in an elastic medium. This problem was analyzed in many books (e.g., Broberg (1956), Hopkins (1960), Nowacki (1970), Kokhmanyuk et al. (1980), Slepyan (1972), Filippov, Egorychev (1977), Kubenko (1979)).

The problem was first solved in Jeffreys (1931). The author analyzed a particular case of the elastic medium, when the Lamé elastic constants were assumed to be equal in value, $\lambda = \mu$. It was also assumed that a uniform pressure was applied to the surface of cavity. The same solution is presented in Jeffreys (1959) and Jeffreys, Swirles (1966). The problem was also solved in Sharpe (1942). The author applied a double Fourier transform with respect to radius r and time t, made inverse transformation using the theory of residue functions, and presented a comparison of the results obtained and the experimental data. For this reason, in many publications the problem of radial expansion of spherical cavity in an elastic medium is named the *Sharpe problem*.

In the case when $\lambda \neq \mu$ and the pressure at the cavity's surface is $p(t) = \exp(-\alpha t) H(t)$, where $H(t)$ is the Heaviside function, the Sharpe problem was solved by employing the Laplace transform (Selberg (1952)) and the Fourier transform with respect to time (Blake (1952). A similar problem for the stepwise and exponential laws of variation of pressure in time was solved in late 1930s (Kawasumi, Yosiyama (1935), Sezawa (1935), Sezawa, Kanai (1936)).

The similar solution and a great body of numerical results and curves were presented in Allen, Goldsmith (1955) and Goldsmith, Allen (1955). A solution of this problem was also presented in Sabodash, Tsurpal (1969).

Besides the methods of integral transforms mentioned above, the ray method (Podilchuk, Rubtsov (1979)) and the representation of the solution in the form of diverging spherical waves (Achenbach (1973)) were applied.

The results of investigation of the influence of the boundary conditions at a cavity's surface, as well as of a shape of pressure pulse, were presented in Vaněk (1955), Chakraborty (1958), Vodicka (1963), Gaek, Shumilo (1965), Bhattacharya (1969), Matsumoto, Nakahara (1971), Plakhotnyi (1971), and Ponomarev (1971). An analysis of energy spectrum in the case of the problem under consideration was presented in Baranov, Gryazev (1979).

When the pressure variation at the cavity's surface is predetermined in the form of the Dirac delta function with respect to time, the solution of the problem was derived in Das Gupta (1954). An analysis of the similar problems, when the displacements are predetermined in the form of the Dirac delta functions, was presented in Verma (1957). The shear waves in the case of centrally symmetric vibrations were analyzed in Pec (1957), but these cannot exist in that case. When the velocity of the boundary of spherical cavity in acoustic space is prescribed, the problem was analyzed in Gladkov (1977) and Stepanischen (1973).

An experimental solution of the problem under consideration for the elastic medium was presented in Kirillov (1947) and Fadeev (1968). The problems of real conditions after explosion in a continuous medium and related problems of determination of the cavity's radius and the loads at the cavity's surface were studied in Morris (1950), Holzer (1966), Batalov, Svidinsky (1971), Kostyuchenko, Rodionov (1974), Rodean (1971), and Rodean (1971).

A number of works is devoted to studies of propagation of elastic waves from spherical cavity under complicated boundary conditions. The presence of a thin-walled elastic shell stiffly joined with an elastic or acoustic medium does not make any principal difficulties. Such problems were studied in Glenn, Kidder (1983), Forrestal, Sagartz (1971), Duffey, Johnson (1981), and Bedrosian, Dimaggio (1972).

In order to match better the results of solution and the real conditions of explosion, the authors of Konovalov (1970) and Kasiak et al. (1977) assumed that the pressure in a cavity is proportional to the volume of products of detonation. At that, they applied a general integral of the wave equation and got a governing equation which turned out to be a nonlinear ordinary differential equation of the second order to be solved numerically.

The case when the boundary conditions are prescribed at the surface of a cavity, and this surface moves following a given law regardless of deformations, was considered in Yanyutin (1980) and Achenbach, Sun (1966). In the first paper, the problem was reduced to the Volterra equation; in the second one, the problem was reduced to an ordinary differential equation with variable coefficients. The numerical results were presented for an evident case when the radius of cavity varies linearly with time.

When studying the problems of radial vibrations, a presence of internal boundaries complicates the wave pattern since, besides the diverging waves, it is necessary to take into account a system of converging waves.

Similarly to other, more complicated problems, the numerical methods turn out to be universal, as they can be applied in the case of interfaces of different shape. Thus, in Chou, Koenig (1966) and Rose et al. (1973), the method of characteristics was applied to the studies of transient radial vibrations of a thick-walled sphere and a cylinder under a given load. The so-called *discrete stepwise method* (in fact, the finite element method) was applied for the same purposes in Mehta, Davids (1966).

When solving the problems of dynamics of spherically shaped or cylindrically shaped bodies, we can consider the method of finite integral transforms to be a general one (Slepyan, Yakovlev (1980)). This method was employed for studying radial vibrations of elastic spheres and cylinders (Cinelli (1966), McKinney (1971), Senitsky (1971), Senitsky, Syromyatnikova (1976)).

The first works devoted to solution of the problem of transient radial vibrations of solid spheres were based on application of the Fourier method of separation of variables. Thus, in Clebsch (1862), this method was used to derive the formulas for determination of components of stress–strain state of a solid elastic sphere when the centrally symmetric initial conditions were prescribed but no surface loads were applied to. A similar method was used in the case of a hollow sphere Baker, Allen (1958) and Courbon (1971). A detailed study of radial vibrations of a hollow elastic sphere subjected to nonstationary pressure applied at the internal surface is presented in Tranter (1942). The author employed the Laplace transform and determined its inversion by means of residue functions. A similar problem in the case when the pressure is applied at the inner and outer surfaces was solved using the same method in Matsumoto, Nakahara (1971) and Suzuki (1967). The same problem was studied by the method of complete separation of variables in Brodacki (1965). Let us note that the method of inversion of Laplace transform by means of residue functions and the method of complete separation of variables are practically equivalent. In both cases, it is necessary to solve the sufficiently complicated transcendental equations, the roots of which are the eigenfrequencies of the system.

Another effective method to solve the problems of radial vibrations of spherically shaped bodies is the method of summation of elementary waves. This method was used for different boundary conditions at the internal surface (stiff restraint and free surface) in Wheeler (1973), Fridman (1976b), Kasiak, Włodarczyk (1975), and Kasiak, Włodarczyk (1980); at that, in the last work, a non-hollow sphere was considered. At the sphere's center $r = 0$, a limiting condition for the components of the stress–strain state was stated. In Slepyan (1972), solving the similar problem, the author employed a series expansion of the Laplace transforms in terms of the exponential functions; this method is equivalent to the one mentioned above. For one-dimensional problems, similar expansions were applied, for example, in Lurie (1950).

The problem of transient radial vibrations of a thick-walled elastic or acoustic sphere was solved in Babaev (1981), Kokhmanyuk et al. (1980), and Yanyutin, Titarev (1979) by application of the inversion of the Laplace transform with respect to time by means of the Volterra integral equations; the cases of immovable and movable boundaries were considered. A comparison of the results obtained when applying the linear theory of elasticity and the theory of thick-walled shells was presented in Shchipitsina (1972).

Now, let us review the works devoted to propagation of radial elastic waves from a spherical cavity stiffly joined to a thick-walled shell. The problem of radial vibrations of a thick-walled sphere immersed into an infinite elastic medium, in the case when a uniform nonstationary pressure is applied to the shell's internal surface, was studied by means of the Laplace transform with respect to time in Datta, Sengupta (1984), Reismann, Gideon (1971), and Sengupta, Roy (1971). A limiting case of the problem of propagation of waves from a cavity was considered. In the first of these works, a variant of random loads was analyzed. An exact solution of this problem in the case of deterministic loads was presented in Grigolyuk et al. (1977a) and Tarlakovsky (1975). The method was based on application of the general integral of the wave equation and representation of the solution in the form of a superposition of the waves reflected from the interfaces. Then, the Laplace transform was used; the algorithm of inversion of the Laplace transform allowed an exact determination of the originals for an arbitrary number of wave.

Solution of a similar problem by means of inversion of the Laplace transform using the method of expansion of the transform into a series was presented in Moskalenko (1965) and Nayfeh (1979). A study of the influence of boundary conditions at the internal surface of a thick-walled shell on the system vibrations was presented in Galazyuk, Gorechko (1980b) and Podstrugach et al. (1981). The authors examined two variants: a free surface and an absolutely stiff core. In the first of these works, it was assumed that the material of a thick-walled sphere was an acoustic medium.

Filling of the cavity in a thick-walled shell with a continuous medium can be regarded to be a particular case of the cavity support. Thus, in

Kheisin (1967), the author studied the problem of radial vibration of a continuous elastic sphere in an acoustic medium under the given initial conditions. A similar problem for the case of an elastic space in accounting for the mass forces was analyzed in Kamen, Ostapenko (1982a). The solution obtained did not take into account the waves reflected from the center of the core.

In the elastic systems under consideration, the disturbances can be produced by a point source of spherical wave placed in the medium, which fills the reservoir. In the case of radial vibrations, the source coincides with the center of spherical interfaces. The first stage of approximate solution of these type of problems is to make an assumption of rigidity of the reservoir's walls. In Babaev (1974a), the author derived a solution of this problem for the case of acoustic filler applying the Laplace transform with respect to time and making a series expansion of the transforms in terms of the exponential functions. A problem of propagation of the centrally symmetric waves from a source located at the center of an elastic sphere placed in an infinite acoustic medium was considered in Shaw (1973). The solution was derived in integral form. The engineering methods for analysis of spherical blasting chambers were presented in Demchuk (1968) and Maltsev et al. (1985).

The methods of solution applied for analysis of the two-layer media when one of the layers is infinite can also be used for analysis of the media both of which are bounded (thick-walled shells, shells filled with continuous deformable media, etc.).

In Chou, Greif (1968), the method of characteristics was applied for analysis of transient radial vibrations of thick-walled elastic spherical or cylindrical shells nested into each other. Representation of the solution in the form of superposition of elementary spherical waves was applied for analysis of a similar problem in Kasiak, Włodarczyk (1976). However, the algorithm developed can be realized in practice for the initial instants only.

The dynamics of two thick-walled spherical shells separated by a clearance was studied in Yanyutin, Titarev (1982). At that, the authors solved sequentially two problems: determining the system vibrations, when the clearance is nonzero, and determining the system motion, when the clearance disappears. In Bhaduri, Kanoria (1982), a problem of forced radial vibrations of a thick-walled shell was solved by means of the Laplace transform with respect to time followed by inversion by residue functions.

When the number of interfaces is relatively high, $n \geq 2$, the methods mentioned above are not always efficient. Thus, in Kamen, Ostapenko (1982b) and Filippov, Bakhramov (1978), the authors proposed to apply the method of summation of elementary waves for studying dynamics of multilayer spherical bodies. However, a practical implementation of the solution becomes possible only when a relatively small number of elementary waves is considered. The expressions for amplitude of discontinuity at the fronts of the first stress waves refracted by the surfaces of contacts were derived in Kheinloo (1979).

Among other methods of solution for the dynamic problems of multilayer media, we can distinguish the following ones:

(1) the method of finite integral transforms (Wankhede, Bhonsle (1980));

(2) the inversion of the Laplace transform by its reduction to the integral equations (Kokhmanyuk et al. (1980), Titarev (1984));

(3) determination of the inverse integrals of the Fourier transforms and of the Laplace transforms using the numerical techniques (Biryulya (1983), Poddubnyak et al. (1985));

(4) the method of characteristics and its modifications (Lepikhin (1981), Yang, Achenbach (1970), Kornilov, Lepikhin (1982));

(5) application of equations of the theory of elasticity and equations of the acoustic medium in the divergent form followed by an application of the Godunov shock-capturing method (Panichkin (1984)).

In Gorshkov, Tarlakovsky (1981), Gorshkov, Tarlakovsky (1983a), and Gorshkov, Tarlakovsky (1983b), the authors reduced the problem of propagation of elastic waves in a piecewise homogeneous medium with a spherical symmetry to a system of integral equations of convolution type applying the influence functions for each layer. A system of differential-difference equations with a retarded argument intended for study of dynamics of a multilayer inhomogeneous hollow sphere was derived in Molodtsov (1981); free vibrations of a thick-walled shell were analyzed.

In Efimova, Stepanenko (1984), the authors applied the finite difference numerical method for solution of the problem of radial vibrations of a system which consisted of the infinitely elastic medium, thin-walled shell, and acoustic medium. A behavior of the systems of concentric thin-walled spherical shells separated by acoustic or elastic media was studied in Babaev (1983), Babaev (1984), and Babaev et al. (1983). The outer shell was surrounded by an acoustic media and the internal one was filled with a similar material. The authors reduced the problem to a system of Volterra integral equations with respect to the radial displacements of the shells.

2.1 Propagation of Disturbances from a Cavity

Let us consider a classical problem when the pressure applied to the surface of a cavity is uniformly distributed over this surface and varies in time as a step function, that is, $p(\tau) = p_0 H(\tau)$ (see Fig. 2.1).

The problem is axially symmetric, consequently, $\psi = 0$ and the functions sought depend only on r and τ. The cavity's radius b is assumed to be a characteristic linear dimension in (1.11). Then, taking into account (1.12)–(1.14), we obtain the following boundary value problem:

$$\gamma^2 \frac{\partial^2 \varphi}{\partial \tau^2} = \Delta \varphi,$$

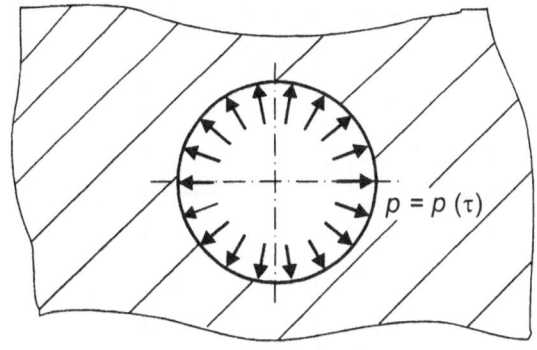

Fig. 2.1. Spherically symmetric problem of propagation of waves from a cavity

$$\Delta = \frac{1}{r^2} \frac{\partial}{\partial r} \left(r^2 \frac{\partial}{\partial r} \right),$$

$$\sigma_{rr}|_{r=1} = -p(\tau),$$

$$u = \frac{\partial \varphi}{\partial r},$$

$$\sigma_{rr} = \beta \left(\frac{\partial u}{\partial r} + 2\kappa \frac{u}{r} \right), \qquad \sigma_{\theta\theta} = \kappa \sigma_{rr} + (1 - \kappa)(1 + 2\kappa) \frac{u}{r}. \qquad (2.1)$$

All the other components of the stress–strain state are equal to zero. It is assumed that as $r \to \infty$, the disturbances are absent.

Assuming the absence of disturbances at the initial time $\tau = 0$, we complete the relationships (2.1) by the initial conditions

$$\varphi|_{\tau=0} = \dot{\varphi}|_{\tau=0} = 0, \qquad (2.2)$$

where dots are used to designate differentiation with respect to time τ.

According to the results of Sect. 1.5, we can write the solution of the problem (2.1)–(2.2) in the form of the diverging wave

$$\varphi(r, \tau) = \frac{1}{r} g[\tau - \gamma(r - 1)] H[\tau - \gamma(r - 1)], \qquad (2.3)$$

where $g(x)$ is an arbitrary function. Under such a choice, the conditions at infinity are satisfied immediately.

If it is assumed $g(0) = \dot{g}(0) = 0$, the initial conditions (2.2) are satisfied as well.

The displacement u and the stress σ_{rr} can be represented via an arbitrary function g as follows:

$$u = -\frac{1}{r^2} [g(\tau - \gamma r + \gamma) + \gamma r g'(\tau - \gamma r + \gamma)] H(\tau - \gamma r + \gamma),$$

$$\sigma_{rr} = \frac{\beta}{r^3} [\gamma^2 r^2 g''(\tau - \gamma r + \gamma) + 2(1 - \kappa) \gamma r g'(\tau - \gamma r + \gamma)$$

$$+ 2(1 - \kappa) g(\tau - \gamma r + \gamma)] H(\tau - \gamma r + \gamma). \qquad (2.4)$$

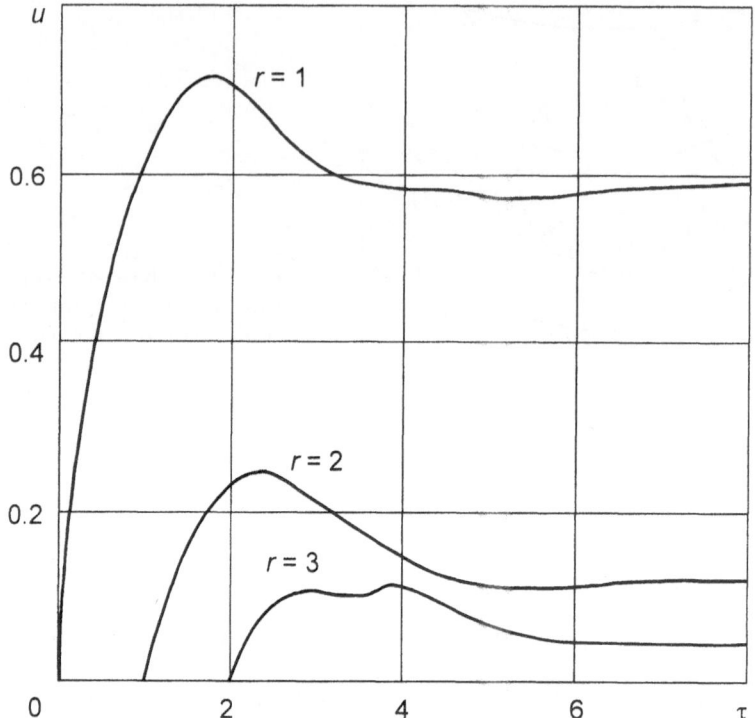

Fig. 2.2. Propagation of radial disturbances from a cavity. The displacements at different points of the elastic medium

Taking into account the boundary condition at the cavity's surface $r = 1$, we obtain the following initial value Cauchy problem for the function g:

$$\gamma^2 g''(\tau) + 2\left(1 - \kappa\right) \gamma\, g'(\tau) + 2\left(1 - \kappa\right) g(\tau) = -\frac{p_0}{\beta}\,,$$

$$g(0) = \dot{g}(0) = 0\,. \tag{2.5}$$

The corresponding characteristic equation is as follows:

$$\gamma^2 s^2 + 2\left(1 - \kappa\right)\gamma s + 2\left(1 - \kappa\right) = 0\,.$$

Because $1 - \kappa > 0$, its roots are equal to

$$s_{1,2} = -\alpha \pm i\omega\,, \qquad \alpha = \frac{1 - \kappa}{\gamma}\,, \qquad \omega = \frac{1}{\gamma}\sqrt{1 - \kappa^2}\,.$$

As a result, we can write the solution of the Cauchy problem (2.5) in the form

$$g(\tau) = \frac{p_0}{2\beta(1 - \kappa)}\left[e^{-\alpha\tau}\left(\cos\omega\tau + \frac{\alpha}{\omega}\sin\omega\tau\right) - 1 + \kappa\right]\,. \tag{2.6}$$

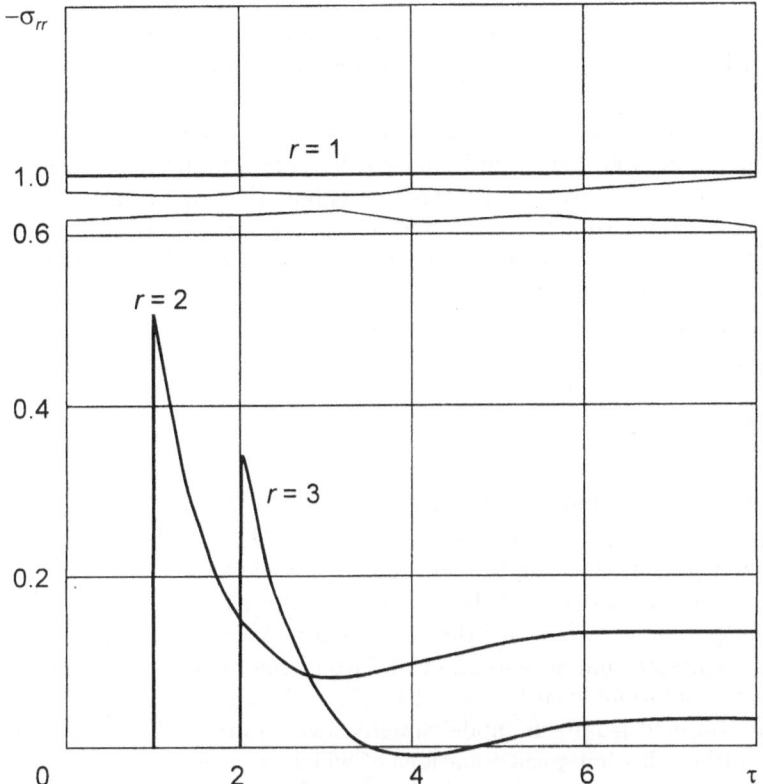

Fig. 2.3. Propagation of radial disturbances from a cavity. The radial stresses at different points of the elastic medium

Finally, substituting (2.6) into (2.4), it is easy to derive explicit expressions for the displacements and stresses.

Similarly, we can obtain a solution of the problem in the case when the displacement $u = u_0(\tau)$ is set at the boundary, and under other variants of the boundary conditions.

If a spherical cavity is placed into an acoustic medium, then, as it was mentioned in Chap. 1, it is necessary to set $\kappa = 1$. From the formulas (2.4) and (2.5), we obtain

$$g''(\tau) = -\frac{p_0}{\beta\gamma^2}, \qquad p = -\sigma_{rr} = -\frac{p_0}{r} H(\tau - \gamma r + \gamma), \qquad (2.7)$$

where p is the pressure in the acoustic medium.

Let us consider a cavity in granite with $\kappa = 0.143$. The radial displacements u and the stresses σ_{rr} versus time τ are presented in Figs. 2.2 and 2.3 for $r = 1, 2, 3$. The value $r = 1$ corresponds to the cavity's surface. The dimensionless parameters are chosen such that $\gamma = 1$, $\beta = 1$, and $p_0 = 1$. The jumps in the radial stress correspond to the instants when the elastic wave arrives at a point of observation. The curves illustrate clearly how the wave is dampened with an increase in the distance between the point of observation and the cavity's surface.

Let us now consider a problem of spherical expansion of a spherical cavity with the complicated boundary conditions which approximate the conditions in the problem of an explosion in an elastic medium (Kasiak et al. (1977)). In that case, we assume that the gas inside the spherical cavity is polytropic and the pressure at the surface of contact of the gas and the elastic medium can be described by the following law:

$$p(\tau) = p_0 \left[\frac{V_0}{V(\tau)} \right]^n H(\tau), \qquad V_0 = \frac{4}{3}\pi,$$

$$V = V_0 \left[1 + u(1, \tau) \right]^3 ; \tag{2.8}$$

here, V_0 is the cavity's volume at the time $\tau = 0$, $V(\tau)$ is the volume of the region occupied by the gas at an arbitrary instant τ ($V(0) = V_0$), and n is the polytropic exponent; u is the displacement of the elastic medium.

The mathematical statement of the problem remains similar to that in the previous case (2.1), unless we regard the pressure $p(\tau)$ as the function defined by (2.8). Let us notice that in Kasiak et al. (1977) the boundary condition was stated at the deformable surface $r = 1 + u(1, \tau)$. However, the refinement of the results obtained from this model should have an order of the error of the linear theory of elasticity, the equations of which were applied.

Searching again for the potential $\varphi(r, \tau)$ in the form of diverging wave (2.3) and taking into account (2.8) and the boundary conditions (2.1), we obtain the nonlinear equation with respect to the function $g(\tau)$:

$$[1 - \gamma g'(\tau) + g(\tau)]^{3n} [\gamma^2 g''(\tau) + 2(1 - \kappa)\gamma g'(\tau) + 2(1 - \kappa) g(\tau)] = -\frac{p_0}{\beta},$$

$$g(0) = \dot{g}(0) = 0. \tag{2.9}$$

This equation was solved numerically in Kasiak et al. (1977). Let us notice that if the exponent of polytrope $n = 0$, the relationship (2.9) transforms into (2.5).

In Figs. 2.4 and 2.5, we present some numerical results obtained in Kasiak et al. (1977). Here, the dimensionless radial stress σ_{rr} and the tangential stress $\sigma_{\theta\theta}$ in an elastic medium at the distance $r = 10$ from the cavity's surface are plotted versus the time parameter $\tau_1 = \tau - r$. The parameters of the material, a calc spar, are: Young's modulus $E = 0.89 \cdot 10^4$ MPa, Poisson's ratio $\nu = 0.243$, and density $\rho = 2.85 \cdot 10^3$ kg/m^3. At that, the dimensionless parameter $\kappa = 0.321$. The units are chosen in such a manner that $\beta = 1$, $\gamma = 1$, and $p_0 = 1$. Each of the figures presents three curves corresponding to different polytropic exponents of the gas in the cavity, $n = 2, 3, 4$. At that, when the exponent n decreases, the maximum values of the tensile stress

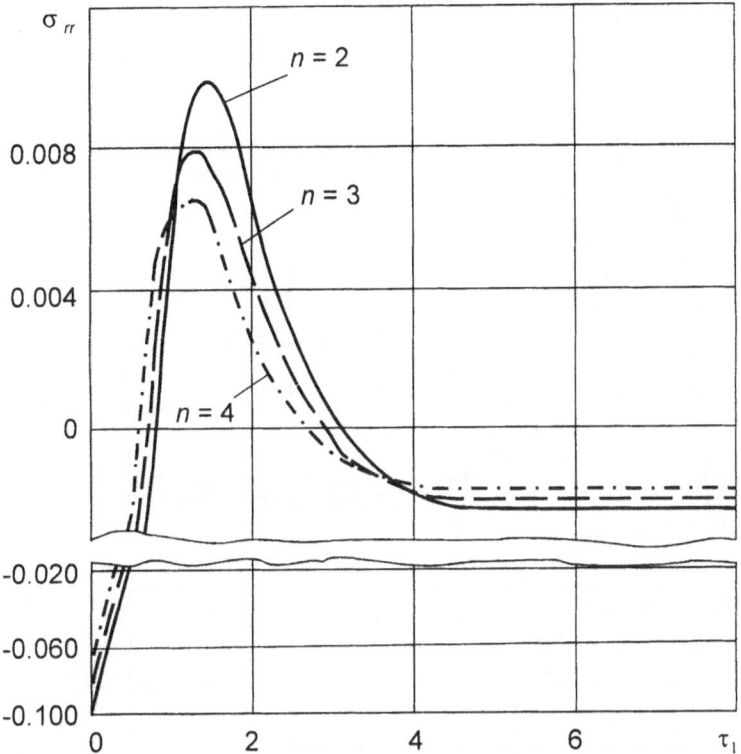

Fig. 2.4. The radial stress at $r = 10$ in the case of a polytropic gas in a cavity (Kasiak et al. (1977))

increase. The absolute value of the stresses reaches its maximum at the front of the elastic wave ($\tau_1 = 0$) and corresponds to compressive stresses. At the instants $\tau_1 > 5$, the stress–strain state of the medium becomes practically time-independent.

Another variant of boundary conditions of the problem of extension of spherical cavity relates to the case when the cavity's surface is reinforced by a thin-walled shell. We assume that an elastic medium, which occupies the whole space, contacts with a thin-walled homogeneous isotropic spherical shell at the surface $r = 1$ (see Fig. 2.1). When defining the dimensionless values, according to (1.11), we set $b = R_1$, where R_1 is the radius of the shell's middle surface; we attribute the dimensionless values with no subscripts to the surrounding medium and we attribute those with the subscript 1 to the shell's material. According to Sect. 1.3, the radial vibrations of the shell can be described by the following equation:

$$\ddot{w} + \omega_1^2 w = \frac{q}{\gamma_1^2 \beta_1 h_1} ,$$

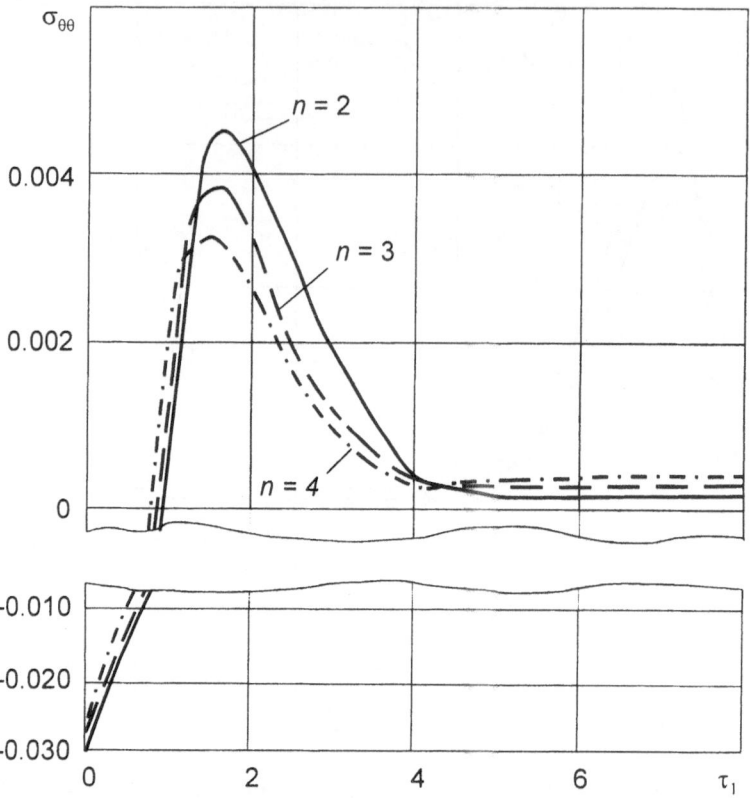

Fig. 2.5. Tangential stresses at $r = 10$ in the case of a polytropic gas in a cavity (Kasiak et al. (1977))

$$\omega_1^2 = \frac{2E_1}{\rho_1 c_*^2 (1 - \nu_1)} = 2 \frac{(1 + 2\kappa_1)(1 - \kappa_1)}{\gamma_1^2}. \tag{2.10}$$

Here, w is the dimensionless radial displacement of the shell, E_1, ν_1, and ρ_1 are Young's modulus, Poisson's ratio, and the density of the shell's material, respectively, q is the pressure at the shell's surface, and $h_1 = \delta/R_1$, where δ is the shell thickness.

Taking into account the contact conditions for the shell and the elastic medium, we arrive at the following boundary value problem:

$$\gamma^2 \frac{\partial^2 \varphi}{\partial \tau^2} = \Delta \varphi, \qquad u\big|_{r=1} = w,$$

$$\ddot{w} + \omega_1^2 w = \frac{q}{\gamma_1^2 \beta_1 h_1}, \qquad q = p + \sigma_{rr}\big|_{r=1} ; \tag{2.11}$$

here, p is the pressure applied at the shell's internal surface. It is assumed that at the initial instant of time the disturbances are absent, that is,

$$\varphi|_{\tau=0} = \dot{\varphi}|_{\tau=0} = w(0) = \dot{w}(0) = 0. \tag{2.12}$$

In order to solve the problem (2.11)–(2.12), let us apply the Laplace transform with respect to time τ (Grigolyuk, Gorshkov (1967), Gelchinsky (1958)). Let s be the transform parameter. The superscript L is used to denote the transform of the corresponding function. Then, assuming $p = p_0 f(\tau) H(\tau)$, in accounting for (2.3) and (2.4) in the space of transforms, we obtain the following analogy to the original problem:

$$\varphi^L(r, s) = \frac{1}{r} g^L(s) e^{-\gamma(r-1)s}, \qquad u^L(1, s) = w^L(s),$$

$$(s^2 + \omega_1^2) w^L(s) = \frac{q^L(s)}{\gamma_1^2 \beta_1 h_1}, \qquad q^L(s) = \frac{p_0}{s} + \sigma_{rr}^L(1, s),$$

$$u^L(r, s) = -\frac{1}{r^2} R_{01}(\gamma r s) g^L(s) e^{-\gamma(r-1)s},$$

$$\sigma^L(r, s) = \frac{\beta_1}{r^3} Q_{01}(\gamma r s) g^L(s) e^{-\gamma(r-1)s},$$

$$R_{01}(s) = s + 1, \qquad Q_{01}(s) = s^2 + 2(1 - \kappa)(s + 1). \tag{2.13}$$

Then, we obtain the transform:

$$g^L(s) = -\frac{p_0 f^L(s)}{\gamma_1^2 h_1 \beta_1 (s^2 + \omega_1^2) R_{01}(\gamma s) + \beta Q_{01}(\gamma s)}. \tag{2.14}$$

The original of the function $g^L(s)$ can be determined applying the inverse integral for the Laplace transform and the theory of residue functions; we have

$$g(\tau) = \sum_{i=0}^{3} \operatorname*{res}_{s=s_i} g^L(s) e^{s\tau}; \tag{2.15}$$

here, $\operatorname*{res}_{s=s_i} f(s)$ is the residue function of the function $f(s)$ at the point $s = s_i$, where s_i, $i = 0, 1, 2, 3$, are the roots of the denominator.

The formulas (2.15), together with the relationships (2.1) and (2.4), allow us to determine the components of the stress–strain state of the surrounding medium. The dimensionless displacement and membrane stress in the shell can be calculated from the following relationships:

$$w(\tau) = u(1, \tau), \qquad \sigma = \frac{1}{2} \omega_1^2 \gamma_1^2 \beta_1 w. \tag{2.16}$$

If the surrounding medium is of an acoustic type, then the denominator structure in (2.16) is simplified (the degree of the polynomial remains the same). At that, it is necessary to set $\kappa = 1$; then, according to (2.13), we get $Q_{01}(\gamma s) = \gamma^2 s^2$.

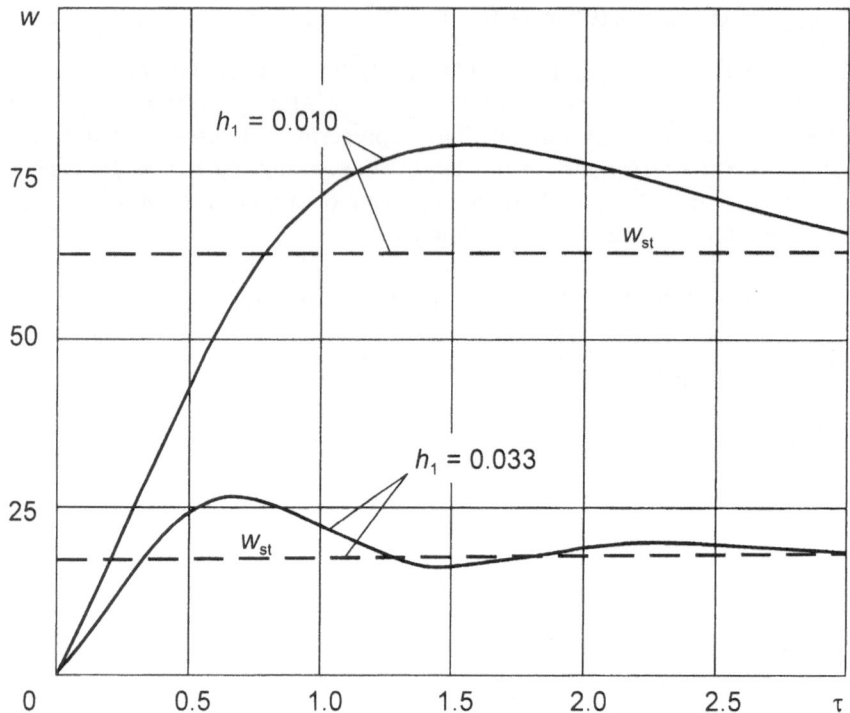

Fig. 2.6. Radial vibrations of a supported cavity ($\alpha = 0$). The displacements of the shell

The solution (2.14) obtained for the particular case when there is no elastic reaction on the cavity's surface (the shell thickness $h_1 = 0$) gives us an analogy to (2.5) in the space of Laplace transforms. Thus, at the limiting case as $h_1 \to 0$, we obtain a solution for the propagation of elastic wave from a spherical cavity under nonstationary pressure.

Let us analyze an example presented in Forrestal, Sagartz (1971). It was assumed that a thin-walled spherical steel shell was placed into an acoustic medium (water). The dimensionless parameters were assumed to be as follows: $\beta = 0.00862$, $\gamma = 1$, $\kappa = 1$, $\beta_1 = 1$, $\gamma_1 = 0.252$, and $\kappa_1 = 0.393$. The load varied in time exponentially, that is, $f(\tau) = \exp(-\alpha\tau)$ and $p_0 = 1$.

The numerical results are presented in Figs. 2.6–2.8. In Figs. 2.6 and 2.7, the radial displacement w and the membrane stresses σ_1 are plotted versus time for two values of shell thickness, $h_1 = 0.010$ and $h_1 = 0.033$. Here, we assume that $f(\tau) \equiv 1$ ($\alpha = 0$). The maximum displacement and the maximum stresses decrease with an increase in the shell thickness. The dashed straight lines correspond to the stationary values of displacements and stresses; using the limiting properties of the Laplace transforms (Dötsch (1974), Lavrentev, Shabat (1987)) at $f(\tau) \equiv 1$, from (2.13)–(2.16)

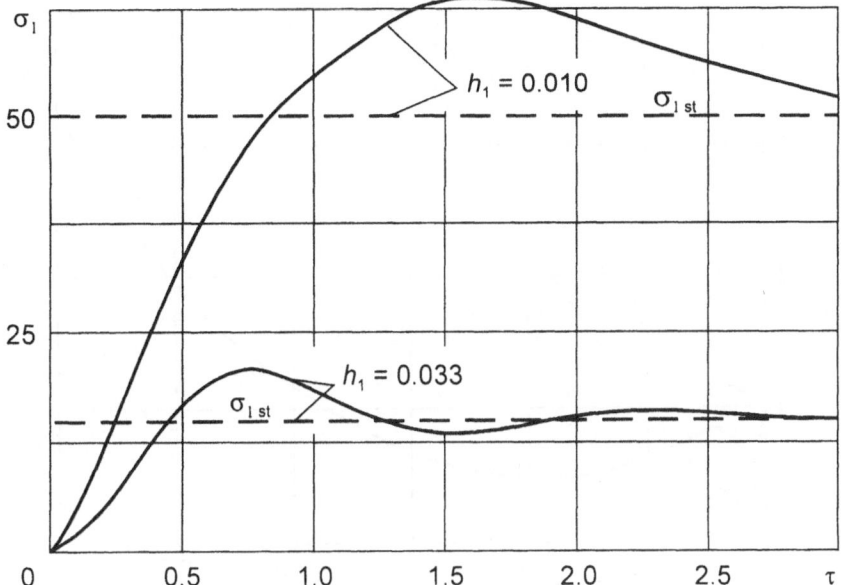

Fig. 2.7. Radial vibrations of a supported cavity ($\alpha = 0$). The stresses in the shell

we get

$$w_{\mathrm{st}} = \lim_{\tau \to \infty} w(\tau) = \lim_{s \to 0} sw^{\mathrm{L}}(s) = \frac{p_0}{\gamma_1^2 h_1 \beta_1 \omega_1^2 + 2\beta(1 - \kappa)},$$

$$\sigma_{\mathrm{st}} = \frac{p_0}{2\left[h_1 + \dfrac{4\beta(1 - \kappa)}{\beta_1 h_1^2 \gamma_1^2}\right]}.$$

In Fig. 2.8, the pressure p for $\alpha = -1$, the pressure $p_0 = -\sigma_{rr}(1, \tau)$ at the surface of a shell of thickness $h_1 = 0.01$ in the surrounding medium, and the total pressure q which acts on the shell (a dashed line) are plotted versus time.

2.2 Vibrations of Thin-Walled Isotropic Shells Contacting Elastic Media

In this section, we shall consider an important problem of propagation of elastic spherical waves from a point source located at the center of a thin-walled isotropic spherical shell surrounded by and filled with the elastic media (Fig. 2.9). We shall designate the external medium by the subscript 0 and

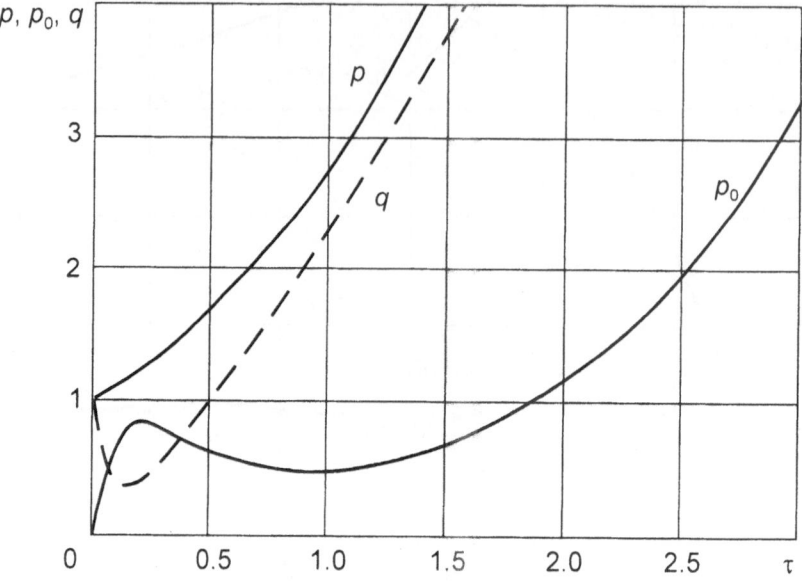

Fig. 2.8. Radial vibrations of a supported cavity ($h_1 = 0.01$). The load at the internal surface of the shell p ($\alpha = 1$), the pressure at the external surface of the shell p_0, and the total pressure q

the internal one by the subscript 2. Similarly to what we did in Sect. 2.1, we shall designate all the parameters attributed to the shell by the subscript 1.

Keeping to the basic notation of Sect. 1.1, let us represent the boundary value problem to be used for studying transient vibrations of the system 'elastic medium – shell – elastic medium' as follows:

$$\gamma_0^2 \frac{\partial^2 \varphi_0}{\partial \tau^2} = \Delta \varphi_0 \,,$$

$$\gamma_2^2 \frac{\partial^2 \varphi_2}{\partial \tau^2} = \Delta \varphi_2 \,,$$

$$u_0|_{r=1} = u_2|_{r=1} + u_{2s}|_{r=1} = w \,,$$

$$\ddot{w} + \omega_1^2 w = \frac{q}{\gamma_1^2 \beta_1 h_1} \,,$$

$$q = \sigma_{rr}^{(0)}\Big|_{r=1} - \left(\sigma_{rr}^{(2)} + \sigma_{rrs}^{(2)} \right)\Big|_{r=1} \,. \tag{2.17}$$

Here, u_{2s} and $\sigma_{rrs}^{(2)}$ are the displacement and the radial stress in a spherical wave propagating from a point source; this wave can be described by the elastic potential

$$\varphi_{2s}(r, \tau) = \frac{1}{r} f[\tau - \gamma_2(r - 1)] \, H[\tau - \gamma_2(r - 1)] \,, \tag{2.18}$$

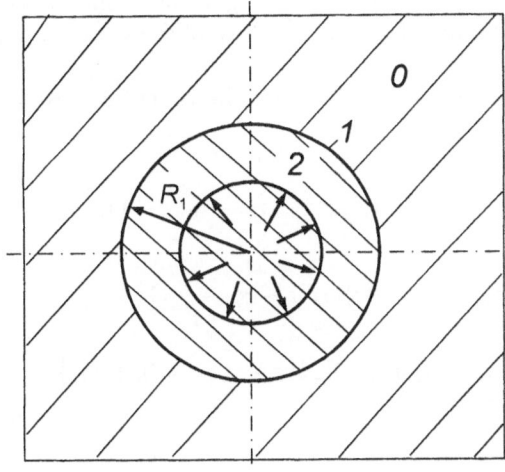

Fig. 2.9. To the problem of vibrations of a thin-walled spherical shell surrounded by and filled with an elastic media

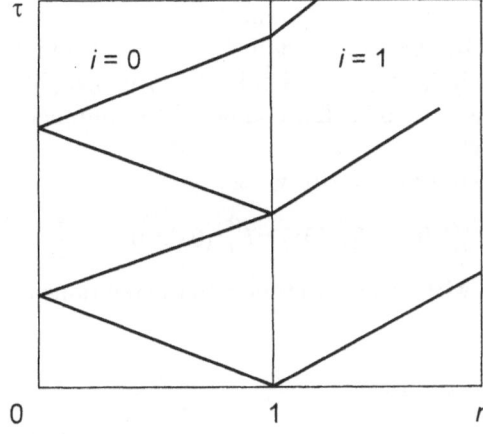

Fig. 2.10. Vibrations of thin-walled shell contacting elastic media. The wave pattern

where f is an arbitrary function, which defines the law of variation of the potential at the wave front. It is necessary to supplement the relationships (2.17) by the condition of absence of the disturbances at infinity in the media designated by the subscript 0, by the boundedness condition for the functions sought at the sphere's center

$$|\varphi_2(r, \tau)| < \infty, \qquad \varphi_2(r, \tau) = O(1), \qquad r \to 0, \tag{2.19}$$

and by the initial conditions,

$$\varphi_0|_{\tau=0} = \dot{\varphi}_0|_{\tau=0} = \varphi_2|_{\tau=0} = \dot{\varphi}_2|_{\tau=0} = w(0) = \dot{w}(0) = 0. \tag{2.20}$$

In order to derive a solution of the problems (2.17)–(2.20), let us apply a general solution of the wave equations (1.114) at $n = 0$. It is necessary to note that both types of wave, converging and diverging, exist in the internal medium because this medium is bounded. A wave pattern of the system under study is presented in Fig. 2.10.

Then, analyzing the wave pattern, we can present the potentials $\varphi_0(r, \tau)$ and $\varphi_2(r, \tau)$ in the following form:

$$\varphi_0(r, \tau) = \frac{1}{r} \sum_{k=0}^{\infty} f_k^{(0)}[\tau - \tau_k - \gamma_0(r - 1)] \, H[\tau - \tau_k - \gamma_0(r - 1)],$$

$$\varphi_2(r, \tau) = \frac{1}{r} \sum_{k=0}^{\infty} \Big\{ f_{0k}^{(2)}[\tau - \tau_k - \gamma_2(1 - r)] \, H[\tau - \tau_k - \gamma_2(1 - r)]$$

$$+ f_{1k}^{(2)}[\tau - \tau_k - \gamma_2(r + 1)] \, H[\tau - \tau_k - \gamma_2(r + 1)] \Big\},$$

$$\tau_k = 2\gamma_2 k. \tag{2.21}$$

Here, $f_k^{(0)}$, $f_{0k}^{(2)}$, and $f_{1k}^{(2)}$ are arbitrary functions characterizing the converging and diverging waves, respectively, the superscripts 0 and 2 correspond to the number of the elastic medium. In the external medium, the converging waves are absent, and that is conditioned by the absence of disturbances at infinity. Let us note that we put the upper limit of summation in the formulas (2.21) to be infinite conventionally. In reality, for a finite time τ, the number of elementary waves will be finite.

In order to satisfy the initial conditions (2.20), we set

$$f_k^{(0)}(0) = \dot{f}_k^{(0)}(0) = f_{0k}^{(2)}(0) = \dot{f}_{0k}^{(2)}(0) = f_{1k}^{(2)}(0) = \dot{f}_{1k}^{(2)}(0) = 0. \tag{2.22}$$

Applying the Laplace transform with respect to time τ to the relationships (2.21), we get

$$\varphi_0^L(r, s) = \frac{1}{r} \sum_{k=0}^{\infty} f_k^{(0)L}(s) \, e^{-[\tau_k + \gamma_0(r-1)]s},$$

$$\varphi_2^L(r, s) = \frac{1}{r} \sum_{k=0}^{\infty} \Big\{ f_{0k}^{(2)L}(s) \, e^{-\gamma_2(1-r)s} + f_{1k}^{(2)L}(s) \, e^{-\gamma_2(r+1)s} \Big\} e^{-\tau_k s}.$$

$$\tag{2.23}$$

In order to satisfy the boundary conditions as $r \to 0$ (2.19), let us take into account that

$$e^r - e^{-r} = O(r), \qquad r \to 0. \tag{2.24}$$

Using (2.24), from the second equality of (2.23) we obtain

$$\varphi_2^L(r, s) = \sum_{k=0}^{\infty} \frac{1}{r} \Big\{ e^{-\gamma_2 r s} \left[f_{0k}^{(2)L}(s) + f_{1k}^{(2)L}(s) \right] + O(\gamma_2 r s) \Big\} e^{(-\gamma_2 - \tau_k)s},$$

$$r \to 0. \tag{2.25}$$

From (2.25) it follows that the function $\varphi_2^L(r, s)$ is bounded as $r \to 0$ only if

$$f_{0k}^{(2)} + f_{1k}^{(2)} = 0. \tag{2.26}$$

Let us note that it is possible to prove that the last equality provides also the boundedness of all of the components of the stress–strain state of the internal medium at the sphere's center $r = 0$.

Let us designate

$$f_k^{(2)} = f_{1k}^{(2)} = -f_{0k}^{(2)}. \tag{2.27}$$

Taking into account (2.27), the formula (2.23), and the relationship of the potential φ with the components of the stress–strain state (2.1), we obtain the following expressions for the Laplace transforms of displacements and stresses:

$$u_0^L(r, s) = -\frac{1}{r^2} \sum_{k=0}^{\infty} R_{01}(\gamma_0 rs) f_k^{(0)L}(s) e^{-[\tau_k + \gamma_0(r-1)]s},$$

$$u_2^L(r, s) = -\frac{1}{r^2} \sum_{k=0}^{\infty} \left[-R_{01}(-\gamma_2 rs) e^{-\gamma_2(1-r)s} \right.$$

$$\left. + R_{01}(\gamma_2 rs) e^{-\gamma_2(1+r)s} \right] f_k^{(2)L}(s) e^{-\tau_k s},$$

$$\sigma_{rr}^{(0)L}(r, s) = \frac{\beta_0}{r^3} \sum_{k=0}^{\infty} Q_{01}^{(0)}(\gamma_0 rs) f_k^{(0)L}(s) e^{-[\tau_k + \gamma_0(r-1)]s},$$

$$\sigma_{rr}^{(2)L}(r, s) = \frac{\beta_2}{r^3} \sum_{k=0}^{\infty} \left[-Q_{01}^{(2)}(-\gamma_2 rs) e^{-\gamma_2(1-r)s} \right.$$

$$\left. + Q_{01}^{(2)}(\gamma_2 rs) e^{-\gamma_2(1+r)s} \right] f_k^{(2)L}(s) e^{-\tau_k s},$$

$$\sigma_{\theta\theta}^{(i)} = \kappa_i \sigma_{rri} + (1 - \kappa_i)(1 + 2\kappa_i) \frac{u_i}{r},$$

$$Q_{01}^{(i)}(s) = s^2 + 2(1 - \kappa_i)(s + 1), \qquad i = 0, 2. \tag{2.28}$$

For the transforms of the potential, displacement, and stress in the disturbing wave, we have

$$\varphi_{2s}^L(r, s) = \frac{f^L(s)}{r} e^{-\gamma_2(r-1)s},$$

$$u_{2s}^L(r, s) = -\frac{f^L(s)}{r^2} R_{01}(\gamma_2 rs) e^{-\gamma_2(r-1)s},$$

$$\sigma_{rrs}^{(2)L}(r, s) = \frac{\beta_2 f^L(s)}{r^3} Q_{01}^{(2)}(\gamma_2 rs) e^{-\gamma_2(r-1)s}. \tag{2.29}$$

Taking into account (2.28) and (2.29), applying the Laplace transform with respect to time τ to the boundary conditions at $r = 1$ in (2.17), we obtain the following equality:

$$-\sum_{k=0}^{\infty} R_{01}(\gamma_0 s)\, f_k^{(0)\mathrm{L}}(s)\, \mathrm{e}^{-\tau_k s} = -\sum_{k=0}^{\infty} R_{01}(-\gamma_2 s)\, f_k^{(2)\mathrm{L}}(s)\, \mathrm{e}^{-\tau_k s}$$

$$+\sum_{k=1}^{\infty} R_{01}(\gamma_2 s)\, f_{k-1}^{(2)\mathrm{L}}(s)\, \mathrm{e}^{-\tau_k s} + f^{\mathrm{L}}(s)\, R_{01}(\gamma_2 s) = w^{\mathrm{L}}(s)\,,$$

$$\beta_1 \gamma_1^2 h_1 (s^2 + \omega_1^2)\, w^{\mathrm{L}}(s) = \beta_0 \sum_{k=0}^{\infty} Q_{01}^{(0)}(\gamma_0 s)\, f_k^{(0)\mathrm{L}}(s)\, \mathrm{e}^{-\tau_k s}$$

$$+\beta_2 \sum_{k=0}^{\infty} Q_{01}^{(2)}(-\gamma_2 s)\, f_k^{(2)\mathrm{L}}(s)\, \mathrm{e}^{-\tau_k s}$$

$$-\beta_2 \sum_{k=1}^{\infty} Q_{01}^{(2)}(\gamma_2 s)\, f_{k-1}^{(2)\mathrm{L}}(s)\, \mathrm{e}^{-\tau_k s} - \beta_2 f^{\mathrm{L}}(s)\, Q_{01}^{(2)}(\gamma_2 s) = q^{\mathrm{L}}(s)\,.$$

$$(2.30)$$

Equating in (2.30) the coefficients at the same powers of retarding (at the exponential functions of equal powers), we arrive at the recursive relationships with respect to the functions $f_k^{(0)\mathrm{L}}(s)$ and $f_k^{(2)\mathrm{L}}(s)$ $(k \geq 0)$, which we represent in the following matrix form:

$$\begin{pmatrix} R_{01}(-\gamma_2 s) & R_{01}(\gamma_0 s) \\ \beta_2 Q_{01}^{(2)}(-\gamma_2 s) & T_0(s) \end{pmatrix} \begin{pmatrix} f_k^{(2)\mathrm{L}}(s) \\ f_k^{(0)\mathrm{L}}(s) \end{pmatrix} = \begin{pmatrix} R_{01}(\gamma_2 s) \\ \beta_2 Q_{01}^{(2)}(\gamma_2 s) \end{pmatrix} f_{k-1}^{(2)\mathrm{L}}(s)\,,$$

$$T_0(s) = \beta_0 Q_{01}^{(0)}(\gamma_0 s) + \beta_1 \gamma_1^2 h_1 (s^2 + \omega_1^2)\, R_{01}(\gamma_0 s)\,,$$

$$f_{-1}^{(2)\mathrm{L}}(s) = f^{\mathrm{L}}(s)\,. \qquad (2.31)$$

The following recursive relationships will be a solution of this matrix equation:

$$f_k^{(2)\mathrm{L}}(s) = H_1^{\mathrm{L}}(s)\, f_{k-1}^{(2)\mathrm{L}}(s)\,, \qquad f_k^{(0)\mathrm{L}}(s) = H_2^{\mathrm{L}}(s)\, f_{k-1}^{(2)\mathrm{L}}(s)\,,$$

$$H_1^{\mathrm{L}}(s) = \frac{Y_1(s)}{X(s)}\,, \qquad H_2^{\mathrm{L}}(s) = \frac{Y_2(s)}{X(s)}\,,$$

$$X(s) = R_{01}(-\gamma_2 s)\, T_0(s) - \beta_2 Q_{01}^{(2)}(-\gamma_2 s)\, R_{01}(\gamma_0 s)\,,$$

$$Y_1(s) = R_{01}(\gamma_2 s)\, T_0(s) - \beta_2 Q_{01}^{(2)}(\gamma_2 s)\, R_{01}(\gamma_0 s)\,,$$

$$Y_2(s) = \beta_2 \left[Q_{01}^{(2)}(\gamma_2 s)\, R_{01}(-\gamma_2 s) - Q_{01}^{(2)}(-\gamma_2 s)\, R_{01}(\gamma_2 s) \right] = -2\beta_2 \gamma_2^3 s^3\,.$$

$$(2.32)$$

Equations (2.32) allow us to obtain the transforms of the functions $f_k^{(0)}$ and $f_k^{(2)}$ for an arbitrary $k \geq 0$ and, thus, to determine the transforms of the components of the stress–strain state of the elastic media and the shell.

Since the functions $X(s)$, $Y_1(s)$, and $Y_2(s)$ are the polynomials of the degrees 4, 4, and 3, respectively, the originals of the rational functions $H_1^L(s)$ and $H_2^L(s)$ can be derived without difficulty applying the theory of residue functions (Grigolyuk, Gorshkov (1967)). However, since $H_1^L(s)$ is an improper fraction, its original is a generalized function (Dötsch (1974)), namely,

$$H_1(\tau) = c\,\delta(\tau) + H_{1*}(\tau)\,,$$

$$H_{1*}(\tau) = \sum_{j=1}^{4} \operatorname*{res}_{s_j} \left[H_1^L(s) - c \right] e^{s\tau}\,, \qquad X(s_j) = 0\,,$$

$$c = \lim_{s \to \infty} H_1^L(s) = -1\,, \tag{2.33}$$

where $\delta(\tau)$ is the Dirac delta function.

Using (2.33), we can represent the analogy to the recursive relationships (2.32) in the space of originals in the form of convolutions (Dötsch (1974))

$$f_k^{(2)}(\tau) = -f_{k-1}^{(2)}(\tau) + H_{1*}(\tau) * f_{k-1}^{(2)}(\tau)\,,$$

$$f_k^{(0)}(\tau) = H_2(\tau) * f_{k-1}^{(2)}(\tau)\,, \qquad f_{-1}^{(2)}(\tau) = f(\tau)\,. \tag{2.34}$$

The relationships (2.34) give us an opportunity to derive all functions $f_k^{(0)}(\tau)$ and $f_k^{(2)}(\tau)$ ($k \geq 0$) and, hence, to determine the displacements and the stresses of the elastic media and the shell. However, though the convolutions in (2.34) can be calculated in closed form, it is difficult to derive analytic expressions for a relatively large value of k (i.e., large value of τ). Hence, it is necessary to apply methods of numerical integration (Babaev (1974a)).

Let us demonstrate a possible way to obtain an exact inversion of the Laplace transform for the relationships of the type (2.32). In order to do this, we must, first, notice that for an arbitrary k, the solution of (2.32) can be represented in the following explicit form:

$$f_k^{(2)L}(s) = \left[H_1^L(s) \right]^{k+1} f^L(s)\,. \tag{2.35}$$

We can see that the singularities of $f_k^{(2)L}(s)$ are the singular points of the function $f^L(s)$ and the poles s_j of the multiplicity $(k+1)$, where s_j, $j = 1, 2, 3, 4$, are the roots of the equation $X(s) = 0$.

In order to determine the originals, it is sufficient to determine the residue functions for the functions $f_k^{(2)L}(s) \exp(s\tau)$. It is evident (see Dötsch (1974), Lavrentev, Shabat (1987)) that

$$\operatorname*{res}_{s_j} f_k^{(2)L}(s)\, e^{s\tau} = \frac{1}{(k-1)!} \frac{d^{k-1}}{ds^{k-1}} \left[(s - s_j)^k f_k^{(2)L}(s)\, e^{s\tau} \right] \Bigg|_{s=s_j}\,.$$

Applying the Leibniz formula, we get

$$\operatorname*{res}_{s_j} f_k^{(2)\mathrm{L}}(s)\, e^{s\tau} = \sum_{p=0}^{k-1} \frac{\tau^p\, e^{s_j\tau}}{p!}\, d_{k-1-p}^k,$$

$$d_l^k = \frac{1}{l!}\left[(s-s_j)^k f_k^{(2)\mathrm{L}}(s)\right]^{(l)}_{s=s_j}. \tag{2.36}$$

Thus, when calculating the residue functions, it is sufficient to know the sequence of the derivatives d_l^k, $l = 0, 1, \ldots, k-1$. They can be determined by deriving the analogy to the recursive relationship (2.32):

$$d_l^k = \sum_{i=0}^{l} h_i d_{l-i}^{k-1} = \{h_i\} * \{d_i^{k-1}\},$$

$$h_i = \frac{1}{i!}\left[(s-s_j)\, H_1^\mathrm{L}(s)\right]^{(i)}_{s=s_j},$$

$$d_i^{-1} = \frac{1}{i!}\left[(s-s_j)\, f^\mathrm{L}(s)\right]^{(i)}_{s=s_j}. \tag{2.37}$$

Here, asterisk '*' is used to designate the convolution of the sequences $\{h_i\}$ and $\{d_i^{k-1}\}$. Let us note that the expression (2.34) is similar by form to the formulas (2.37). In the first case, we deal with the integral convolutions of functions, in the second one, we deal with the convolution product of sequences.

In order to close the algorithm, it is sufficient to obtain a method of calculation of the sequence of the derivatives r_i at the point $s = s_j$ for the function

$$R(s) = \frac{Q_1(s)}{Q_2(s)}(s-s_j),$$

where $Q_1(s)$ and $Q_2(s)$ are the polynomials; here, s_j is either a simple zero of the function $Q_2(s)$ or not the zero. The sequence of the derivatives of the function $R(s)$ at the points $s = s_j$ can be calculated using the following recursive formulas, which follows from the Leibniz formula:

$$r_i = \frac{q_{1i}}{q_{21}} - \frac{1}{q_{21}}\sum_{l=1}^{i} r_{i-l}\, q_{2l},$$

$$r_i = \frac{1}{i!}\, R^{(i)}(s_j), \qquad q_{1i} = \frac{1}{i!}\, Q_1^{(i)}(s_j), \qquad q_{2i} = \frac{1}{i!}\, Q_2^{(i)}(s_j). \tag{2.38}$$

Thus, the algorithm presented gives us an opportunity to determine the functions $f_k^{(0)}(\tau)$ and $f_k^{(2)}(\tau)$ exactly. In order to determine the originals of displacements and stresses in the elastic media and in the shell by the formulas (2.28), (2.17), and (2.16), it is also necessary to know the originals of these functions multiplied by s and s^2. Following this purpose, it is necessary to multiply the relationship (2.32) sequentially by s and s^2, and to apply the algorithm described above.

The function $f(\tau)$ characterizes the variation of the components of the stress–strain state at the wave front produced by a point source. In the future, we shall choose the function $f(\tau)$ considering the condition of equality of the radial stress $\sigma_{rr\,s}^{(2)}$ at the surface $r = 1$ at $\tau = 0$ to some constant p_0. As a result, we get

$$\sigma_{rr\,s}^{(2)}(1,\,0) = \lim_{s \to \infty} s\sigma_{rr\,s}^{(2)}(1,\,s) = \lim_{s \to \infty} \beta_2 s f^{L}(s)\, Q_{01}^{(2)}(\gamma_2 s) = p_0\,. \qquad (2.39)$$

It is evident that these conditions will be satisfied by the following equalities:

$$f^{L}(s) = \frac{p_0}{\beta_2 \gamma_2^2 (s + \alpha)^3}\,,$$
$$f(\tau) = \frac{p_0}{2\gamma_2^2 \beta_2}\, \tau^2\, e^{-\alpha\tau}\, H(\tau)\,. \qquad (2.40)$$

Let us demonstrate that from the solution obtained the solution of the problem of propagation of a disturbance from a cavity supported by a spherical shell solved in Sect. 2.1 follows as a particular case. In fact, as we can see from Sect. 1.4, in order to make a passage to the limit related to the absence of medium, it is necessary to set $\beta_2 = 0$ and $\gamma_2 \to \infty$. Taking into account that

$$\lim_{\gamma_2 \to \infty} e^{-\tau_k s} = 0\,, \qquad k > 0\,,$$

from (2.34) we obtain

$$u_0^{L}(r,\,s) = -\frac{1}{r^2}\, R_{01}(\gamma_0 r s)\, f_0^{(0)L}(s)\, e^{-\gamma_0(r-1)s}\,,$$

$$\sigma_{rr}^{(0)L}(r,\,s) = -\frac{\beta_0}{r^3}\, Q_{01}(\gamma_0 r s)\, f_0^{(0)L}(s)\, e^{-\gamma_0(r-1)s}\,,$$

$$u_2^{L}(r,\,s) \equiv 0\,, \qquad \sigma_{rr}^{(2)L}(r,\,s) \equiv 0\,. \qquad (2.41)$$

Making a passage to the limit as $\gamma_2 \to \infty$ and $\beta_2 \to 0$ in the recursive relationships (2.32) in accounting for (2.40) and considering $f(\tau) = 1$, we arrive at

$$f_0^{(0)L}(s) = \frac{2p_0}{sT_0(s)}\,. \qquad (2.42)$$

Except for the notation, the formulas (2.14) and (2.42) are identical. The presence of the factor 2 in the numerator of (2.42) can be explained by the fact that the load at the shell's surface is produced by the incoming wave, and even when the passage to the limit is provided, the effect of the pressure doubling caused by reflection takes place. The difference in signs in (2.42) and (2.14) is caused by the fact that the positive stress in the elastic medium

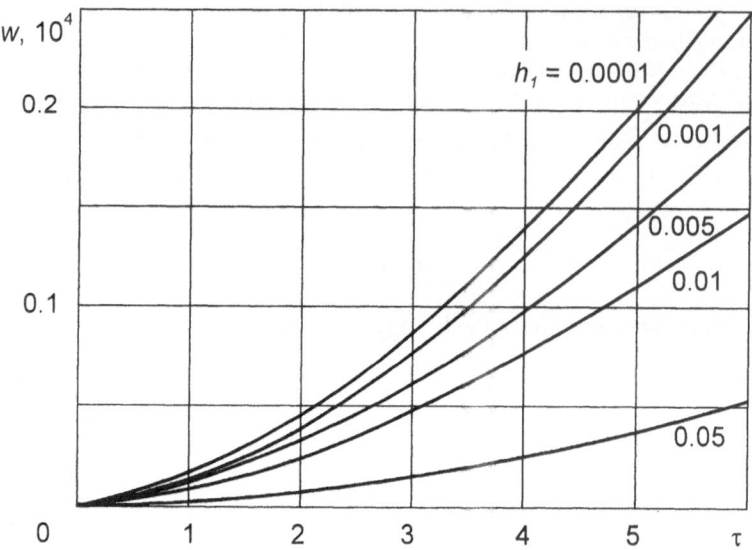

Fig. 2.11. The displacements of a thin-walled shell calculated under different values of the thickness h_1

corresponds to the distributed load applied to the internal surface of the shell and directed towards the sphere's center.

Let us point out a particular case of the problem considered, when the media which fill and surround the shell are of acoustic type. As we have already mentioned, in this variant it is necessary to set $\kappa_0 = \kappa_2 = 1$. Let p_0 and p_2 be the pressures in the external and internal media. Then, the formulas (2.28) and (2.32) remain valid under the following simplifications:

$$Q_{01}^{(0)}(s) = Q_{01}^{(2)}(s) = s^2, \qquad p_0 = -\sigma_{rr}^{(0)}, \qquad p_2 = -\sigma_{rr}^{(2)},$$
$$X(s) = R_{01}(-\gamma_2 s)\, T_0(s) - \beta_2 s^2 \gamma_2^2 R_{01}(\gamma_0 s),$$
$$Y_1(s) = R_{01}(\gamma_2 s)\, T_0(s) - \beta_2 \gamma_2^2 s^2 R_{01}(\gamma_0 s),$$
$$T_0(s) = \beta_1 \gamma_1^2 h_1 (s^2 + \omega_1^2)\, R_{01}(\gamma_0 s) + \beta_0 \gamma_0^2 s^2. \tag{2.43}$$

Except for the notation and the choice of dimensionless parameters, these formulas are identical to the known solution (Babaev, Kubenko (1977b)).

Let us now discuss some numerical results obtained by means of the algorithm presented. We set $p_0 = 1$ and $\alpha = 0$; these parameters characterize the variation of stress in a disturbing wave. The media, which surround and fill the shell, are assumed to be of acoustic type, hence, $\kappa_0 = \kappa_1 = 1$.

The curves in Figs 2.11 and 2.12 illustrate the influence of the thickness of a spherical shell on the radial displacement and stress in the shell. When performing calculations, it was assumed that the steel shell was surrounded by and filled with water ($\kappa_1 = 0.393$, $\gamma_0 = \gamma_2 = 1$, $\gamma_1 = 0.252$, $\beta_1 = 1$,

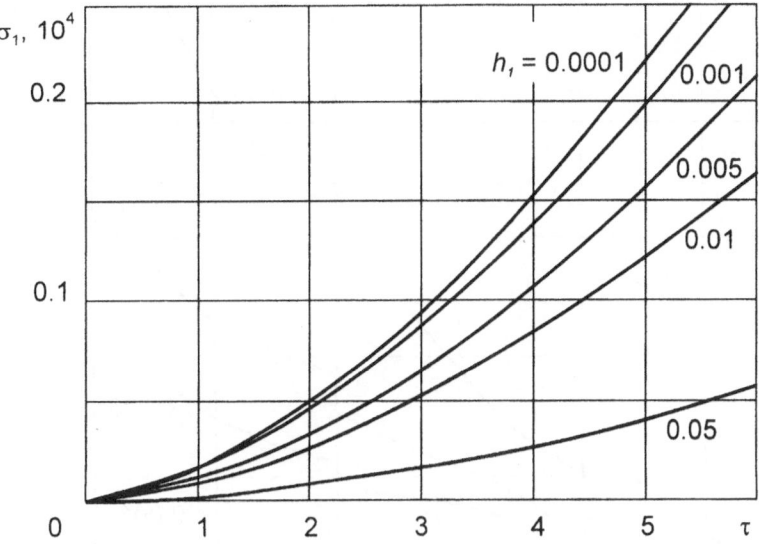

Fig. 2.12. The stresses in a thin-walled shell calculated under different values of the thickness h_1

$\beta_0 = \beta_2 = 0.00862$). Calculations were performed under the following values of thickness: $h_1 = 0.0001, 0.001, 0.005, 0.01, 0.05$.

The results of calculation of the pressure p_2 in the medium at the inner surface of the shell are presented in Fig. 2.13; the results of calculation of the pressure p in the fluid at the outer surface are presented in Fig. 2.14. The curves are plotted for different values of the thickness h_1. The discontinuities of the first kind of the function $p_2(\tau)$ and the discontinuities of the derivative of the function $p(\tau)$ correspond to the instants of time when the reflected waves' fronts arrive at the inner shell's surface. It follows from Fig. 2.14 that as $\tau \to \infty$, the pressure in the internal medium approaches the stationary value p_{st}, which depends on the shell thickness. The smaller is the shell thickness, the more 'acoustically soft' it gets, that is, the distortions of the signal, which passes through the shell, become lower following the Heaviside function $H(\tau)$. At $h_1 = 0.0001$, the pressure p at the external surface of the shell does not differ significantly from the function $H(\tau)$.

The influence of the surrounding and filling materials, as well as of the shell's material, on the displacement of the shell is illustrated in Fig. 2.15. Curve *1* corresponds to the system 'water – steel – water'; curve *4* corresponds to the system 'water – aluminum – water' ($\kappa_1 = 0.493$; $\gamma_0 = \gamma_2 = 1$; $\gamma_1 = 0.236$; $\beta_0 = \beta_2 = 0.0222$); curve *5* corresponds to the system 'glycerin – steel – water' ($\kappa_1 = 0.393$; $\gamma_0 = 1.32$; $\gamma_2 = 1$; $\gamma_1 = 0.333$; $\beta_0 = 0.00862$; $\beta_2 = 0.0177$); curve *6* corresponds to the system 'mercury – steel – water' ($\kappa_1 = 0.393$; $\gamma_2 = 1$; $\gamma_0 = 1$; $\gamma_1 = 0.252$; $\beta_0 = 0.00862$; $\beta_2 = 0.109$); (at the

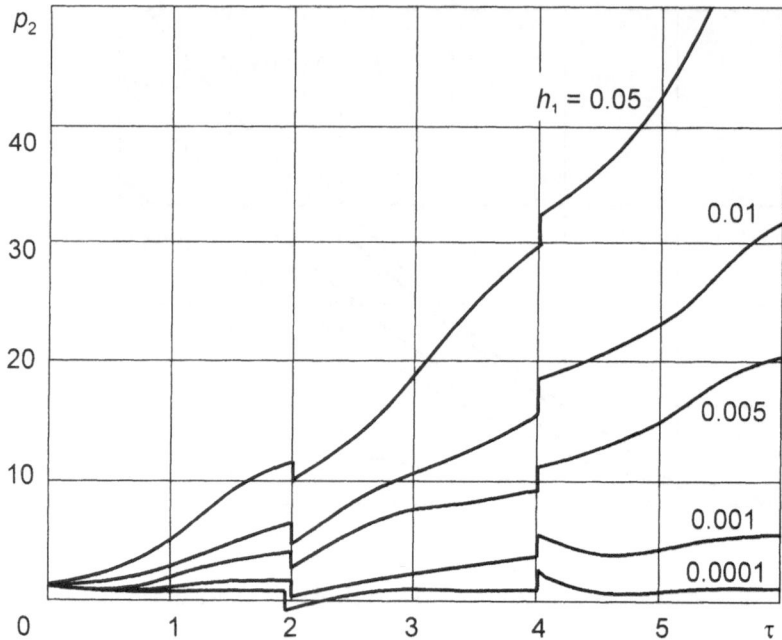

Fig. 2.13. The pressure at the internal surface of a thin-walled shell

first place, we put the materials of filling medium, at the third one, we put the materials of surrounding medium). At that, since the pressure in a disturbing wave is attributed to the parameters of the shell, then, in order to make a comparison of curve *4* with the variant of a steel shell (curve *1*), it is necessary to set $p_0 = (\lambda + 2\mu)_{\text{steel}}/(\lambda + 2\mu)_{\text{alum.}} \approx 2.58$. Curves *2* and *3* characterize the influence of the accounting for the waves irradiated into the surrounding media for the systems 'water – steel – water'. Curve *2* corresponds to the absence of internal medium ($\beta_0 = 0$) and ignoring the interaction of the internal medium and the shell ($\beta_2 = 0$); curve *3* corresponds to the case when $\beta_0 \neq 0$ and $\beta_2 = 0$. From the comparison of curves *1*, *2*, and *3* it follows that ignoring the pressure of interaction brings an increase in the displacements.

2.3 Vibrations of a Thick-Walled Sphere in an Elastic Medium

In this section, we shall analyze the problem of transient radial vibrations of a thick-walled elastic spherical shell surrounded by and filled with the elastic media (Fig. 2.16). This problem presumes generalization of the problem considered in Sect. 2.2.

As it follows from the solution of the relatively simple problem presented in Sect. 2.2, a presence of the elastic medium, which fills the bounded re-

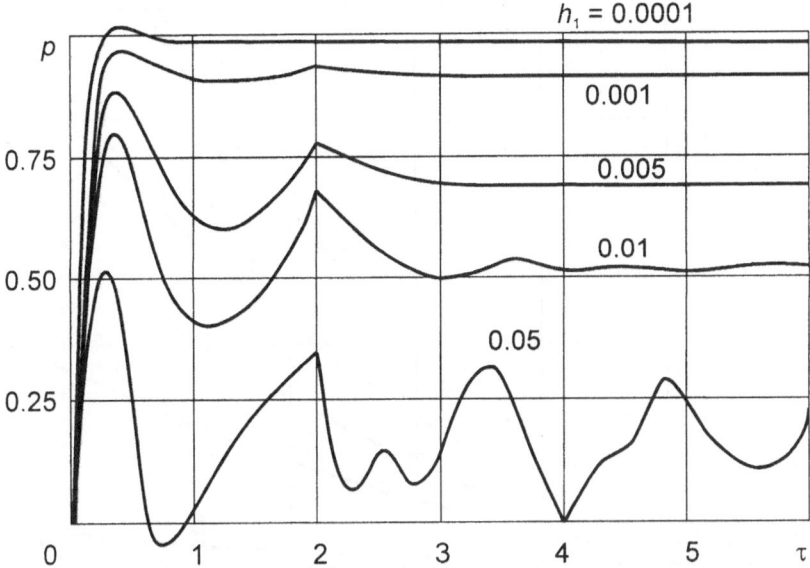

Fig. 2.14. The pressure at the external surface of a thin-walled shell

gion, brings us to a more complicated problem, as compared to the case of unbounded regions. This is caused by presence of two systems of waves in a bounded region: the diverging waves and the converging waves. When there is more than two interfaces, the problem is complicated by adding of the refracted waves passing through the boundaries of different media.

Besides the sufficiently general problem of radial vibrations to be discussed here, the more simple problems of radial vibrations of homogeneous and hollow elastic spheres under dynamic loads, as well as the case, when the internal and external media are absent, are also of interest to us. However, we do not discuss these problems separately since, on the one hand, their solution can be derived as a particular case of the problem under study, and, on the other hand, the arguments and derivations to be developed for the solutions are similar.

Following Grigolyuk et al. (1977a) and Tarlakovsky (1975), let us consider the problem of radial transient vibrations of a thick-walled elastic spherical shell surrounded by and filled with an elastic media. We shall assume that all three media are homogeneous and isotropic, and designate them by the subscripts $i = 0, 1, 2$ (Fig. 2.16). We shall distinguish two types of the problem, internal and external. Considering an external problem, we assume that at the initial time $\tau = 0$, the uniformly distributed nonstationary pressure $p(\tau)$ is applied to the external surface of the shell $r = 1$. Considering an internal problem, we assume that the front of the spherical compression wave produced by a point source located at the sphere's center O reaches the inter-

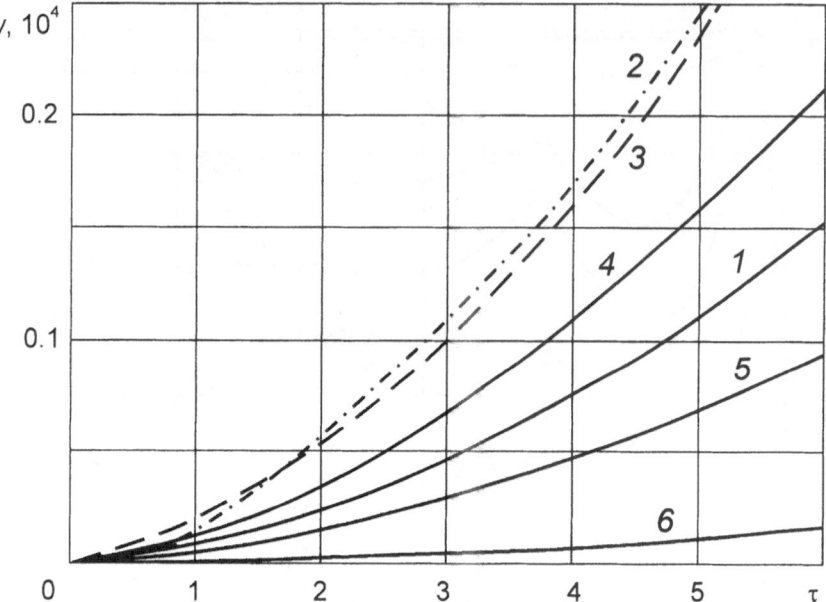

Fig. 2.15. The influence of the material of the surrounding and filling media, as well as of the shell's material, on the displacements

nal surface. Similarly to Sect. 2.2, the intensity of the point source is defined by the elastic potential $\varphi_{2s}(r, \tau)$ determined by (2.18).

Using the notation introduced before, we obtain the following boundary value problem:

$$\gamma_i^2 \frac{\partial^2 \varphi_i}{\partial \tau^2} = \Delta \varphi_i, \qquad i = 0, 1, 2,$$

$$u_1|_{r=1} = u_0|_{r=1}, \qquad \sigma_{rr}^{(1)}\Big|_{r=1} = \sigma_{rr}^{(0)}\Big|_{r=1} - p(\tau),$$

$$u_1|_{r=r_1} = u_2|_{r=r_1} + u_{2s}|_{r=r_1}, \qquad p(\tau) = f(\tau)\, H(\tau),$$

$$\sigma_{rr}^{(1)}\Big|_{r=1} = \sigma_{rr}^{(2)}\Big|_{r=1} + \sigma_{rrs}^{(2)}\Big|_{r=1}. \tag{2.44}$$

Here, the subscript s corresponds to the components of the stress–strain state in the incoming wave. With that, for the external problem, we have $u_{2s} = 0$ and $\sigma_{rrs}^{(2)} = 0$, and for the internal one, we have $p(\tau) \equiv 0$. When defining the dimensionless parameters (1.11), we state: the dimension b is equal to the radius of shell's external surface; $\lambda_* = \lambda_1$, and $\mu_* = \mu_1$, where λ_1 and μ_1 are the Lamé elastic constants for the shell's material; $c_* = c_1^{(0)}$ is the speed of propagation of compression waves in the external medium. This case corresponds to $\gamma_0 = 1$ and $\beta_1 = 1$.

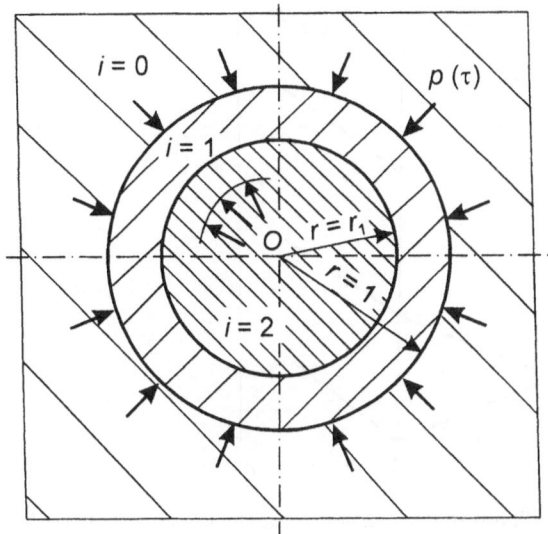

Fig. 2.16. To the problem of radial vibrations of a thick-walled spherical shell

At infinity, as $r \to \infty$ and $i = 0$, we still assume the condition of absence of disturbances, and at the center of the filler, as $r \to \infty$ and $i = 2$, we assume the condition of boundedness of the elastic potential (2.19).

We assumed that before the disturbances were applied to the shell's surface, the whole system was at rest. Thus, the initial conditions follow:

$$\varphi_i(r, 0) = \dot{\varphi}_i(r, 0) = 0, \qquad i = 0, 1, 2. \tag{2.45}$$

The patterns of wave formation for the external and internal problems are illustrated in Figs. 2.17 and 2.18. Considering the corresponding characteristics, the general integral of the wave equation (1.114) at $n = 0$, and the results of the analysis of the boundary conditions as $r \to 0$ obtained in Sect. 2.2 (the formulas (2.23)–(2.27)), we can write the potentials $\varphi_i(r, \tau)$ as follows:

$$\varphi_0(r, \tau) = \frac{1}{r} \sum_{k,\,m=0}^{\infty} f_{km}^{(0)}[\tau - \tau_{km} - \tau_0^* - (r-1)]\, H[\tau - \tau_{km} - \tau_0^* - (r-1)],$$

$$\varphi_1(r, \tau) = \frac{1}{r} \sum_{k,\,m=0}^{\infty} \left\{ f_{0km}^{(1)}[\tau - \tau_{km} - \tau_0^* - \gamma_1(1-r)] \right.$$

$$\times H[\tau - \tau_{km} - \tau_0^* - \gamma_1(1-r)]$$

$$\left. + f_{1km}^{(1)}[\tau - \tau_{km} - \tau_1^* - \gamma_1(r-r_1)]\, H[\tau - \tau_{km} - \tau_1^* - \gamma_1(r-r_1)] \right\},$$

$$\varphi_2(r, \tau) = \frac{1}{r} \sum_{k,\,m=0}^{\infty} \left\{ -f_{km}^{(2)}[\tau - \tau_{km} - \tau_1^* - \gamma_2(r_1-r)] \right.$$

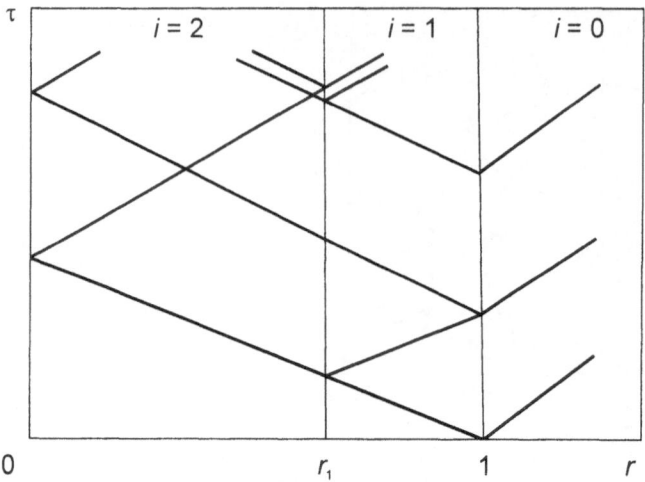

Fig. 2.17. Formation of waves. The radial vibrations of a thick-walled sphere. The external problem

$$\times H[\tau - \tau_{km} - \tau_1^* - \gamma_2(r_1 - r)]$$

$$+ f_{km}^{(2)}[\tau - \tau_{km} - \tau_1^* - \gamma_2(r + r_1)]\, H[\tau - \tau_{km} - \tau_1^* - \gamma_2(r + r_1)]\Big\}\,,$$

$$\tau_{km} = 2k\gamma_2 r_1 + 2m\gamma_1\delta\,, \qquad \delta = 1 - r_1\,. \tag{2.46}$$

Here, for the external problem, we have $\tau_0^* = 0$ and $\tau_1^* = \gamma_1\delta$, and, for the internal one, we have $\tau_0^* = \gamma_1\delta$ and $\tau_1^* = 0$. The functions $f_{km}^{(i)}$, $i = 0, 2$, and $f_{jkm}^{(1)}$, $j = 0, 1$ correspond to the converging and diverging waves in the corresponding media. The representation in the form of a sum of the potentials (2.46) is the representation of the elementary spherical waves reflected from the boundary surfaces and refracted at the boundaries at the corresponding instants of time (Fig. 2.17 and 2.18). The presence of the Heaviside function $H(\tau)$ as the multiples in the terms proves that each of the waves appears only at a definite moment (the moment of reflection or refraction) determined by the argument of the Heaviside function. The presence of double sums in (2.46) can be explained by the presence of two bounded regions $i = 1$ and $i = 2$, that is, by the presence of two internal boundaries $r = r_1$ and $r = 1$.

The functions $\varphi_i(r, \tau)$, $i = 0, 1, 2$, defined by (2.46) satisfy the wave equations (2.44) and the conditions at infinity and at the sphere's center for arbitrary functions $f_{km}^{(i)}$, $i = 0, 2$, and $f_{jkm}^{(1)}$, $j = 0, 1$. The former functions should be determined from the boundary conditions at the external surface $r = 1$ and at the internal surface $r = r_1$ of the spherical shell. At that, in order to follow the initial conditions (2.45), it is sufficient to set

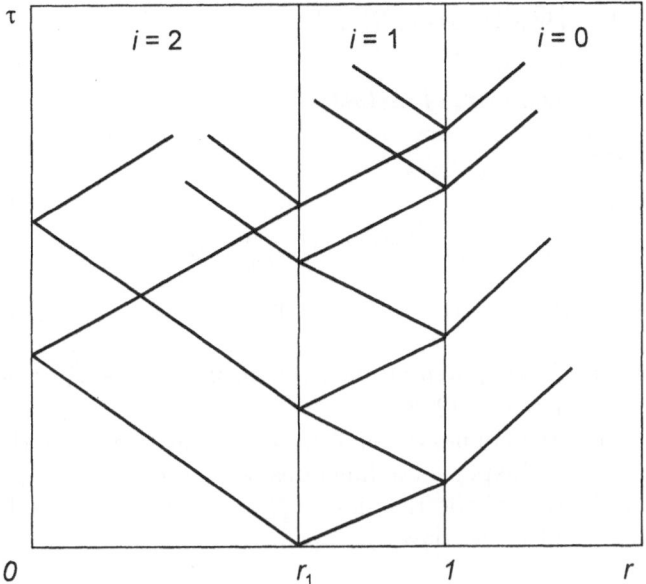

Fig. 2.18. Formation of waves. The radial vibrations of a thick-walled sphere. The internal problem

$$f_{km}^{(i)}(0) = \dot{f}_{km}^{(i)}(0) = f_{jkm}^{(1)}(0) = \dot{f}_{jkm}^{(1)}(0) = 0, \qquad i = 0, 2, \qquad j = 0, 1.$$

(2.47)

Using the relationship between the components of the stress–strain state of the elastic media and the corresponding potentials, as well as the representations (2.46), we obtain the following expressions for the displacement and stress images in the Laplace transform space with respect to time τ:

$$u_0^L(r, s) = -\frac{1}{r^2} \sum_{k, m=0}^{\infty} R_{01}(rs) f_{km}^{(0)L}(s) e^{-(r-1+\tau_{km}+\tau_0^*)s},$$

$$u_1^L(r, s) = -\frac{1}{r^2} \sum_{k, m=0}^{\infty} \left\{ R_{01}(-\gamma_1 rs) f_{0km}^{(1)L}(s) e^{-[\gamma_1(1-r)+\tau_0^*]s} \right.$$

$$\left. + R_{01}(\gamma_1 rs) f_{1km}^{(1)L}(s) e^{-[\gamma_1(r-r_1)+\tau_1^*]s} \right\} e^{-\tau_{km}s},$$

$$u_2^L(r, s) = -\frac{1}{r^2} \sum_{k, m=0}^{\infty} \left[-R_{01}(-\gamma_2 rs) e^{-\gamma_2(r_1-r)s} \right.$$

$$\left. + R_{01}(\gamma_2 rs) e^{-\gamma_2(r+r_1)s} \right] f_{km}^{(2)L}(s) e^{-(\tau_1^*+\tau_{km})s},$$

$$\sigma_{rr0}^L(r, s) = \frac{\beta_0}{r^3} \sum_{k, m=0}^{\infty} Q_{01}^{(0)}(rs) f_{km}^{(0)L}(s) e^{-(r-1+\tau_{km}+\tau_0^*)s},$$

$$\sigma_{rr1}^{L}(r,\,s) = \frac{1}{r^3} \sum_{k,\,m=0}^{\infty} \left\{ Q_{01}^{(1)}(-\gamma_1 rs)\, f_{0km}^{(1)L}(s)\, e^{-[\gamma_1(1-r)+\tau_0^*]\,s} \right.$$

$$\left. + Q_{01}^{(1)}(\gamma_1 rs)\, f_{1km}^{(1)L}(s)\, e^{-[\gamma_1(r-r_1)+\tau_1^*]\,s} \right\} e^{-\tau_{km}s}\,,$$

$$\sigma_{rr2}^{L}(r,\,s) = \frac{\beta_2}{r^3} \sum_{k,\,m=0}^{\infty} \left[-Q_{01}^{(2)}(-\gamma_2 rs)\, e^{-\gamma_2(r_1-r)\,s} \right.$$

$$\left. + Q_{01}^{(2)}(\gamma_2 r_1 s)\, e^{-\gamma_2(r+r_1)\,s} \right] f_{km}^{(2)L}(s)\, e^{-(\tau_1^*+\tau_{km})\,s}\,,$$

$$\sigma_{\theta\theta i} = \kappa_i \sigma_{rri} + (1-\kappa_i)(1+2\kappa_i)\frac{u_i}{r}\,, \qquad i = 0,\,1,\,2\,. \tag{2.48}$$

The transforms of the elastic potential φ_{2s}, the displacements u_{2s}, and the stresses $\sigma_{rrs}^{(2)}$ are determined by (2.29).

Substituting (2.48) into the boundary conditions (2.44) and equating the coefficients at similar powers of exponential functions, we obtain the recursive relationships for determination of the transforms of arbitrary functions. In the case of the external problem, we have

$$f_{0km}^{(1)L} = \frac{Y_{011}}{X_{00}} f_{1k,\,m-1}^{(1)L}\,, \qquad f_{km}^{(0)L} = \frac{Y_{031}}{X_{00}} f_{1k,\,m-1}^{(1)L}\,,$$

$$f_{1km}^{(1)L} = \frac{Z_{011}}{X_{01}} f_{0km}^{(1)L} + \frac{Z_{013}}{X_{01}} f_{k-1,\,m}^{(2)L}\,,$$

$$f_{km}^{(2)L} = \frac{Z_{031}}{X_{01}} f_{0km}^{(1)L} + \frac{Z_{033}}{X_{01}} f_{k-1,\,m}^{(2)L}\,, \qquad k,\,m > 0\,,$$

$$f_{0km}^{(1)L} = f_{1km}^{(1)L} = f_{km}^{(2)L} \equiv 0\,, \qquad k,\,m < 0\,. \tag{2.49}$$

In the case of the internal problem, we have the similar relationships

$$f_{0km}^{(1)L} = \frac{Y_{011}}{X_{00}} f_{1km}^{(1)L}\,, \qquad f_{km}^{(0)L} = \frac{Y_{031}}{X_{00}} f_{1km}^{(1)L}\,,$$

$$f_{1km}^{(1)L} = \frac{Z_{011}}{X_{01}} f_{0k,\,m-1}^{(1)L} + \frac{Z_{013}}{X_{01}} f_{k-1,\,m}^{(2)L}\,,$$

$$f_{km}^{(2)L} = \frac{Z_{031}}{X_{01}} f_{0k,\,m-1}^{(1)L} + \frac{Z_{033}}{X_{01}} f_{k-1,\,m}^{(2)L}\,, \qquad k,\,m > 0\,,$$

$$f_{0km}^{(1)L} = f_{1km}^{(1)L} = f_{km}^{(2)L} \equiv 0\,, \qquad k,\,m < 0\,. \tag{2.50}$$

The polynomials $Y_{0ij}(s)$, $Z_{0ij}(s)$, $i,\,j = 1,\,3$, $X_{00}(s)$, and $X_{01}(s)$ are determined by the following matrix relationships:

$$Y_0 = \begin{pmatrix} Y_{011} & Y_{013} \\ Y_{031} & Y_{033} \end{pmatrix}\,, \qquad Z_0 = \begin{pmatrix} Z_{011} & Z_{013} \\ Z_{031} & Z_{033} \end{pmatrix}\,,$$

$$Y_0 = -X_{00}\, M_0^{-1}\, N_0\,, \qquad Z_0 = -X_{01}\, K_0^{-1}\, L_0\,,$$

$$X_{00} = \det M_0\,, \qquad X_{01} = \det K_0\,,$$

$$\mathrm{M}_0 = \begin{pmatrix} Q_{01}^{(1)}(-\gamma_1 s) & -\beta Q_{01}^{(0)}(s) \\ R_{01}(-\gamma_1 s) & -R_{01}(s) \end{pmatrix}, \qquad \mathrm{N}_0 = \begin{pmatrix} Q_{01}^{(1)}(\gamma_1 s) & 0 \\ R_{01}(\gamma_1 s) & 0 \end{pmatrix},$$

$$\mathrm{K}_0 = \begin{pmatrix} Q_{01}^{(1)}(\gamma_1 r_1 s) & \beta_2 Q_{01}^{(2)}(-\gamma_2 r_1 s) \\ R_{01}(\gamma_1 r_1 s) & R_{01}(-\gamma_2 r_1 s) \end{pmatrix},$$

$$\mathrm{L}_0 = \begin{pmatrix} Q_{01}^{(1)}(-\gamma_1 r_1 s) & -\beta_2 Q_{01}^{(2)}(\gamma_2 r_1 s) \\ R_{01}(-\gamma_1 r_1 s) & R_{01}(\gamma_2 r_1 s) \end{pmatrix}. \tag{2.51}$$

Here, M_0^{-1} is the matrix inverse to M_0 and $\det \mathrm{M}_0$ is the determinant of the matrix M_0. The polynomials $Q_{01}^{(i)}(s)$ and $R_{01}(s)$ are determined in Sects. 2.1 and 2.2. The usage of subscripts of elements of the polynomial matrices Y_0 and Z_0 is caused by the fact that these matrices can be obtained as a particular case of the corresponding matrices from the solution of a more general axially symmetric problem to be considered in Chap. 3.

Making some manipulations in (2.51), we can obtain the following explicit expressions for the polynomials presented in the recursive relationships (2.49) and (2.50):

$$X_{00}(s) = -Q_{01}^{(1)}(-\gamma_1 s)\, R_{01}(s) + \beta_0 Q_{01}^{(0)}(s)\, R_{01}(-\gamma_1 s)\,,$$

$$Y_{011}(s) = Q_{01}^{(1)}(\gamma_1 s)\, R_{01}(s) - \beta_0 Q_{01}^{(2)}(s)\, R_{01}(\gamma_1 s)\,,$$

$$Y_{031}(s) = Q_{01}^{(1)}(\gamma_1 s)\, R_{01}(-\gamma_1 s) - Q_{01}^{(1)}(-\gamma_1 s)\, R_{01}(\gamma_1 s) = -2\,(\gamma_1 s)^3\,,$$

$$X_{01}(s) = R_{01}(-\gamma_2 r_1 s)\, Q_{01}^{(1)}(\gamma_1 r_1 s) - \beta_2 R_{01}(\gamma_1 r_1 s)\, Q_{01}^{(2)}(-\gamma_2 r_1 s)\,,$$

$$Z_{011}(s) = -R_{01}(-\gamma_2 r_1 s)\, Q_{01}^{(1)}(-\gamma_1 r_1 s) + \beta_2 R_{01}(-\gamma_1 r_1 s)\, Q_{01}^{(2)}(-\gamma_2 r_1 s)\,,$$

$$Z_{013}(s) = R_{01}(-\gamma_2 r_1 s)\, \beta_2\, Q_{01}^{(2)}(\gamma_2 r_1 s) - \beta_2 Q_{01}^{(2)}(-\gamma_2 r_1 s)\, R_{01}(\gamma_2 r_1 s)$$
$$= -2\beta_2(\gamma_2 r_1 s)^3\,,$$

$$Y_{013} = Y_{033} = 0\,,$$

$$Z_{031}(s) = R_{01}(\gamma_1 r_1 s)\, Q_{01}^{(1)}(-\gamma_1 r_1 s) - R_{01}(-\gamma_1 r_1 s)\, Q_{01}^{(1)}(\gamma_1 r_1 s)$$
$$= 2\,(\gamma_1 r_1 s)^3\,,$$

$$Z_{033}(s) = R_{01}(\gamma_2 r_1 s)\, Q_{01}^{(1)}(\gamma_1 r_1 s) - \beta_2 R_{01}(\gamma_1 r_1 s)\, Q_{01}^{(2)}(\gamma_2 r_1 s)$$
$$= -Z_{011}(-s)\,. \tag{2.52}$$

Analyzing the boundary conditions (2.44), we get the initial conditions for the recursive relationships (2.49) and (2.50) ($k = m = 0$); in the case of the external problem, we have

$$f_{000}^{(1)\mathrm{L}} = \frac{R_{01}(s)}{X_{00}(s)}\, f^{\mathrm{L}}(s)\,, \qquad f_{00}^{(0)\mathrm{L}} = -\frac{R_{01}(-\gamma_1 s)}{X_{00}(s)}\, f^{\mathrm{L}}(s)\,; \tag{2.53}$$

in the case of the internal problem, we have

$$f_{100}^{(1)\mathrm{L}} = f^{\mathrm{L}}(s)\, \frac{Z_{013}(s)}{X_{01}(s)}\,, \qquad f_{00}^{(2)\mathrm{L}} = f^{\mathrm{L}}(s)\, \frac{Z_{031}(s)}{X_{01}(s)}\,. \tag{2.54}$$

Table 2.1. Multiplicities of the poles of images. Radial vibrations

Function	s_{0j}	s_{1j}	s_α
External problem			
$f_{km}^{(0)L}$, $f_{0km}^{(1)L}$	$m+1$	$m+k$	1
$f_{1km}^{(1)L}$, $f_{km}^{(2)L}$	$m+1$	$m+k+1$	1
Internal problem			
$f_{km}^{(0)L}$, $f_{0km}^{(1)L}$	$m+1$	$m+k+1$	1
$f_{1km}^{(1)L}$, $f_{km}^{(2)L}$	m	$m+k+1$	1

Thus, when the function $f(\tau)$ is known, the transforms of arbitrary functions can be obtained by means of the relationships (2.49) and (2.53) or (2.50) and (2.54). Hence, the components of the stress–strain state for all three elastic media can be determined using (2.48). The algorithm for the exact solution of the problem under study should be supplemented by the method of calculation of the originals of the Laplace transform; this technique is presented in formulas (2.35)–(2.38). It is only necessary to know the multiplicities of poles for the transforms of arbitrary functions.

Let us assume that s_α is the pole of the function $f^L(s)$ and s_{0j} and s_{1j}, $j = 1, 2, 3, 4$, are the poles of the polynomials $X_{00}(s)$ and $X_{01}(s)$, respectively. Then, by analysis of the recursive relationships obtained, we can derive the multiplicities of the poles s_α, s_{0j}, and s_{1j} presented in Table 2.1.

Let us note that though the recursive relationships (2.49) and (2.50) are two-dimensional, it is easy to derive their solutions, that is, to obtain the explicit expressions for the transforms of arbitrary functions. In order to do this, let us apply the discrete Laurent transform (Dötsch (1974)) for analysis of the external problem.

We shall designate the transform of some sequence $\{f_k^L\}$, $k = 0, 1, \ldots$, by a bar. We have

$$\bar{f}(z) = \sum_{k=0}^{\infty} f_k^L z^{-k} , \qquad (2.55)$$

where z is the parameter of the Laurent transform.

The function $\bar{f}(z)$ is analytic outside of some circle of radius r_α. The formula for inversion is of the form (Dötsch (1974))

$$f_k^L = \frac{1}{2\pi i} \int_C \bar{f}(z) z^{k-1} \, dz , \qquad (2.56)$$

where C is an arbitrary closed contour which encloses the circle $|z| = r_\alpha$.

Applying this transform with respect to m to the system of equations (2.49), we can obtain

$$\bar{f}_{k+1}^{(2)} - A\,\bar{f}_k^{(2)} = 0\,, \qquad A = \frac{Z_{033}}{X_{01}}\frac{z-a}{z-b}\,,$$

$$a = \frac{Y_{011}}{X_{00}}\frac{Z_{033}Z_{011} - Z_{013}Z_{031}}{X_{01}Z_{033}}\,, \qquad b = \frac{Y_{011}}{X_{00}}\frac{Z_{011}}{X_{01}}\,. \tag{2.57}$$

From (2.49) it follows that

$$f_{0m}^{(2)\mathrm{L}} = \frac{R_{01}(s)}{X_{00}}\frac{Z_{031}}{X_{01}}b^m f^{\mathrm{L}}(s)$$

or (Dötsch (1974))

$$\bar{f}_0^{(2)} = f^{\mathrm{L}}(s)\frac{R_{01}(s)}{X_{00}}\frac{Z_{031}}{X_{01}}\frac{z}{z-b}\,.$$

Then, the solution of the difference equation (2.57) takes the following form:

$$\bar{f}_k^{(2)}(z) = \bar{f}_0^{(2)}A^k = f^{\mathrm{L}}(s)\frac{R_{01}(s)}{X_{00}}\frac{Z_{031}}{X_{01}}\left(\frac{Z_{033}}{X_{01}}\right)^k\frac{z(z-a)^k}{(z-b)^{k+1}}\,. \tag{2.58}$$

The function $\bar{f}_k^{(2)}(z)$ has a singularity at the pole $z = b$ of the multiplicity $k+1$. Hence, from (2.56) and (2.57), we obtain

$$f_{km}^{(2)\mathrm{L}} = \operatorname*{res}_{z=b}\bar{f}_k^{(2)}(z)\,z^{m-1}$$

$$= f^{\mathrm{L}}(s)\frac{R_{01}(s)}{X_{00}}\frac{Z_{031}}{X_{01}}\left(\frac{Z_{033}}{X_{01}}\right)^k\frac{1}{k!}\frac{\mathrm{d}^{(k)}}{\mathrm{d}z^k}\left[z^m(z-a)^k\right]\Big|_{z=b}$$

$$= f^{\mathrm{L}}(s)\frac{R_{01}(s)}{X_{00}}\left(\frac{Y_{011}}{X_{00}}\right)^m\frac{Z_{031}}{X_{01}^{m+k+1}}$$

$$\times \sum_{i=0}^{\min(k,\,m)}\binom{m}{i}\binom{k}{i}Z_{011}^{m-i}Z_{033}^{k-i}\left(Z_{013}Z_{031}\right)^i\,, \qquad k,\,m \ge 0\,, \tag{2.59}$$

where

$$\binom{n}{m} = \frac{n!}{m!\,(n-m)!}$$

is the binomial coefficient.

The functions $f_{1km}^{(1)L}$ and $f_{0km}^{(1)L}$ can be obtained similarly:

$$f_{1km}^{(1)L} = f^L(s) \frac{R_{01}(s)}{X_{00}} \left(\frac{Y_{011}}{X_{00}}\right)^m \frac{Z_{013}Z_{031}}{X_{01}^{m+k+1}} \sum_{i=0}^{\min(k-1,\,m)} \binom{m+1}{i}$$

$$\times \binom{k-1}{i} Z_{011}^{m-i} Z_{033}^{k-1-i} (Z_{013}Z_{031})^i, \qquad k \geq 1,\, m \geq 0,$$

$$f_{0km}^{(1)L} = f^L(s) \frac{R_{01}(s)}{X_{00}} \left(\frac{Y_{011}}{X_{00}}\right)^m \frac{Z_{013}Z_{031}}{X_{01}^{m+k}} \sum_{i=0}^{\min(k-1,\,m-1)} \binom{m}{i}$$

$$\times \binom{k-1}{i} Z_{011}^{m-i-1} Z_{033}^{k-i-1} (Z_{013}Z_{031})^i, \qquad k,\, m \geq 1,$$

$$f_{10m}^{(1)L} = f^L(s) \frac{R_{01}(s)}{X_{00}} \left(\frac{Z_{011}}{X_{01}}\right)^{m+1} \left(\frac{Y_{011}}{X_{00}}\right)^m, \qquad m \geq 0,$$

$$f_{0k0}^{(1)L} \equiv 0, \qquad k \geq 1,$$

$$f_{00m}^{(1)L} = f^L(s) \frac{R_{01}(s)}{X_{00}} \left(\frac{Z_{011}}{X_{01}}\right)^m \left(\frac{Y_{011}}{X_{00}}\right)^m, \qquad m \geq .0. \qquad (2.60)$$

Using (2.59), (2.60), and (2.48), we can derive the corresponding explicit formulas for the Laplace images of displacements and stresses.

The similar formulas can be derived for the internal problem:

$$f_{km}^{(2)L} = \frac{f^L(s)}{X_{01}^{m+k+1}} \left(\frac{Y_{011}}{X_{00}}\right)^m \sum_{i=0}^{\min(k,\,m-1)} \binom{m-1}{i}$$

$$\times \binom{k+1}{i+1} Z_{011}^{m-1-i} Z_{033}^{k-i} (Z_{013}Z_{031})^{i+1}, \qquad m \geq 1,$$

$$f_{1km}^{(1)L} = f^L(s) \frac{Z_{013}}{X_{01}^{m+k+1}} \left(\frac{Y_{011}}{X_{00}}\right)^m$$

$$\times \sum_{i=0}^{\min(k,\,m)} \binom{m}{i}\binom{k}{i} Z_{011}^{m-i} Z_{033}^{k-i} (Z_{013}Z_{031})^i,$$

$$f_{0km}^{(1)L} = \frac{Y_{011}}{X_{00}} f_{1km}^{(1)L}, \qquad k,\, m \geq 0,$$

$$f_{k0}^{(2)L} = f^L(s) \left(\frac{Z_{033}}{X_{01}}\right)^{k+1}, \qquad k \geq 0. \qquad (2.61)$$

2.4 Limiting Cases in Problems of Vibrations of Thick-Walled Sphere

Solutions of the problems presented in this section can be obtained as the particular cases of the problem of transient radial vibrations of a thick-walled spherical shell contacting an elastic media.

2.4.1 External and Internal Acoustic Media ($\kappa_0 = \kappa_1 = 1$)

In this case, the recursive systems of equations (2.49) and (2.50) retain their representations for the external and for the internal problems, respectively. We should only change the signs in the initial conditions (2.53) and (2.54). The expressions for the polynomials X_{00}, X_{01}, Y_{0ij}, and Z_{0ij}, $i, j = 1, 2$, are represented by the formulas (2.52) in accounting for the simplification

$$Q_{01}^{(0)}(s) = Q_{01}^{(2)}(s) = s^2. \tag{2.62}$$

2.4.2 The Absence of Internal Medium ($\beta_0 = 0$)

Here, for the internal and external problems, we can write the polynomials X_{00} and Y_{011} as follows:

$$X_{00}(s) = -Q_{01}^{(1)}(-\gamma_1 s), \qquad X_{011}(s) = Q_{01}^{(1)}(\gamma_1 s). \tag{2.63}$$

We obtain the exact solution which was derived for the case of axially symmetric vibrations in Molodtsov (1981) by means of a direct inversion of the Laplace transform applying residue functions.

2.4.3 A Solid Sphere in an Elastic Medium

We shall obtain the solution of this problem by applying the results of Sect. 2.2.

Let us set $\gamma_1 = \gamma_2 = \gamma$, $\kappa_1 = \kappa_2 = \kappa$, and $\beta_2 = 1$. Then, from (2.52), we obtain

$$Q_{01}^{(1)}(s) = Q_{01}^{(2)}(s) = Q_{01}(s),$$
$$X_{01}(s) = Q_{01}(\gamma r_1 s) R_{01}(-\gamma r_1 s) - R_{01}(\gamma r_1 s) Q_{01}(-\gamma r_1 s) = -2(\gamma r_1 s)^3,$$
$$Z_{011}(s) = Z_{033}(s) \equiv 0,$$
$$Z_{013}(s) = X_{01}(s) = -Z_{031}(s), \tag{2.64}$$

We can rewrite the recursive equations (2.49) and (2.50) in the following form:

$$f_{1km}^{(1)L} = f_{k-1,m}^{(2)L}, \qquad f_{km}^{(2)L} = -\frac{Y_{011}}{X_{00}} f_{1k,m-1}^{(1)L}. \tag{2.65}$$

The relationships supplementary to (2.65) can be obtained from the initial conditions (2.53) and (2.54); for the external problem, we get

$$f_{0km}^{(1)L} = \frac{Y_{011}}{X_{00}} f_{1k,m-1}^{(1)L}, \qquad f_{000}^{(1)L} = \frac{R_{01}(s)}{X_{00}} f^{L}(s), \tag{2.66}$$

for the internal problem, we get

$$f_{0km}^{(1)L} = \frac{Y_{011}}{X_{00}} f_{1km}^{(1)L}, \qquad f_{100}^{(1)L} = f^{L}(s), \qquad f_{00}^{(2)L} \equiv 0. \tag{2.67}$$

From (2.65), we have

$$f_{km}^{(2)L} = -\frac{Y_{011}}{X_{00}} f_{k-1,m-1}^{(2)L}. \tag{2.68}$$

Performing the Laurent transform with respect to m in (2.68), we obtain the following difference equation:

$$\bar{f}_{k+1}^{(2)} + \frac{Y_{011}}{z X_{00}} \bar{f}_{k}^{(2)} = f_{k+1,0}^{(2)L}. \tag{2.69}$$

However, from (2.68) it follows that $f_{k+1,0}^{(2)L} = f_{0,m+1}^{(2)L} \equiv 0$, $k, m = 0, 1, \dots$. Thus, we get

$$\bar{f}_{k}^{(2)} = \left(-\frac{Y_{011}}{z X_{00}}\right)^{k} \bar{f}_{0}^{(2)} = \left(-\frac{Y_{011}}{X_{00}}\right)^{k} \frac{1}{z^{k}} f_{00}^{(2)L},$$

$$f_{km}^{(2)L} = \left(-\frac{Y_{011}}{X_{00}}\right)^{k} f_{00}^{(2)L}, \qquad k = m,$$

$$f_{km}^{(2)L} \equiv 0, \qquad k \neq m. \tag{2.70}$$

Designating $f_{kk}^{(2)L} = f_{1,k-1,k}^{(1)L} = f_{k}^{L}$, from (2.48) we obtain formulas for the Laplace images of the displacements and stresses in a solid sphere in the case of the external problem:

$$u^{L}(r, s) = -\frac{1}{r^{2}} \sum_{k=0}^{\infty} \left[-R_{01}(-\gamma rs) e^{-\gamma(1-r)s} \right.$$

$$\left. + R_{01}(\gamma rs) e^{-\gamma(1+r)s} \right] f_{k}^{L}(s) e^{-2k\gamma s},$$

$$\sigma_{rr}^{L}(r, s) = \frac{1}{r^{3}} \sum_{k=0}^{\infty} \left[-Q_{01}(-\gamma rs) e^{-\gamma(1-r)s} \right.$$

$$\left. + Q_{01}(\gamma rs) e^{-\gamma(1+r)s} \right] f_{k}^{L}(s) e^{-2k\gamma s},$$

$$f_{k}^{L}(s) = -\frac{Y_{011}}{X_{00}} f_{k-1}^{L}, \qquad f_{0}^{L}(s) = -\frac{R_{01}(s)}{X_{00}} f^{L}(s). \tag{2.71}$$

A similar analysis leads us to the formulas for the internal problem. However, if in this case assuming the initial instant of time to be the first contact of the internal wave and the surface $r = 1$ (but not $r = r_1$, as it was assumed before), then the stresses and displacements will be determined by (2.71) and the initial conditions for an arbitrary function f_k^L will take the form $f_0^L(s) = f^L(s)$.

Let us note that in the absence of external medium ($\beta_0 = 0$), the solution of the external problem coincides with the solution presented in Wheeler (1973).

In order to test validity of (2.71), let us derive a steady-state value of the radial stresses $\sigma_r(r, \infty)$ in the sphere under the unit external pressure $f(\tau) = H(\tau)$. Then, $f^L(s) = s^{-1}$. We shall take into account the following limiting relationship between originals and images (Dötsch (1974)):

$$\sigma_{rr}(r, \infty) = \lim_{\tau \to \infty} \sigma_{rr}(r, \tau) = \lim_{s \to 0} s\, \sigma_{rr}^L(r, s);$$

note that because of the load discontinuity, the convergence of the series (2.71) on a closed set, which includes the point $s = 0$, is nonuniform. For this reason, in order to make a passage to the limit, we shall, first, sum up (2.71) as a series which corresponds to the infinitely decreasing geometric sequence

$$\sigma_{rr}^L(r, s) = -\frac{1}{r^3} \frac{R_{01}(s)}{s} \frac{Q_{01}(\gamma rs)\, e^{-\gamma rs} - Q_{01}(-\gamma rs)\, e^{\gamma rs}}{X_{00}(s)\, e^{\gamma s} + Y_{011}(s)\, e^{-\gamma s}}. \tag{2.72}$$

Applying the expansion of e^s into the Tailor series in the point $s = 0$ and the relationships (2.52), we can arrive at the following asymptotic formulas as $s \to 0$:

$$Q_{01}(s)\, e^{-s} - Q_{01}(-s)\, e^s = -\frac{2}{3}(1 + 2\kappa)\, s^3 + O(s^4),$$

$$R_{01}(s)\, e^{-s} - R_{01}(-s)\, e^s = \frac{2}{3} s^3 + O(s^4), \qquad s \to 0,$$

$$X_{00}(s)\, e^{\gamma s} + Y_{011}(s)\, e^{-\gamma s} = -\frac{2}{3}(\gamma s)^3 [1 + 2\kappa + 2\beta_0(1 - \kappa)] + O(s^4). \tag{2.73}$$

Then, from (2.72), we obtain

$$\sigma_{rr}(r, \infty) = -\frac{1}{1 + 2\beta_0 \dfrac{1 - \kappa}{1 + 2\kappa}}. \tag{2.74}$$

In the absence of external medium, $\beta_0 = 0$, we obtain a result which corresponds to the stress–strain state under uniform compression $\sigma_{rr}(r, \infty) = -1$.

2.4.4 Propagation of Disturbances from a Cavity

This case follows from a solution of the problem for a solid sphere with $\lambda = \mu = 0$. Taking into account that all arbitrary functions $f_k \equiv 0$ for $k \neq 0$, from (2.52) and (2.53) we get

$$u_0^{\mathrm{L}}(r, s) = -\frac{1}{r^2} R_{01}(rs) f^{(0)\mathrm{L}}(s) e^{-(r-1)s},$$

$$f^{(0)\mathrm{L}}(s) = -\frac{f^{\mathrm{L}}(s)}{\beta_0 Q_{01}^{(0)'}(s)},$$

$$\sigma_{rr0}^{\mathrm{L}}(r, s) = \frac{\beta_0}{r^3} Q_{01}^{(0)}(rs) f^{(0)\mathrm{L}}(s) e^{-(r-1)s}, \tag{2.75}$$

where, when determining the coefficient β_0, we assume that λ_1 and μ_1 are some constants, which are not related to the medium 1.

It is evident that the formulas (2.75) give us the known solution of the problem on propagation of a pressure pulse from a spherical cavity in infinite elastic space, which was studied in Sect. 2.1.

2.4.5 Reflection of an Elastic Wave from Walls of an Absolutely Rigid Reservoir

Let us assume in the solution of the problem for a solid sphere that $\beta_0 = \infty$; it corresponds to the absence of displacements at the surface $r = 1$. In this case, the displacements and stresses will be determined by (2.71); at that, when $\beta_0 = \infty$, we obtain

$$f_k^{\mathrm{L}} = -\frac{Y_{01}}{X_{00}} f_{k-1}^{\mathrm{L}} = \frac{R_{01}(\gamma s)}{R_{01}(-\gamma s)} f_{k-1}^{\mathrm{L}} = \frac{1 + \gamma s}{1 - \gamma s} f_{k-1}^{\mathrm{L}},$$

$$f_0^{\mathrm{L}} = f^{\mathrm{L}}(s). \tag{2.76}$$

Assuming $\gamma = 1$, we get a solution, presented in Babaev (1974a).

2.4.6 Free Internal Surface $r = r_1$

It is necessary to assume that $f_{km}^{(2)} \equiv 0$ (i.e., the absence of internal medium, $\beta_2 = 0$) (Grigolyuk et al. (1977a), Tarlakovsky (1975)). The arbitrary functions $f_{1km}^{(1)}$ and $f_{0km}^{(2)}$ do not depend on k. Let us designate $f_{1km}^{(1)} = f_{1m}$ and $f_{0km}^{(1)} = f_{0m}$.

Then, for the internal problem, the recursive relationships (2.49) and (2.50) will take the following form:

$$f_{0m}^{\mathrm{L}} = \frac{Y_{011}}{X_{00}} f_{1, m-1}^{\mathrm{L}}, \qquad f_{1m}^{\mathrm{L}} = \frac{Z_{011}}{X_{01}} f_{0m}^{\mathrm{L}},$$

$$f_{00}^{\mathrm{L}} = \frac{R_{01}(s)}{X_{00}} f^{\mathrm{L}}(s). \tag{2.77}$$

The displacements and stresses are determined by (2.48) (summation should be made over the index m only), the polynomials Y_{011}, X_{01}, and R_{01} are determined by (2.52), and the polynomials Z_{011} and X_{01} are as follows:

$$Z_{011}(s) = -Q_{01}^{(1)}(-\gamma_1 r_1 s), \qquad X_{01}(s) = Q_{01}^{(1)}(\gamma_1 r_1 s).$$

Let us note that when $\beta_0 = 0$ (i.e., when the external medium is absent), we obtain an exact solution of the problem discussed, for instance, in Fridman (1976b).

Then, we determine a steady-state value of the radial stresses $\sigma_{rr}^{(1)}(r, \infty)$ in a thick-walled sphere under the unit load $f(\tau) = H(\tau)$, $f^{L}(s) = s^{-1}$, and write the solution (2.77) as follows:

$$f_{0m}^{L} = \left(\frac{Y_{011}}{X_{00}}\right)^m \left(\frac{Z_{011}}{X_{01}}\right)^m \frac{R_{01}(s)}{X_{00}} f^{L}(s),$$

$$f_{1m}^{L} = \frac{Z_{011}}{X_{01}} f_{0m}^{L}. \tag{2.78}$$

Summing up the appropriate series in (2.71), we obtain

$$\sigma_{rr}^{(1)L}(r, s) = f^{L}(s) \frac{R_{01}(s)}{r^3} \frac{A(s)}{B(s)},$$

$$A(s) = Q_{01}^{(1)}(-\gamma_1 rs) X_{01}(s) e^{\gamma_1(r-r_1)s} + Q_{01}^{(1)}(\gamma_1 rs) Z_{011}(s) e^{-\gamma_1(r-r_1)s},$$

$$B(s) = X_{00}(s) X_{01}(s) e^{\gamma_1 \delta s} - Y_{011}(s) Z_{011}(s) e^{-\gamma_1 \delta s}. \tag{2.79}$$

Similarly to (2.73), it is possible to demonstrate that as $s \to 0$, the following asymptotic relationships are valid:

$$R_{01}(-\gamma_1 rs) Q_{01}^{(1)}(\gamma_1 r_1 s) e^{\gamma_1(r-r_1)s} - R_{01}(\gamma_1 rs) Q_{01}^{(1)}(-\gamma_1 r_1 s) e^{-\gamma_1(r-r_1)s}$$
$$= -\frac{2}{3} \gamma_1^3 [(1 + 2\kappa_1) r_1^3 + 2(1 - \kappa_1) r^3] + O(s^4),$$

$$R_{01}(-\gamma_1 rs) R_{01}(\gamma_1 r_1 s) e^{\gamma_1(r-r_1)s} - R_{01}(\gamma_1 rs) R_{01}(-\gamma_1 r_1 s) e^{-\gamma_1(r-r_1)s}$$
$$= -\frac{2}{3} \gamma_1^3 (r^3 - r_1^3) s^3 + O(s^4),$$

$$A(s) = \frac{4}{3} \gamma_1^3 (1 - \kappa_1)(1 + 2\kappa_1)(r^3 - r_1^3) s^3 + O(s^4),$$

$$B(s) = -\frac{4}{3} \gamma_1^3 \{(1 - \kappa_1)(1 + 2\kappa_1)(1 - r_1^3)$$
$$+ \beta_0(1 - \kappa_0)[3r_1^3 + 2(1 - \kappa_1)(1 - r_1^3)]\} s^3 + O(s^4), \quad s \to 0. \tag{2.80}$$

Passing in (2.79) to the limit as $s \to 0$, we obtain

$$\sigma_{rr}^{(1)}(r, \infty)$$
$$= -\frac{1}{r^3} \frac{(1 - \kappa_1)(1 + 2\kappa_1)(r^3 - r_1^3)}{(1 - \kappa_1)(1 + 2\kappa_1)(1 - r_1^3) + \beta_0(1 - \kappa_0)[(1 + 2\kappa_1)r_1^3 + 2(1 - \kappa_1)]}. \tag{2.81}$$

When $\beta_0 = 0$, we get the solution of the Lamé problem for a hollow sphere in the absence of internal pressure (Love (1959)):

$$\sigma_{rr}^{(1)} = -\frac{1}{1 - r_1^3}\left(1 - \frac{r_1^3}{r^3}\right).$$

A presence of the free surface $r = r_1$ in the internal problem brings us to the study of a problem of dynamic load applied to the internal surface of a thick-walled sphere located in an infinite elastic medium. At that, we ignore the wave processes inside the sphere. This problem was studied in Reismann, Gideon (1971) and Rose et al. (1973).

The transforms of displacements and stresses are determined by (2.48) at $\tau_0^* = \gamma_1 \delta$ and $\tau_1^* \equiv 0$ (summation should be made over the index m only). From (2.50), we obtain the following recursive relationships for $f_{0m}^{L} = f_{0km}^{(1)L}$ and $f_{1m}^{L} = f_{1km}^{(1)L}$:

$$f_{0m}^{L} = \frac{Y_{011}}{X_{00}} f_{1m}^{L}, \qquad f_{1m}^{L} = \frac{Z_{011}}{X_{01}} f_{0, m-1}^{L}, \tag{2.82}$$

Here, Z_{011} and X_{01} are of the same meaning as in the case of the external problem. The initial conditions in that case are determined by the relationship $\sigma_{rr}^{(1)}(r_1, \tau) = -p(\tau)$, where $p(\tau) = f(\tau)H(\tau)$ is the pressure at the surface $r = r_1$. Finally, we get the initial conditions in the form

$$f_{10}^{L}(s) = \frac{r_1^3 f^{L}(s)}{Q_{01}^{(1)}(\gamma_1 r_1 s)}. \tag{2.83}$$

2.4.7 Absolutely Rigid Fixture at the Internal Surface $r = r_1$. External Problem ($\beta_2 = \infty$, $\gamma_2 = 0$)

In this case, $f_{km}^{(2)} \equiv 0$ and other arbitrary functions do not depend on the index k. Let us designate $f_{0m}^{L} = f_{0km}^{(1)L}$ and $f_{1m}^{L} = f_{1km}^{(1)L}$. The images of the displacements and stresses are determined by (2.52) (summation should be made over the index m only). Passing in (2.49) and (2.53) to the limit as $\beta_2 \to \infty$, we obtain the recursive relationships (2.77); at that, the polynomials Z_{011} and X_{01} have the following form:

$$X_{01}(s) = -R_{01}(\gamma_1 r_1 s), \qquad Z_{011}(s) = R_{01}(-\gamma_1 r_1 s). \tag{2.84}$$

Similarly to the previous case, we shall obtain the steady-state values of the components of the stress–strain state of the sphere. For example, in accounting for (2.80), as $\tau \to \infty$, the stress $\sigma_{rr}^{(1)}(r, \tau)$ is equal to

$$\sigma_{rr}^{(1)}(r, \infty) = -\frac{1}{r^3} \frac{2(1-\kappa_1)r_1^3 + (1+2\kappa_1)r^3}{(1+2\kappa_1) + 2(1-\kappa_1)r_1^3 + 2\beta_0(1-\kappa_0)(1-r_1^3)}.$$

(2.85)

In the absence of external medium, the static solution takes the following form:

$$\sigma_{rr}^{(1)} = -\frac{1 + 2\kappa_1 + 2(1-\kappa_1)\dfrac{r_1^3}{r^3}}{(1+2\kappa_1) + 2(1-\kappa_1)r_1^3}.$$

2.5 Separation of Discontinuities in Solutions

In the problems under consideration in this chapter, we characterize the variations of load in time by the function $f(\tau)$. According to the method applied, this function may be chosen arbitrarily but it should satisfy the conditions of existence of the Laplace transform (Dötsch (1974), Lavrentev, Shabat (1987)). The loads considered are of a smooth character; that means that the function $f(\tau)$ is continuous. However, in developing a mathematical model, it is convenient to employ step functions to simulate rapidly varying loads in the vicinity of some instant τ_0. Thus, it is necessary to derive the solutions for the cases when $f(\tau)$ is a piecewise continuous function in any finite interval. Besides, when we know a solution of the dynamic problem for the load presented in the form of the Heaviside function $H(\tau)$, we can obtain a solution for the load of any type applying the Duhamel method (Arsenin (1974), Dötsch (1974)).

Let us study the problems under consideration in the case when a piecewise continuous load is applied. In order to do this, we shall use the more general problem of vibrations of a thick-walled sphere in elastic media (see Sect. 2.3).

When the load has the discontinuities in time of the first order, the same discontinuities will be inherent to the components of the stress tensor, and the displacements will be continuous in time; at that, the discontinuities for the ith medium, $i = 0, 1, 2$, will propagate along the characteristics $\tau_{0km}^{(i)} + \gamma_i r = \text{const}$ and $\tau_{1km}^{(i)} - \gamma_i r = \text{const}$, namely,

$$\tau_{1km}^{(0)} = r - 1 + \tau_{km} + \tau_0^*, \qquad \tau_{0km}^{(1)} = \gamma_1(1 - r) + \tau_{km} + \tau_0^*,$$

$$\tau_{1km}^{(1)} = \gamma_1(r - r_1) + \tau_{km} + \tau_1^*, \qquad \tau_{0km}^{(2)} = \gamma_2(r_1 - r) + \tau_{km} + \tau_1^*,$$

$$\tau_{1km}^{(2)} = \gamma_2(r + r_1) + \tau_{km} + \tau_1^*, \qquad k, m = 0, 1, \dots.$$

(2.86)

Let us designate by $[\sigma_{ri}]_{0km}$ and $[\sigma_{ri}]_{1km}$ the values of discontinuities of the radial stresses in the converging and diverging waves in the ith medium at the characteristics, which correspond to the values of time $\tau_{0km}^{(i)}$ and $\tau_{1km}^{(i)}$. Then, we can write

$$[\sigma_{ri}]_{jkm} = \lim_{\tau \to \tau_{jkm}^{(i)}+0} \sigma_{ri}(r, \tau) - \lim_{\tau \to \tau_{jkm}^{(i)}-0} \sigma_{ri}(r, \tau),$$

$$i = 0, 1, 2, \qquad j = 0, 1. \quad (2.87)$$

We shall take into account the limiting equality which relates the original $f(\tau)$ and its transform $f^L(s)$ (Dötsch (1974)),

$$\lim_{\tau \to +0} f(\tau) = \lim_{s \to \infty} s f^L(s),$$

and the following conditions, which should be satisfied by all arbitrary functions:

$$f_{km}^{(0)}(\tau) = f_{0km}^{(1)}(\tau) = f_{1km}^{(1)}(\tau) = f_{km}^{(2)}(\tau) \equiv 0, \qquad \tau < 0.$$

Then, from (2.48), for example, for $[\sigma_{r1}]_{0km}$, we obtain

$$[\sigma_{r1}]_{0km}$$

$$= \lim_{\tau \to \tau_{0km}^{(1)}+0} \frac{1}{r^3} L^{-1} \left[Q_{01}^{(1)}(-\gamma_1 r s) f_{0km}^{(1)L}(s) e^{-\gamma_1(1-r) s - (\tau_0^* + \tau_{km}) s} \right]$$

$$- \lim_{\tau \to \tau_{0km}^{(1)}-0} \frac{1}{r^3} L^{-1} \left[Q_{01}^{(1)}(-\gamma_1 r s) f_{0km}^{(1)L}(s) e^{-\gamma_1(1-r) s - (\tau_0^* + \tau_{km}) s} \right]$$

$$= \lim_{\tau \to +0} \frac{1}{r^3} L^{-1} \left[Q_{01}^{(1)}(-\gamma_1 r s) f_{0km}^{(1)L}(s) \right]$$

$$= \frac{1}{r^3} \lim_{s \to \infty} s Q_{01}^{(1)}(-\gamma_1 r s) f_{0km}^{(1)L}(s).$$

Here, $L^{-1}[f^L(s)]$ is the original of the transform $f^L(s)$. Similarly, we obtain

$$[\sigma_{r0}]_{1km} = \frac{\beta_0}{r^3} \lim_{s \to \infty} s Q_{01}^{(0)}(rs) f_{km}^{(0)L}(s),$$

$$[\sigma_{r1}]_{1km} = \frac{1}{r^3} \lim_{s \to \infty} s Q_{01}^{(1)}(\gamma_1 rs) f_{1km}^{(1)L}(s),$$

$$[\sigma_{r2}]_{0km} = -\frac{\beta_2}{r^3} \lim_{s \to \infty} s Q_{01}^{(2)}(-\gamma_2 rs) f_{km}^{(2)L}(s),$$

$$[\sigma_{r2}]_{1km} = \frac{\beta_2}{r^3} \lim_{s \to \infty} s Q_{01}^{(2)}(\gamma_2 rs) f_{km}^{(2)L}(s). \quad (2.88)$$

Making the necessary rearrangements and passages to the limits as $s \to \infty$ in (2.49), (2.53) and (2.50), (2.54), we obtain the following recursive relationships for the discontinuities.

The external problem:

$$[\sigma_{r1}]_{000} = \lim_{s\to\infty} \frac{s\,f^{L}(s)\,R_{01}(s)}{r^3 X_{00}(s)}\,Q_{01}^{(1)}(-\gamma_1 rs) = -\frac{1}{r}\,\frac{\gamma_1}{\gamma_1 + \beta_0}\,,$$

$$[\sigma_{r0}]_{100} = -\lim_{s\to\infty} \frac{\beta_0 s\,f^{L}(s)\,R_{01}(-\gamma_1 s)}{r^3 X_{00}(s)}\,Q_{01}^{(0)}(rs) = \frac{1}{r}\,\frac{\beta_0}{\gamma_1 + \beta_0}\,,$$

$$[\sigma_{r1}]_{0km} = a_{00}[\sigma_{r1}]_{1k,\,m-1}\,,$$

$$[\sigma_{r2}]_{0km} = -[\sigma_{r2}]_{1km}\,,$$

$$[\sigma_{r1}]_{1km} = a_{11}[\sigma_{r1}]_{0km} + a_{12}[\sigma_{r2}]_{1,\,k-1,\,m}\,,$$

$$[\sigma_{r2}]_{0km} = a_{21}[\sigma_{r1}]_{0km} + a_{22}[\sigma_{r2}]_{1,\,k-1,\,m}\,. \tag{2.89}$$

The internal problem:

$$[\sigma_{r1}]_{100} = \lim_{s\to\infty} \frac{s\,f^{L}(s)\,Z_{013}(s)}{r^3 X_{01}(s)}\,Q_{01}^{(1)}(\gamma_1 rs) = \frac{1}{r}\,\frac{2r_1\gamma_1}{\gamma_1 + \beta_2\gamma_2}\,,$$

$$[\sigma_{r2}]_{000} = -\lim_{s\to\infty} \frac{\beta_2 s\,f^{L}(s)\,Z_{033}(s)}{r^3 X_{01}(s)}\,Q_{02}^{(2)}(-\gamma_2 rs) = -\frac{1}{r}\,\frac{r_1(\gamma_1 - \beta_2\gamma_2)}{\gamma_1 + \beta_2\gamma_2}\,,$$

$$[\sigma_{r1}]_{0km} = a_{00}[\sigma_{r1}]_{1km}\,,$$

$$[\sigma_{r2}]_{0km} = -[\sigma_{r2}]_{1km}\,,$$

$$[\sigma_{r1}]_{1km} = a_{11}[\sigma_{r1}]_{0k,\,m-1} + a_{12}[\sigma_{r2}]_{1,\,k-1,\,m}\,,$$

$$[\sigma_{r2}]_{0km} = a_{21}[\sigma_{r1}]_{0k,\,m-1} + a_{22}[\sigma_{r2}]_{1,\,k-1,\,m}\,. \tag{2.90}$$

Deriving (2.89) and (2.90), we have assumed that the function $f(\tau)$, which determines the law of variation of the load, is chosen in such a manner that, for the external problem,

$$\lim_{s\to\infty} s\,f^{L}(s) = 1\,,$$

and for the internal one,

$$\lim_{s\to\infty} \beta_2 r_1^{-3} s\,f^{L}(s)\,Q_{01}^{(2)}(\gamma_2 r_1 s) = 1\,.$$

The coefficients a_{00} and a_{ij}, $i, j = 1, 2$, can be determined by the following formulas:

$$a_{00} = \lim_{s\to\infty} \frac{Y_{011}(s)}{X_{00}(s)}\,\frac{Q_{01}^{(1)}(-\gamma_1 rs)}{Q_{01}^{(1)}(\gamma_1 rs)} = -\frac{\gamma_1 - \beta_0}{\gamma_1 + \beta_0}\,,$$

$$a_{11} = \lim_{s\to\infty} \frac{Z_{011}(s)}{X_{01}(s)}\,\frac{Q_{01}^{(1)}(\gamma_1 rs)}{Q_{01}^{(1)}(-\gamma_1 rs)} = -\frac{\gamma_1 - \beta_2\gamma_2}{\gamma_1 + \beta_2\gamma_2}\,,$$

$$a_{12} = \lim_{s\to\infty} \frac{Z_{013}(s)}{X_{01}(s)}\,\frac{Q_{01}^{(1)}(\gamma_1 rs)}{\beta_2 Q_{01}^{(2)}(\gamma_2 rs)} = \frac{2\gamma_1}{\gamma_1 + \beta_2\gamma_2}\,,$$

$$a_{21} = -\lim_{s \to \infty} \frac{Z_{031}(s)}{X_{01}(s)} \frac{\beta_2 Q_{01}^{(2)}(-\gamma_2 rs)}{Q_{01}^{(1)}(-\gamma_1 rs)} = \frac{2\gamma_2 \beta_2}{\gamma_1 + \beta_2 \gamma_2},$$

$$a_{22} = -\lim_{s \to \infty} \frac{Z_{033}(s)}{X_{01}(s)} \frac{Q_{01}^{(2)}(-\gamma_2 rs)}{Q_{01}^{(2)}(\gamma_2 rs)} = \frac{\gamma_1 - \beta_2 \gamma_2}{\gamma_1 + \beta_2 \gamma_2}. \tag{2.91}$$

The recursive relationships (2.89) and (2.90) allow us to calculate successively all the discontinuities of the radial stresses. From (2.48) it follows that, because of continuity of the displacements u_i, the hoop stresses $\sigma_{\theta i}$, $i = 0, 1, 2$, have discontinuities which differ from those of the radial stresses by the multiple κ_i.

It is easy to derive the general solution (2.89) or (2.90). However, we can also derive it when applying the general expressions for arbitrary functions (2.59), (2.60) or (2.61). For example, let us consider the external problem; then, assuming $k > 1$ and $m > 1$ and making a passage to the limit as $s \to \infty$, from (2.48) and (2.59), (2.60) we can derive $[\sigma_{r1}]_{0km}$ as follows:

$$[\sigma_{r1}]_{0km} = \frac{1}{r} \frac{\gamma_1(\gamma_1 - \beta_0)^m}{(\gamma_1 + \beta_0)^{m+1}(\gamma_1 + \beta_2 \gamma_2)^{m+k}}$$

$$\times \sum_{i=0}^{\min(k-1,\, m-1)} \binom{m}{i}\binom{k-1}{i} (4\gamma_1 \gamma_2)^{i+1} (\gamma_1 - \beta_2 \gamma_2)^{m+k-2(i+1)}.$$

It should be pointed out that the effect of thick-walled spherical shell is evident; in particular, the pressure in the surrounding medium at the sphere's surface at $\tau = 0$ differs from those at the thin-walled or absolutely rigid obstacles. Correspondingly, we have: (1) for the external problem, $\gamma_1/(\gamma_1 + \beta_0)$ and 1 (Grigolyuk, Gorshkov (1976)); (2) for the internal problem, $2\gamma_1/(\gamma_1 + \beta_2 \gamma_2)$ and 2 (Babaev et al. (1980), Babaev (1974a)). The same result was obtained in Veksler (1975) for acoustic media and axially symmetric external problem.

Some numerical results for the problem of vibrations of a thick-walled spherical shell contacting the elastic media are presented in Figs. 2.19–2.26 (external problem) and in Figs. 2.27–2.31 (internal problem). In all these cases it was assumed that the shell's material is steel ($\kappa_1 = 0.393$).

Let us assume the following law of variation of load for the external problem:

$$f(\tau) = e^{-\alpha \tau} H(\tau), \qquad f^{\mathrm{L}}(s) = \frac{1}{s + \alpha}; \tag{2.92}$$

for the internal problem, we assume

$$f(\tau) = \frac{r_1}{2\gamma_2^2 \beta_2} \tau^2 e^{-\alpha \tau} H(\tau), \qquad f^{\mathrm{L}}(s) = \frac{r_1}{\beta_2 \gamma_2^2} \frac{1}{(s + \alpha)^3}. \tag{2.93}$$

If we take the function $f(\tau)$ in the form of (2.93), then the radial stress in the incoming wave at the surface $r = r_1$ at $\tau = 0$ is equal to unit, that is,

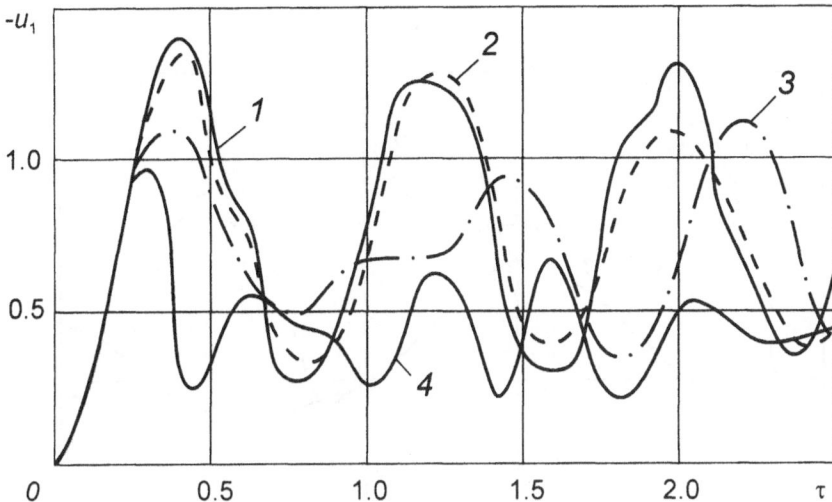

Fig. 2.19. The radial displacement at the external surface of a thick-walled shell in the case of acoustic surrounding media

$$\sigma^{(2)}_{rrs}(r, 0) = \lim_{s\to\infty} s\sigma^{(2)L}_{rrs}(r, s) = \lim_{s\to\infty} \beta_2 s\, f^L(s)\, \frac{Q^{(2)}_{01}(\gamma_2 r_1 s)}{r_1^3} = 1.$$

In Figs. 2.19 and 2.20, we can see the influence of the surrounding and filling media on the radial displacement of the shell $u_1(1, \tau)$ for its relative thickness $\delta = 0.5$ ($\alpha = 0$). The curves presented correspond to vibrations of various systems (the external medium is on the first position and the external one is on the last). In Fig. 2.19, curve *1* corresponds to the system 'water – steel – free surface' ($\gamma_1 = 0.252$; $\beta_0 = 0.00862$), curve *2* corresponds to the system 'water – steel – water' ($\gamma_2 = 1$; $\beta_2 = 0.00862$), curve *3* corresponds to the system 'water – steel – mercury' ($\gamma_2 = 1$; $\beta_2 = 0.109$), and curve *4* corresponds to the system 'water – steel – restraint'. In Fig. 2.20, curve *1* corresponds to the system 'granite – steel – aluminum' ($\gamma_1 = 0.691$; $\beta_0 = 0.167$; $\gamma_2 = 0.648$, $\beta_2 = 0.388$) and curve *2* corresponds to the system 'water – steel – water'.

The curves in Fig. 2.19 coincide until the arrival of a reflected wave at the external surface. Before that, the influence of the shell's filler is not noticeable. It follows from the curves in Fig. 2.20 that the character of the displacements in the elastic and acoustic media which surround the shell is very different.

In Fig. 2.21, we show the influence of the thickness of obstacle on the radial displacements of the shell $u_1(1, \tau)$ (the internal surface is free). The curves were obtained for the following values of thickness: curve *1* corresponds to $\delta = 0.5$, curve *2* corresponds to $\delta = 0.3$, curve *3* corresponds to $\delta = 0.1$, curve *4* corresponds to $\delta = 0.05$, and curve *5* corresponds to $\delta = 0.01$. The dashed straight lines correspond to the steady-state values of displace-

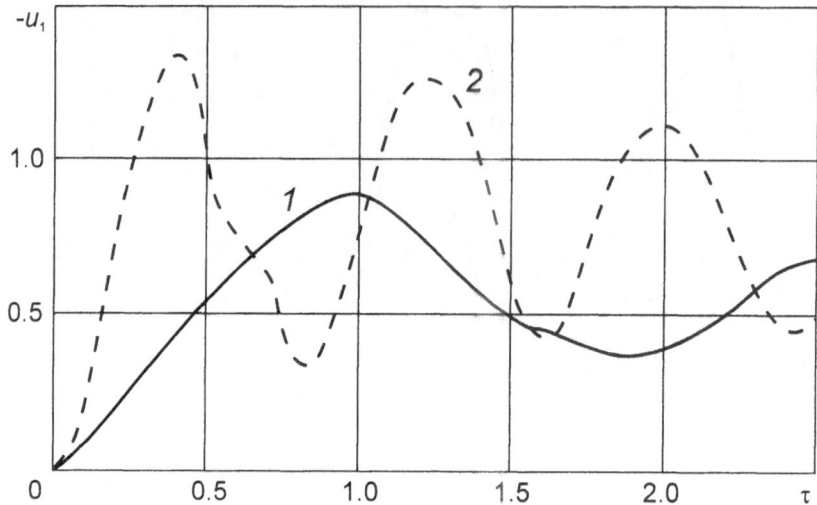

Fig. 2.20. The radial displacement at the external surface of a thick-walled shell in the case of elastic surrounding media

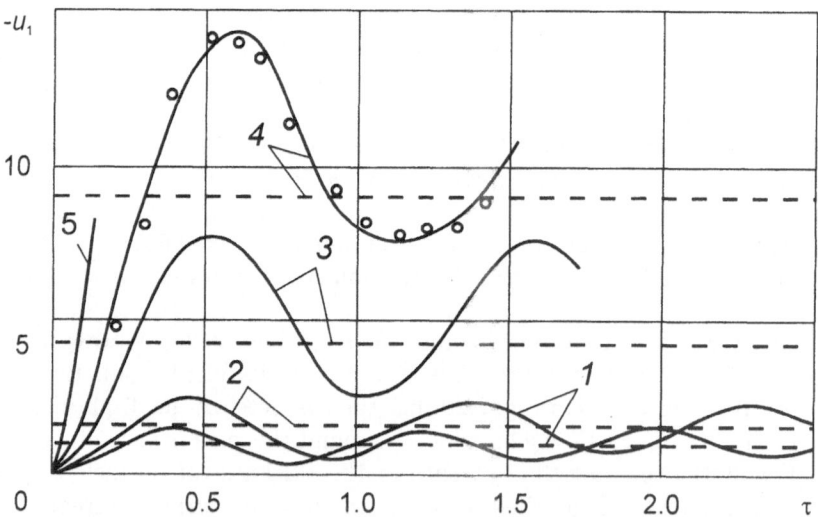

Fig. 2.21. The influence of the shell thickness on the displacements at the external surface (the internal surface is free)

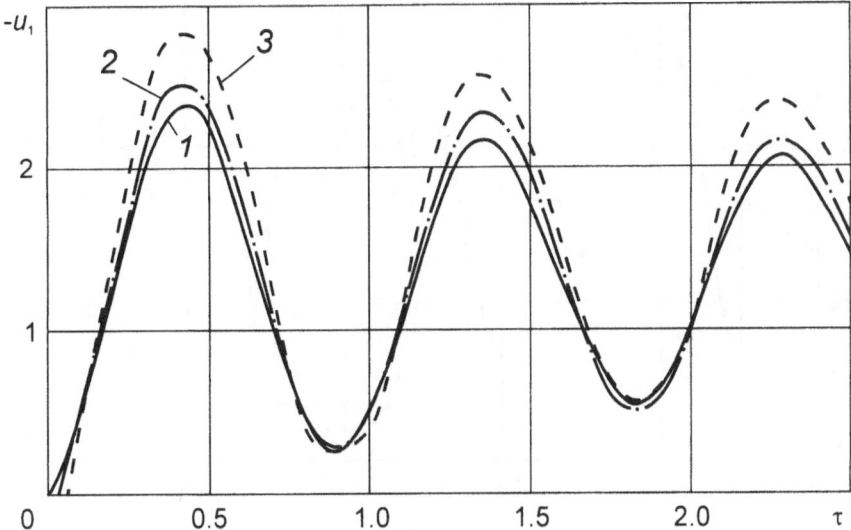

Fig. 2.22. The vibrations of some points of the shell

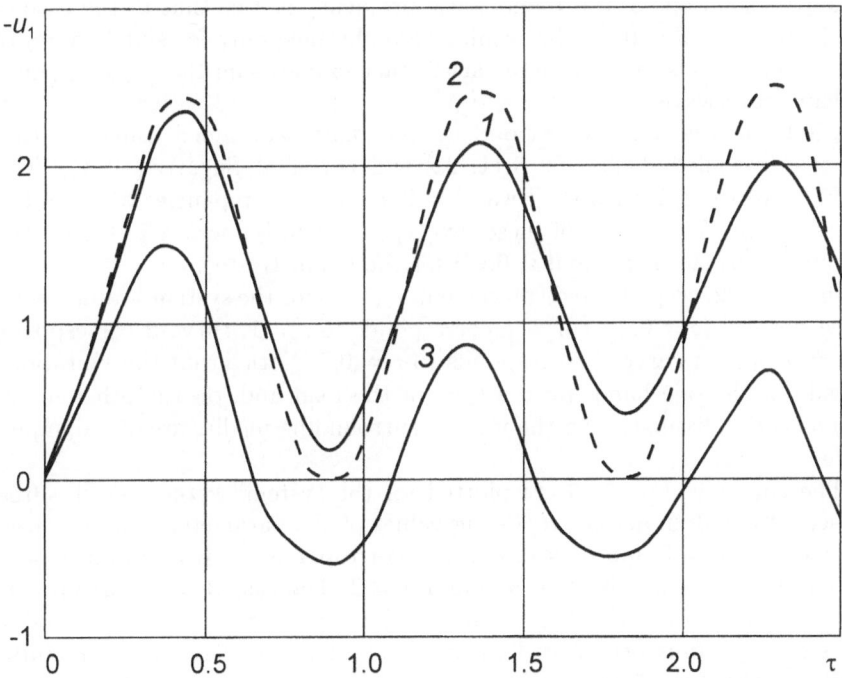

Fig. 2.23. The influence of presence of an external medium and of the load damping coefficient on the displacements at the surface $r = 1$

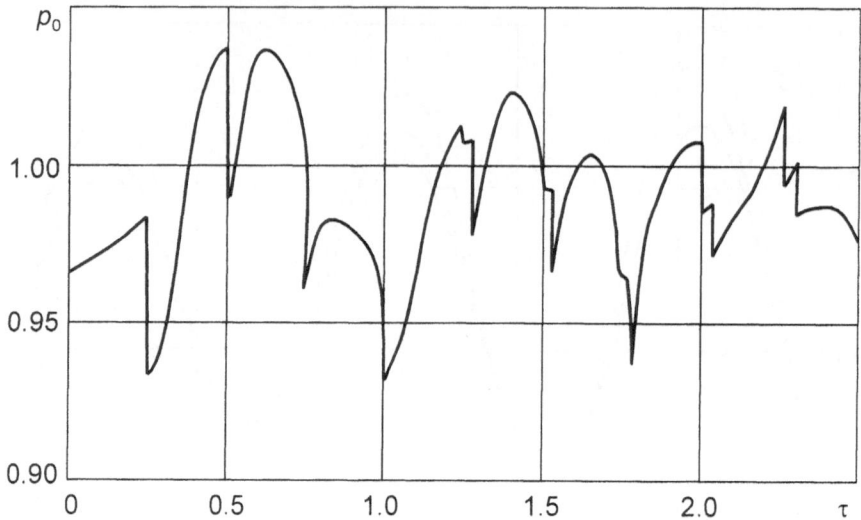

Fig. 2.24. The pressure at the external surface in the acoustic medium (water) surrounding the shell

ment (the solutions of the Lamé static problem). Notice that in the case of small thickness, $\delta = 0.01$, the results were obtained only for small values of time τ since in this case it is necessary to take into account the higher number of elementary waves.

Circles in Fig. 2.21 correspond to the solution obtained using the theory of thin-walled shells (see Sect. 2.2 and Forrestal, Sagartz (1971)). The thickness of the spherical shell was $\delta = 0.05$. The discrepancies between the results obtained by means of these two types of solution was $\approx 11.8\%$ for the maximum displacements and $\approx 9.4\%$ for the hoop stresses.

In Fig. 2.22, we plot the displacement $u_1(1, \tau)$ for the system 'water – steel – free surface' ($\delta = 0.3$); curve 1 corresponds to $r = 1$, curve 2 corresponds to $r = 0.85$, and curve 3 corresponds to $r = 0.7$. Notice that the vibrations caused, on the one hand, by the type of the load and, on the other hand, by the energy dissipation in the infinite surrounding media, are of a damped mode.

The curves in Fig. 2.23 are plotted for the system 'water – steel – free surface', $\delta = 0.3$, under the following values of the parameters: curve 1 corresponds to $\beta_0 \neq 0$ and $\alpha = 0$, curve 2 corresponds to $\beta_0 = 0$ and $\alpha = 0$, and curve 3 corresponds to $\beta_0 \neq 0$ and $\alpha = 2$. The case $\beta_0 = 0$ corresponds to the absence of surrounding medium. The comparison of curves 1 and 2 shows that the presence of media produces a damping effect. From 1 and 3 it follows that if the load decreases by exponential law, the vibrations are of a rapidly damped mode.

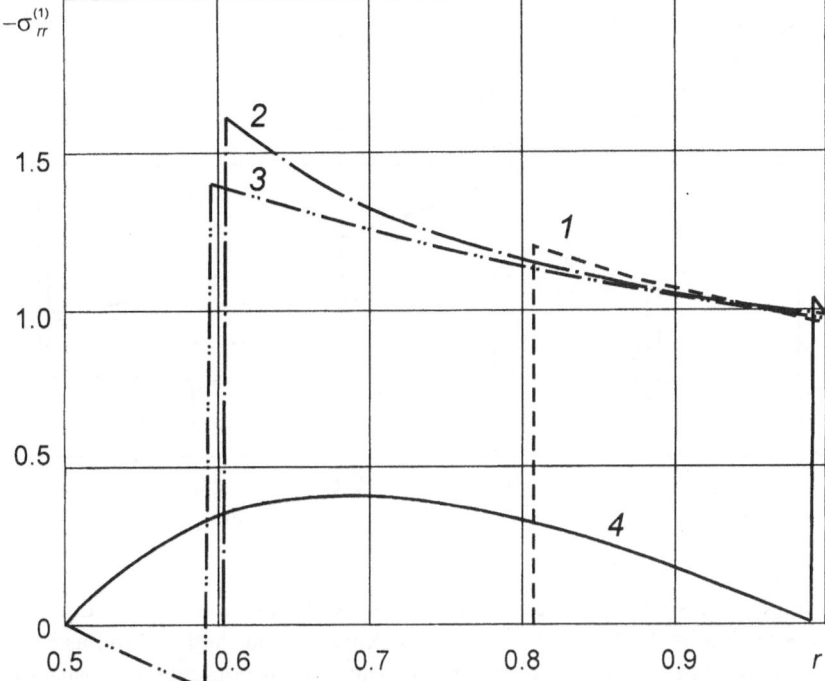

Fig. 2.25. Propagation of a wave of radial stresses in a shell towards the radius

In Fig. 2.24 we illustrate the discontinuous character of the pressure variations at the external shell's surface $p_0 = \sigma_{r1}$ (the system 'water – steel – water', $\delta = 0.5$). Here, the static solution (the Lamé problem) is $p_0 = 1$. The instants of arrival of the discontinuities correspond to the time of arrival at the given point of the reflected and refracted elementary waves. Contrary to thin-walled shells, as we have stated before, the pressure differs from the unit at the initial time $\tau = 0$.

The character of propagation of the stresses in radial waves $\sigma_{rr}^{(1)}$ and in circular waves $\sigma_{\theta\theta}^{(1)}$ through the shell thickness is presented in Figs. 2.25 and 2.26 (the system 'water – steel – free surface', $\delta = 0.5$); at that, curves 1 ($\tau = 0.05$) and 2 ($\tau = 0.1$) correspond to the first converging wave, and curves 3 ($\tau = 0.15$) and 4 ($\tau = 0.25$) correspond to the first diverging wave.

The curves in Figs. 2.27–2.31 are obtained for the internal problem. In Fig. 2.27, we demonstrate the influence of different media on the displacement $u_1(r_1, \tau)$ ($\delta = 0.5$); curve 1 corresponds to the system 'water – steel – water', curve 2 corresponds to the system 'water – steel – mercury', and curve 3 corresponds to the system 'granite – steel – aluminum'. Curve 4 corresponds to the load with the damping coefficient $\alpha = 2$.

The circles correspond to $\beta_2 = 0$, and the crosses correspond to $\beta_0 = 0$ (thus, we do not account for the internal and for the external media, corre-

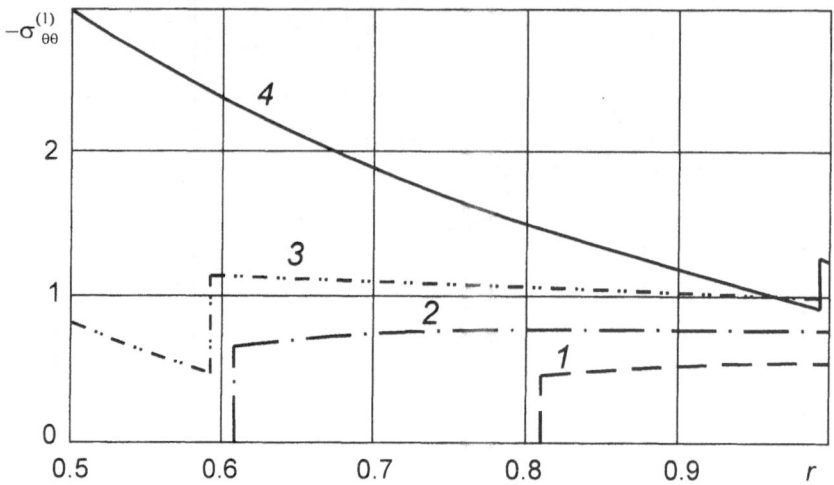

Fig. 2.26. Propagation of a wave of hoop stresses in a shell in radial direction

spondingly). A slight discrepancy with the exact solution, $\approx 5\%$, described by curve *1* can be explained by a large thickness of the shell, as well as by the fact that we analyze radial vibrations. In the case of non-central positioning of the source, we shall expect more influence of the surrounding media.

Figure 2.28 illustrates the variation of the radial displacements u_1 in time for different values of thickness of the shell for the system 'water – steel – water'. Curve *1* corresponds to $\delta = 0.5$, curve *2* corresponds to $\delta = 0.3$, curve *3* corresponds to $\delta = 0.1$, and curve *4* corresponds to $\delta = 0.05$. At that, the solid curves correspond to the internal surface, $r = r_1$, and the dashed ones correspond to the external one, $r = 1$. It is evident that the difference of the displacements at the external and internal surfaces of the shell significantly decreases with a decrease in the thickness.

From Fig. 2.29 it is seen that the pressure at the sphere's internal surface $p_2 = -\sigma_{rr}^{(1)}(r_1, \tau)$ depend on time τ discontinuously. The results for the system 'water – steel – water' are presented for the following values of thickness of the shell: curve *1* corresponds to $\delta = 0.5$, curve *2* corresponds to $\delta = 0.3$, and curve *3* corresponds to $\delta = 0.1$. Let us note that the presence of discontinuities is caused not only by the discontinuities of the external load at the front of the incoming wave, but by the accounting for the elementary waves inside the sphere, as well. The abscissas of discontinuities correspond to the arrival at the internal surface of the waves of both types: the waves reflected from the center in the medium $i = 2$ (i.e., the diverging waves) and the waves in the shell. Curve *4* was plotted for $\beta_2 = 0$.

Consider the initial time $\tau = 0$. Contrary to thin-walled shells, the pressure p_2 at the internal surface of a thick-walled shell is not equal to two; as we have mentioned earlier, that is a peculiarity of thick-walled obstacles.

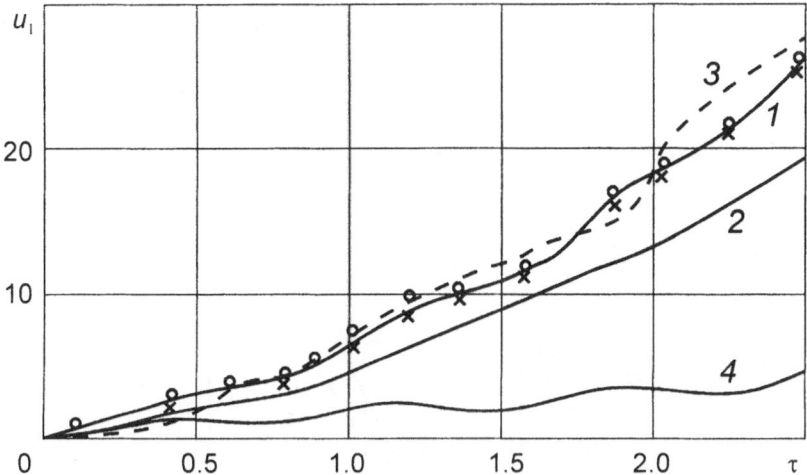

Fig. 2.27. The influence of the characteristics of the media surrounding a shell on the displacement of the internal surface

It should also be mentioned that an undamped mode of displacements and stresses in the case of internal problems is caused by the fact that an integral pulse of the shock wave is not equal to zero (Babaev (1974b)).

In Fig. 2.30, we plot the pressure at the external surface of the sphere $p_0 = -\sigma_{rr}^{(1)}(1, \tau)$ versus time for the system 'water – steel – water'. Curves *1* and *2* correspond to $\delta = 0.3$ and $\delta = 0.1$, respectively. It follows that, first, the variation of pressure is equal to zero until the arrival of the first elementary elastic wave in the shell, and, second, the values of peaks of pressure increase with a decrease in the obstacle thickness.

In Fig. 2.31, we present the distribution of the hoop stresses $\sigma_{\theta\theta}^{(1)}$ through the shell thickness at $\delta = 0.5$ for the system 'water – steel – water'. Curve *1* corresponds to the first elastic wave in the sphere ($\tau = 0.1$), curve *2* corresponds to the wave reflected from the external surface ($\tau = 0.2$), curve *3* corresponds to the arrival of the first wave reflected from the sphere's center at the internal surface ($\tau = 1.2$), and curve *4* corresponds to the presence of two waves reflected from the center ($\tau = 2.2$).

Applying the results of Sect. 2.2, we have solved an internal problem of transient radial vibrations of the system 'water – thin-walled spherical shell – water'. For acoustic media, the similar results were first obtained in Babaev, Kubenko (1977b). For $\delta = 0.05$, the numerical results are plotted in Fig. 2.28 by circles. The maximal discrepancy on the time interval $0 \le \tau \le 1.2$ is $\approx 6\%$ for the displacements and $\approx 10\%$ for the stresses (when making comparison of the stresses, we did not consider the initial stage of interaction since the behavior of a thin-walled shell and the behavior of a three-dimensional elastic body differ significantly). The proximity of the re-

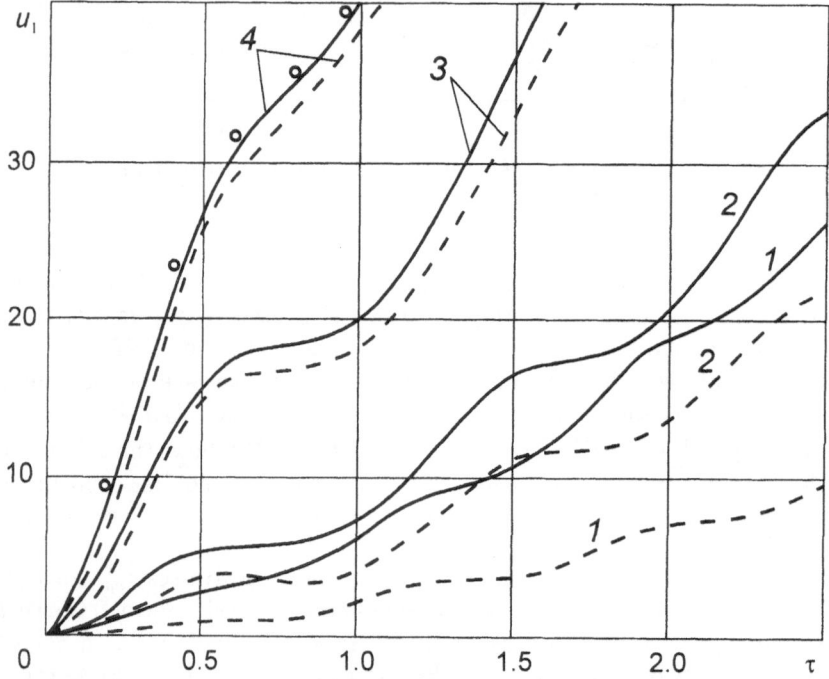

Fig. 2.28. The displacements of the external surface of a shell (solid lines) and of its internal surface (dashed lines) for the shells of various thickness

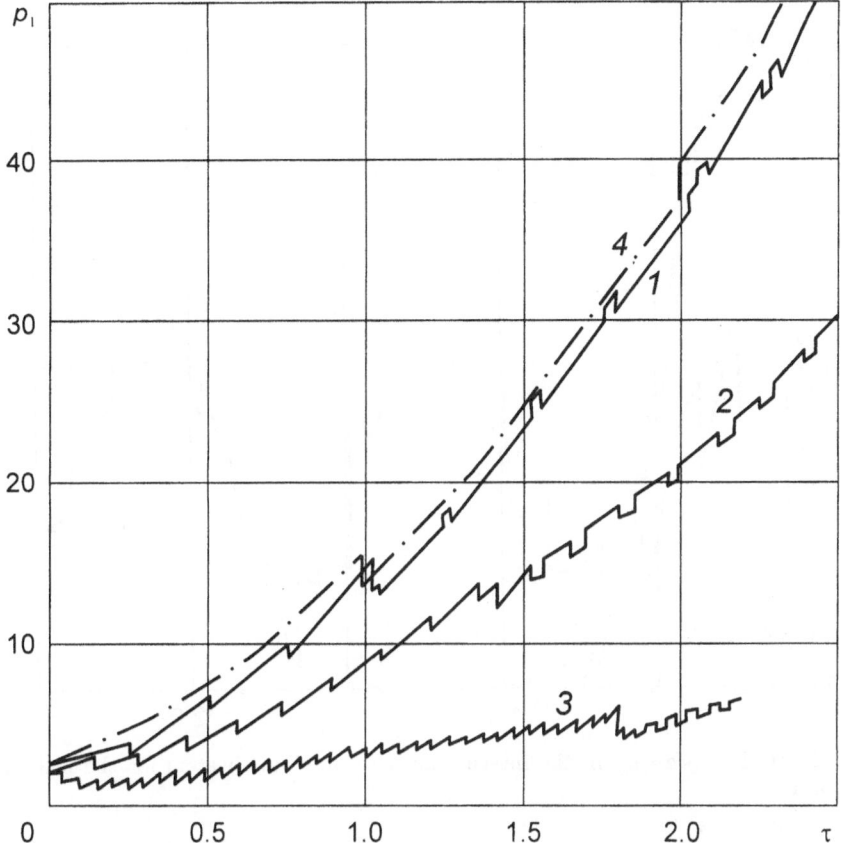

Fig. 2.29. The character of variation of the pressure in the internal acoustic medium at the surface $r = r_1$ for the shells of various thickness

sults obtained by application of these two models, first, witnesses in favor of application of the equations for thin-walled shells on the time intervals which do not include the initial stage and, second, can be explained by the fact that we have analyzed radial vibrations only.

2.6 Vibrations of a Piecewise Homogeneous Elastic Space with Concentric Spherical Interfaces

Considering the problems of explosion in the ground, it makes interest to study propagation of elastic spherical waves in inhomogeneous space from a point source.

In the case of an inhomogeneous medium, its properties can be approximated by piecewise functions. The problems of such a kind can be regarded as the transient ones; these problems are intermediate between the problems

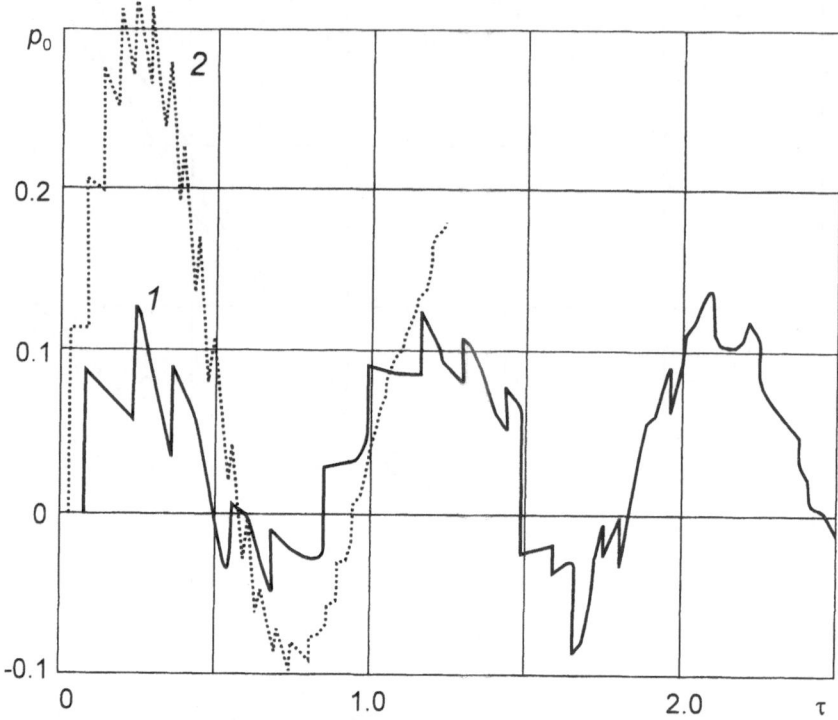

Fig. 2.30. The pressure in the internal acoustic medium (water) for the internal problem

of interaction of homogeneous elastic media and the problems of propagation of waves in inhomogeneous media. In most of the works published until now, only radial vibrations of the media with an arbitrary number of concentric spherical interfaces are analyzed.

From here on, following the work Gorshkov, Tarlakovsky (1981), we shall study the problems of propagation of elastic spherical waves generated by a point source placed in an infinite isotropic space separated by concentric spherical interfaces into $n + 2$ homogeneous media of different properties.

Let us consider an infinite linearly elastic isotropic space separated into $n + 2$ layers (media) by the concentric spherical interfaces of radii $R = R_i$, $i = 0, 1, \ldots, n$, $R_0 > R_1 > \ldots > R_n$, with the center at the point O (see Fig. 2.32). Consequently, the layer designated by the symbol 'i' is bounded by the internal surface of radius $R = R_i$ and by the external surface of radius $R = R_{i-1}$. Thus, the external infinite medium is designated by the subscript $i = 0$, and the internal one (solid sphere) is designated by $i = n + 1$. Each ith medium is homogeneous and is characterized by the Lamé elastic constants λ_i and μ_i and the density ρ_i.

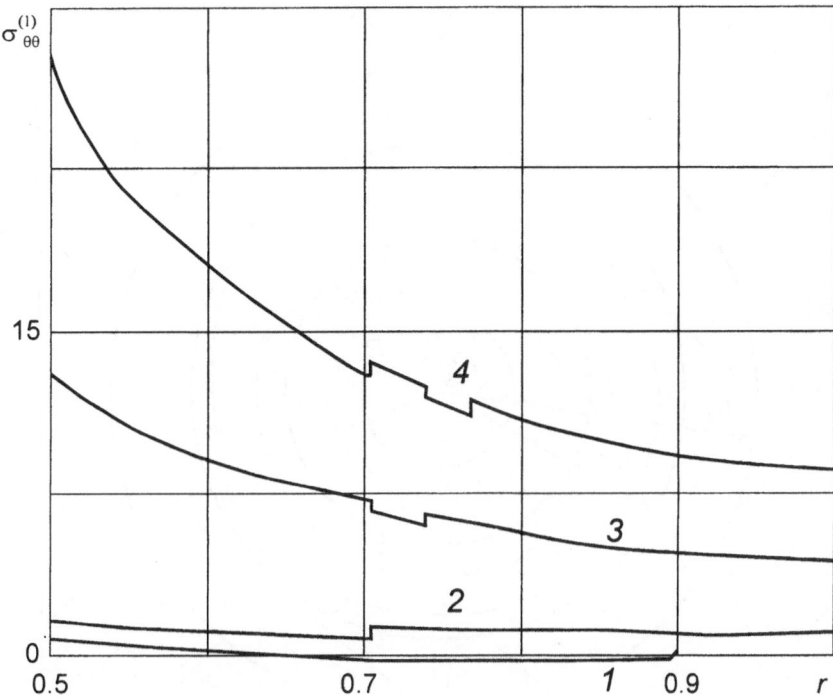

Fig. 2.31. The distribution of the hoop stresses in a shell along the radius at different instants of time

At the initial time $t = 0$, the surface $R = R_n$ is contacted by a front of an elastic spherical wave generated by a source located at the point O; the source is characterized by the potential

$$\varphi_s(r, \tau) = \frac{K r_n f[\tau - \gamma_{n+1}(r - r_n)]}{r}, \tag{2.94}$$

where f is the function which determines the law of variation of the potential in time.

From here on, similarly to (1.11), the following dimensionless values are introduced:

$$r = \frac{R}{R_0}, \qquad \tau = \frac{c_0 t}{R_0}, \qquad r_i = \frac{R_i}{R_0}, \qquad \tilde{u}_i = \frac{u_i}{R_0},$$

$$\tilde{\varphi}_i = \frac{\varphi_i}{R_0^2}, \qquad \tilde{\sigma}_{ri} = \frac{\sigma_{ri}}{\lambda_0 + 2\mu_0}, \qquad \kappa_i = \frac{\lambda_i}{\lambda_i + 2\mu_i}, \qquad \gamma_i = \frac{c_0}{c_i},$$

$$\beta_i = \frac{\lambda_i + 2\mu_i}{\lambda_0 + 2\mu_0}, \qquad c_i = \sqrt{\frac{\lambda_i + 2\mu_i}{\rho_i}}, \qquad i = 0, 1, \ldots, n + 1; \tag{2.95}$$

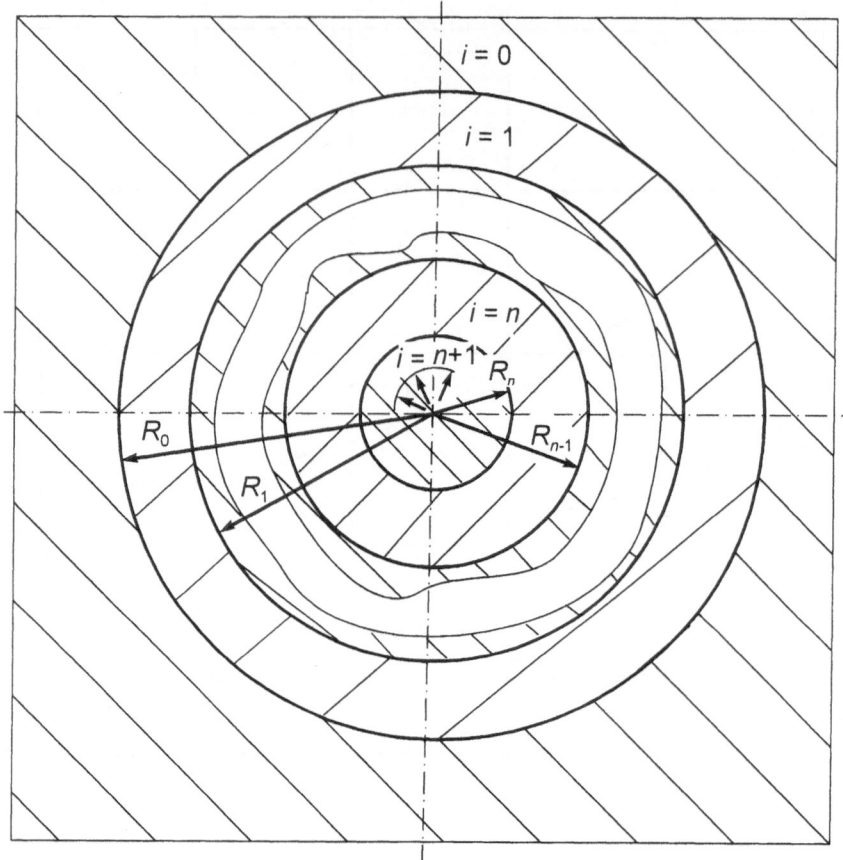

Fig. 2.32. To the problem of propagation of a spherical wave in a piecewise homogeneous elastic space

here, a tilde designates dimensionless values (recall that we omit tildes in dimensionless values), u_i and σ_{ri} are the radial displacement and stress, φ_i and c_i are the elastic potential and the speed of propagation of compression waves.

The functions u_i, σ_{ri}, and φ_i are related by the following dependences:

$$u_i = \frac{\partial \varphi_i}{\partial r}, \qquad \sigma_{ri} = \beta_i \left(\frac{\partial u_i}{\partial r} + \frac{2\kappa_i}{r} u_i \right). \tag{2.96}$$

The radial vibrations of the ith medium can be described by the equation

$$\gamma_i^2 \frac{\partial^2 \varphi_i}{\partial \tau^2} = \frac{\partial^2 \varphi_i}{\partial r^2} + \frac{2}{r} \frac{\partial \varphi_i}{\partial r},$$

$$\varphi_i(r, 0) = \frac{\partial \varphi_i}{\partial \tau}(r, 0) = 0, \qquad i = 0, 1, \dots, n+1. \tag{2.97}$$

Let us represent the conditions of media contact, which presume continuity of the displacements and radial stresses, in the following form:

$$u_i(r_i, \tau) = u_{i+1}(r_i, \tau),$$
$$\sigma_{ri}(r_i, \tau) = \sigma_{r, i+1}(r_i, \tau), \qquad i = 0, 1, \ldots, n-1,$$
$$u_n(r_n, \tau) = u_{n+1}(r_n, \tau) + u_s(r_n, \tau),$$
$$\sigma_{rn}(r_n, \tau) = \sigma_{r, n+1}(r_n, \tau) + \sigma_{rs}(r_n, \tau); \tag{2.98}$$

here, u_s and σ_{rs} are the radial displacement and stress in the incoming wave.

At infinity and at the center of space O, the corresponding potentials of disturbed motion should be finite. Thus, we get the conditions

$$\lim_{r \to \infty} \varphi_0(r, \tau) = 0, \qquad |\varphi_{n+1}(0, \tau)| < C < \infty. \tag{2.99}$$

In order to reduce the problem (2.97)–(2.99) to the system of integral equations, we shall introduce the elastic potentials $\xi_{0i}(r, \tau)$ and $\xi_{1i}(r, \tau)$, $i = 0, 1, \ldots, n+1$, corresponding to the ith medium; each of these potentials satisfies the equation of motion (2.97) and the zero initial conditions; at that, the boundary conditions are as follows:

$$\left. \frac{\partial \xi_{0i}}{\partial r} \right|_{r=r_{i-1}} = \delta(\tau), \qquad \left. \frac{\partial \xi_{0i}}{\partial r} \right|_{r=r_i} = 0, \tag{2.100}$$

$$\left. \frac{\partial \xi_{1i}}{\partial r} \right|_{r=r_{i-1}} = 0, \qquad \left. \frac{\partial \xi_{1i}}{\partial r} \right|_{r=r_i} = \delta(\tau); \tag{2.101}$$

here, $\delta(\tau)$ is the generalized Dirac delta function and $\xi_{00} = \xi_{1,n+1} \equiv 0$.

Let us also define the transition functions $\chi_{m0}^{(i)}(\tau)$ and $\chi_{m1}^{(i)}(\tau)$, $m = 0, 1$, $i = 0, 1, \ldots, n+1$, as the radial stresses which correspond to the potentials ξ_{mi} at the boundaries $r = r_{i-1}$ and $r = r_i$. Thus, from (2.96), we obtain

$$\chi_{m0}^{(i)}(\tau) = \beta_i \left(\frac{\partial^2 \xi_{mi}}{\partial r^2} + \frac{2\kappa_i}{r} \frac{\partial \xi_{mi}}{\partial r} \right) \Bigg|_{r=r_{i-1}},$$

$$\chi_{m1}^{(i)}(\tau) = \beta_i \left(\frac{\partial^2 \xi_{mi}}{\partial r^2} + \frac{2\kappa_i}{r} \frac{\partial \xi_{mi}}{\partial r} \right) \Bigg|_{r=r_i}. \tag{2.102}$$

It can be noticed that the transition functions conventionally applied in the theory of hydroelasticity correspond to the potential, but not to the stresses (pressure).

Using (2.100)–(2.102), as well as the Duhamel integral for the radial stresses in the ith and $(i+1)$th media, we obtain the following conditions at the boundary:

$$\sigma_{ri}(r_i, \tau) = \chi_{01}^{(i)}(\tau) * u_i(r_{i-1}, \tau) + \chi_{11}^{(i)}(\tau) * u_i(r_i, \tau),$$
$$i = 1, 2, \ldots, n,$$
$$\sigma_{r, i+1}(r_i, \tau) = \chi_{00}^{(i+1)}(\tau) * u_{i+1}(r_i, \tau) + \chi_{10}^{(i+1)}(\tau) * u_{i+1}(r_{i+1}, \tau),$$
$$i = 0, 1, \ldots, n-1. \tag{2.103}$$

Here, an asterisk is used to designate the convolution.

Similarly, on the surfaces $r = r_0$ and $r = r_n$ for the media $i = 0$ and $i = n + 1$, we obtain

$$\sigma_{r0}(r_0, \tau) = \chi_{11}^{(0)}(\tau) * u_0(r_0, \tau),$$
$$\sigma_{r, n+1}(r_n, \tau) = \chi_{00}^{(n+1)}(\tau) * u_{n+1}(r_n, \tau). \tag{2.104}$$

Designating $w_i(\tau) = u_i(r_i, \tau), i = 0, 1, \ldots, n$, from the conditions of contact (2.98) and the relationships (2.103) and (2.104) we obtain the following system of integral equations of convolution type in the space of generalized functions:

$$\chi_{01}^{(i)} * w_{i-1} + (\chi_{11}^{(i)} - \chi_{00}^{(i+1)}) * w_i - \chi_{10}^{(i+1)} * w_{i+1} = 0,$$
$$i = 1, \ldots, n-1,$$
$$(\chi_{11}^{(0)} - \chi_{00}^{(1)}) * w_0 - \chi_{10}^{(1)} * w_1 = 0,$$
$$\chi_{01}^{(n)} * w_{n-1} + (\chi_{11}^{(n)} - \chi_{00}^{(n+1)}) * w_n = -\chi_{00}^{(n+1)} * u_s(r_n, \tau) + \sigma_{rs}(r_n, \tau). \tag{2.105}$$

Then, we should find the transition functions $\chi_{mn}^{(i)}(\tau)$. Since their calculation is similar, we shall confine ourselves by derivation of $\chi_{00}^{(i)}(\tau), i = 1, \ldots, n$. The potential $\xi_{0i}(r, \tau)$ satisfies the boundary value problem (2.97) and (2.100). We shall write it in the form of a superposition of converging and diverging spherical waves (see (1.114) for $n = 0$)

$$\xi_{0i}(r, \tau) = \frac{1}{r} \sum_{k=0}^{\infty} \left\{ f_{0k}^{(0i)}[\tau + \gamma_i(r - r_{i-1}) - 2k\gamma_i\delta_i] \right.$$
$$\left. + f_{1k}^{(0i)}[\tau - \gamma_i(r - r_i) - (2k+1)\gamma_i\delta_i] \right\}, \tag{2.106}$$

where $f_{0k}^{(0i)}(\tau)$ and $f_{1k}^{(0i)}(\tau)$ are arbitrary functions corresponding to the converging and diverging waves; here, δ_i is the relative thickness of the layer:

$$\delta_0 = \infty,$$
$$\delta_i = r_{i-1} - r_i, \quad i = 1, 2, \ldots, n,$$
$$\delta_{n+1} = r_n.$$

If $\dot{f}_{0k}^{(0i)}(0) = \dot{f}_{1k}^{(0i)}(0) = f_{0k}^{(0i)}(0) = f_{1k}^{(0i)}(0) = 0$ (here, a dot is used to denote differentiation with respect to time τ), then the function $\xi_{0i}(r, \tau)$ satisfies the wave equation (2.97) and the zero initial conditions. The arbitrary functions can be determined from the boundary conditions (2.100).

In the space of Laplace transform with respect to time τ, the displacement corresponding to the potential $\xi_{0i}^L(r, s)$ has the following form:

$$u_i^L(r, s) = \frac{\partial \xi_{0i}^L}{\partial r} = -\frac{r_{i-1}^2}{r^2} \sum_{k=0}^{\infty} \left\{ R_{01}(-\gamma_i r s) f_{0k}^{(0i)L}(s) e^{-\gamma_i(r_{i-1}-r)s} \right.$$

$$\left. + R_{01}(\gamma_i r s) f_{1k}^{(0i)L}(s) e^{-\gamma_i(r-r_i+\delta_i)s} \right\} e^{-2k\gamma_i \delta_i s} ; \quad (2.107)$$

here, s is a transform parameter.

In order to satisfy the boundary conditions (2.100), we can determine the arbitrary functions $f_{0k}^{(0i)}$ and $f_{1k}^{(0i)}$ by the recursive system of equations

$$f_{00}^{(0i)L} = -\frac{1}{R_{01}(-\gamma_i r_{i-1} s)} ,$$

$$f_{1k}^{(0i)L} = -A_{1i} f_{0k}^{(0i)L} , \qquad f_{0k}^{(0i)L} = -A_{0i} f_{1, k-1}^{(0i)L} ,$$

$$A_{0i}(s) = \frac{R_{01}(\gamma_i r_{i-1} s)}{R_{01}(-\gamma_i r_{i-1} s)} , \qquad A_{1i}(s) = \frac{R_{01}(-\gamma_i r_i s)}{R_{01}(-\gamma_i r_i s)} . \quad (2.108)$$

In accounting for (2.102), the transition function $\chi_{00}^{(i)L}(s)$ has the following form:

$$\chi_{00}^{(i)L}(s) = \sigma_{ri}^L(r_{i-1}, s) = \frac{\beta_i}{r_{i-1}} \left\{ Q_{01}^{(i)}(-\gamma_i r_{i-1} s) f_{00}^{(0i)L}(s) \right.$$

$$+ \sum_{k=1}^{\infty} \left[Q_{01}^{(i)}(-\gamma_i r_{i-1} s) f_{0k}^{(0i)L}(s) \right.$$

$$\left. + Q_{01}^{(i)}(\gamma_i r_{i-1} s) f_{1, k-1}^{(0i)L}(s) \right] e^{-2k\gamma_i \delta_i s} \right\} . \quad (2.109)$$

Let us designate

$$g_{00k}^{(i)L}(s) = Q_{01}^{(i)}(-\gamma_i r_{i-1} s) f_{0k}^{(0i)L}(s) + Q_{01}^{(i)}(\gamma_i r_{i-1} s) f_{1, k-1}^{(0i)L}(s) , \qquad k \geq 1,$$

$$g_{000}^{(i)L}(s) = Q_{01}^{(i)}(-\gamma_i r_{i-1} s) f_{00}^{(0i)L}(s) .$$

Then, from (2.108) and (2.109), we get

$$\chi_{00}^{(i)L}(s) = \frac{\beta_i}{r_{i-1}} \sum_{k=0}^{\infty} g_{00k}^{(i)L}(s) e^{-2k\gamma_i \delta_i s} ,$$

$$g_{001}^{(i)L}(s) = -\frac{2(\gamma_i r_{i-1} s)^3}{R_{01}^2(-\gamma_i r_{i-1} s)} A_{1k}(s) ,$$

$$g_{00k}^{(i)L}(s) = A_{0i} A_{1i} g_{00, k-1}^{(i)L} , \qquad k \geq 2 . \quad (2.110)$$

Turning back to the originals, we obtain the following representation for $\chi_{00}^{(0)}(\tau)$:

$$\chi_{00}^{(i)}(\tau) = \frac{\beta_i}{r_{i-1}} \left\{ \gamma_i r_{i-1} \delta'(\tau) - (1 - 2\kappa_i)\,\delta(\tau) \right.$$

$$\left. + 2 \sum_{k=1}^{\infty} \left[\gamma_i r_{i-1} \delta'(\tau_{0ik}) - 2k \frac{\delta_i}{r_i} \delta(\tau_{0ik}) \right] \right\} + \Phi_{00}^{(i)}(\tau) ,$$

$$\Phi_{00}^{(i)L}(s) = \frac{\beta_i}{r_{i-1}} \sum_{k=0}^{\infty} G_{00k}^{(i)L}(s)\, e^{-2k\gamma_i\delta_i s} ,$$

$$\tau_{0ik} = \tau - 2k\gamma_i\delta_i , \qquad i = 1, \dots, n ; \tag{2.111}$$

here, the transforms $G_{00k}^{(i)L}(s)$ satisfy the following nonhomogeneous recursive relationships:

$$G_{000}^{(i)L} = -\frac{1}{R_{01}}(-\gamma_i r_{i-1} s) , \qquad G_{001}^{(i)L} = P_{00i}(s)\, B_i(s) ,$$

$$G_{00k}^{(i)L} = A_{0i} A_{1i} G_{00,\,k-1}^{(i)L} + 4 \frac{\delta_i}{r_i} R_{01}[-(2k-1)\gamma_i\delta_i s]\, B_i(s) , \qquad k \geq 2 ,$$

$$P_{00i}(s) = -2 \left[\gamma_i^2 r_i r_{i-1} \left(1 + 2\frac{\delta_i^2}{r_i^2}\right) s^2 \right.$$

$$\left. - \gamma_i r_{i-1} \left(1 + \frac{2\delta_i^2}{r_i r_{i-1}} + \frac{2\delta_i}{r_i}\right) s + \frac{2\delta_i}{r_i} \right] ,$$

$$B_i(s) = \frac{1}{R_{01}(-\gamma_i r_{i-1} s)\, R_{01}(\gamma_i r_i s)} . \tag{2.112}$$

The expressions for the other transition functions can be obtained similarly:

$$\chi_{01}^{(i)}(\tau) = \frac{\beta_i r_{i-1}^2}{r_i^3} \sum_{k=0}^{\infty} \left[2\gamma_i \frac{r_i^2}{r_{i-1}} \delta'(\tau_{1ik}) - 2\frac{\delta_i r_i}{r_{i-1}^2}(2k+1)\delta(\tau_{1ik}) \right]$$

$$+ \Phi_{01}^{(i)}(\tau) ,$$

$$\tau_{1ik} = \tau_{0ik} - \gamma_i\delta_i ,$$

$$\Phi_{01}^{(i)L}(s) = \frac{\beta_i r_{i-1}^2}{r_i^3} \sum_{k=0}^{\infty} G_{01k}^{(i)L}(s)\, e^{-(2k+1)\gamma_i\delta_i s} , \qquad i = 1, 2, \dots, n+1 ,$$

$$G_{01k}^{(i)L} = A_{0i} A_{1i} G_{01,\,k-1}^{(i)L} + 4 \frac{\delta_i r_i}{r_{i-1}^2} R_{01}(-2k\gamma_i\delta_i s)\, B_i(s) , \qquad k \geq 1 ,$$

$$G_{010}^{(i)L} = 2 \frac{r_i}{r_{i-1}} \left[-\gamma_i \left(r_i + \frac{\delta_i^2}{r_{i-1}}\right) s + \frac{\delta_i}{r_{i-1}} \right] B_i(s) ; \tag{2.113}$$

$$\chi_{10}^{(i)}(\tau) = \frac{\beta_i r_i^2}{r_{i-1}^3} \sum_{k=0}^{\infty} \left[-2\gamma_i \frac{r_{i-1}^2}{r_i} \delta'(\tau_{1ik}) + 2 \frac{\delta_i r_{i-1}}{r_i^2} (2k+1) \delta(\tau_{1ik}) \right]$$

$$+ \Phi_{10}^{(i)}(\tau),$$

$$\Phi_{10}^{(i)L}(s) = \frac{\beta_i r_i^2}{r_{i-1}^3} \sum_{k=0}^{\infty} G_{10k}^{(i)L} e^{-(2k+1)\gamma_i \delta_i s}, \qquad i = 1, 2, \ldots, n,$$

$$G_{10k}^{(i)L} = A_{0i} A_{1i} G_{10,\,k-1}^{(i)L} - 4 \frac{\delta_i r_{i-1}}{r_i^2} R_{01}(-2k\gamma_i\delta_i s) B_i(s), \qquad k \geq 1,$$

$$G_{100}^{(i)L} = 2 \frac{r_{i-1}}{r_i} \left[\gamma_i \left(r_{i-1} + \frac{\delta_i^2}{r_i} \right) s - \frac{\delta_i}{r_{i-1}} \right] B_i(s); \qquad (2.114)$$

$$\chi_{11}^{(i)}(\tau) = \frac{\beta_i}{r_i} \left\{ -\gamma_i r_i \delta'(\tau) - (1 - 2\kappa_i) \delta(\tau) \right.$$

$$\left. + 2 \sum_{k=1}^{\infty} \left[-\gamma_i r_i \delta'(\tau_{0ik}) + 2k \frac{\delta_i}{r_{i-1}} \delta(\tau_{0ik}) \right] \right\} + \Phi_{11}^{(i)}(\tau),$$

$$\Phi_{11}^{(i)L}(s) = \frac{\beta_i}{r_i} \sum_{k=0}^{\infty} G_{11k}^{(i)L}(s) e^{-2k\gamma_i\delta_i s},$$

$$G_{11k}^{(i)L} = A_{0i} A_{1i} G_{11,\,k-1}^{(i)L} - 4 \frac{\delta_i}{r_{i-1}} R_{01}[-(2k-1)\gamma_i\delta_i s] B_i(s), \qquad k \geq 2,$$

$$G_{110}^{(i)L} = -\frac{1}{R_{01}(\gamma_i r_i s)}, \qquad G_{111}^{(i)L} = P_{11i}(s) B_i(s),$$

$$P_{11i}(s) = 2 \left[\gamma_i^2 r_i r_{i-1} \left(1 + \frac{2\delta_i^2}{r_i^2} \right) s^2 \right.$$

$$\left. - \gamma_i r_{i-1} \left(1 + \frac{2\delta_i^2}{r_i r_{i-1}} + \frac{2\delta_i}{r_i} \right) s + \frac{2\delta_i}{r_i} \right]; \quad (2.115)$$

$$\chi_{00}^{(n+1)}(\tau) = \frac{\beta_{n+1}}{r_n} \sum_{k=0}^{\infty} (-1)^k \left[\gamma_{n+1} r_n \delta'(\tau_{0i,\,n+1}) \right.$$

$$\left. + (2k - 1 + 2\kappa_{n+1}) \delta(\tau_{0i,\,n+1}) \right] + \Phi_{00}^{(n+1)}(\tau),$$

$$\Phi_{00}^{(n+1)L}(s) = \frac{\beta_{n+1}}{r_n} \sum_{k=0}^{\infty} G_{00k}^{(n+1)L}(s) e^{-2k\gamma_{n+1} r_n s},$$

$$G_{00k}^{(n+1)L} = A_{0,\,n+1} G_{00,\,k-1}^{(n+1)L} + (-1)^{k-1} \frac{2k - 1 + 2\kappa_{n+1}}{R_{01}(-\gamma_{n+1} r_n s)}, \qquad k \geq 2,$$

$$G_{000}^{(n+1)L} = -\frac{1}{R_{01}(-\gamma_{n+1} r_n s)},$$

$$G_{001}^{(n+1)L} = -\frac{(1 - 4\kappa_{n+1}) A_{0,\,n+1}}{R_{01}(-\gamma_{n+1} r_n s)}; \qquad (2.116)$$

$$\Phi_{11}^{(0)L}(s) = -\frac{\beta_0}{r_0} \frac{1}{R_{01}(\gamma_0 r_0 s)}, \qquad \Phi_{11}^{(0)}(\tau) = -\frac{\beta_0}{\gamma_0 r_0^2} e^{-\frac{\tau}{\gamma_0 r_0}}. \qquad (2.117)$$

Designating $v_i(\tau) = \dot{w}(\tau)$ and taking into account the properties of the Dirac delta function and its derivative, from (2.105) we obtain the following system of the Volterra integral equations of convolution type of the second kind:

$$w_i = v_i * 1, \qquad i = 0, 1, \dots, n,$$

$$\Phi_{01}^{(i)} * w_{i-1} + (\Phi_{11}^{(i)} - \Phi_{00}^{(i+1)}) * w_i - \Phi_{10}^{(i+1)} * w_{i+1}$$

$$+ \frac{2\beta_i}{r_i} \sum_{k=0}^{\infty} \left[\gamma_i r_{i-1} v_{i-1}(\tau_{1ik}) - \frac{\delta_i}{r_i}(2k+1) w_{i-1}(\tau_{1ik}) \right]$$

$$- (\beta_i \gamma_i + \beta_{i+1} \gamma_{i+1}) v_i(\tau) + \frac{1}{r_i} \left[-\beta_i(1 - 2\kappa_i) + \beta_{i+1}(1 - 2\kappa_{i+1}) \right] w_i(\tau)$$

$$+ \frac{2\beta_i}{r_i} \sum_{k=1}^{\infty} \left[-\gamma_i r_i v_i(\tau_{0ik}) + 2k \frac{\delta_i}{r_{i-1}} w_i(\tau_{0ik}) \right]$$

$$+ \frac{2\beta_{i+1}}{r_i} \sum_{k=1}^{\infty} \left[-\gamma_{i+1} r_i v_i(\tau_{0,i+1,k}) + 2k \frac{\delta_{i+1}}{r_{i+1}} w_i(\tau_{0,i+1,k}) \right]$$

$$+ \frac{2\beta_{i+1}}{r_i} \sum_{k=0}^{\infty} \left[\gamma_{i+1} r_{i+1} v_{i+1}(\tau_{1,i+1,k}) \right.$$

$$\left. - (2k+1) \frac{\delta_{i+1}}{r_i} w_{i+1}(\tau_{1,i+1,k}) \right] = 0,$$

$$\tau_{1ik} = \tau_{0ik} - \gamma_i \delta_i, \qquad i = 1, \dots, n-1,$$

$$\left(\Phi_{11}^{(0)} - \Phi_{00}^{(1)} \right) * w_0 - \Phi_{10}^{(1)} * w_1 - (1 + \beta_1 \gamma_1) v_0(\tau)$$

$$+ \frac{1}{r_0} \left[-(1 - 2\kappa_0) + \beta_1(1 - 2\kappa_1) \right] w_0(\tau)$$

$$+ \frac{2\beta_1}{r_0} \sum_{k=1}^{\infty} \left[-\gamma_1 r_0 v_0(\tau_{01k}) + 2k \frac{\delta_1}{r_1} w_0(\tau_{01k}) \right]$$

$$+ \frac{2\beta_1}{r_0} \sum_{k=0}^{\infty} \left[\gamma_1 r_1 v_1(\tau_{11k}) - (2k+1) \frac{\delta_1}{r_0} w_1(\tau_{11k}) \right] = 0,$$

$$\Phi_{01}^{(n)} * w_{n-1} + \left(\Phi_{11}^{(n)} - \Phi_{00}^{(n+1)} \right) * w_n$$

$$+ \frac{2\beta_n}{r_n} \sum_{k=0}^{\infty} \left[\gamma_n r_{n-1} v_{n-1}(\tau_{1nk}) - (2k+1) \frac{\delta_n}{r_n} w_n(\tau_{1nk}) \right]$$

$$- (\beta_n \gamma_n + \beta_{n+1} \gamma_{n+1}) v_n(\tau)$$

$$+ \frac{1}{r_n} \left[-\beta_n(1 - 2\kappa_n) + \beta_{n+1}(1 - 2\kappa_{n+1}) \right] w_n(\tau)$$

$$+ \frac{2\beta_n}{r_n} \sum_{k=1}^{\infty} \left[-\gamma_n r_n v_n(\tau_{0nk}) + 2k \frac{\delta_n}{r_{n-1}} w_n(\tau_{0nk}) \right]$$

$$- \frac{\beta_{n+1}}{r_n} \sum_{k=1}^{\infty} (-1)^k \left[\gamma_{n+1} r_n v_n(\tau_{0,\,n+1,\,k}) \right.$$

$$\left. + (2k - 1 + 2\kappa_{n+1}) w_n(\tau_{0,\,n+1,\,k}) \right]$$

$$= -\Phi_{00}^{(n+1)} * u_s(r_n, \tau)$$

$$- \frac{\beta_{n+1}}{r_n} \sum_{k=0}^{\infty} (-1)^k \left[\gamma_{n+1} r_n u_s(r_n, \tau_{0,\,n+1,\,k}) \right.$$

$$\left. + (2k - 1 + 2\kappa_{n+1}) u_s(r_n, \tau_{0,\,n+1,\,k}) \right] + \sigma_{rs}(r_n, \tau). \quad (2.118)$$

The system of $2(n + 1)$ equations (2.118) allows us to obtain the displacements and speeds, and, hence, the stresses on the surfaces of contact of spherical layers. Applying the conditions at the surfaces, using the known potentials $\xi_{0i}(r, \tau)$ and $\xi_{1i}(r, \tau)$ and employing the Duhamel integral, it is possible to calculate all the components of the stress–strain state at any point of ith layer.

Let us notice that the upper limit of summation in (2.118) was introduced conventionally. When calculating the sums, it is necessary to take into account that $w_i(\tau) = v_i(\tau) = 0$ at $\tau < 0$. Assuming that the function $f(\tau)$ in (2.94) is of such a kind that the radial stress $\sigma_{rs}(r_n, \tau)$ is a continuous function for any $\tau \geq 0$, it is possible to prove that the unknown functions $w_i(\tau)$ and $v_i(\tau)$ are continuous. Hence, it is possible to search for a solution of (2.118) in a space of continuous functions applying any numerical method known. It is easy to calculate the functions $\Phi_{mn}^{(i)}(\tau)$ applying, for example, the algorithm presented in Sect. 2.2.

It is theoretically possible to derive the solutions for the loads which have discontinuities of the first kind in time τ and to account for these discontinuities in the functions sought. However, a number of discontinuities in the given interval of time increases radically with an increase in the number of interfaces, and accounting for them becomes pointless.

Let us also notice that in the problem considered the spherical layers can be acoustical. To obtain this, it is sufficient to set $\kappa_i = 0$ for the ith medium.

In order to solve the system of equations (2.118) approximately, let us replace the integrals by the finite sums applying the trapezoid formula.

Let d be a time step. Let us consider the sequence of instants $\{\tau_i\}$ such that

$$\tau_j = jd, \qquad j = 0, 1, 2, \ldots . \tag{2.119}$$

We shall search for the mesh functions $\{w_i^{(j)}\}$ and $\{v_i^{(j)}\}$ instead of the functions $w_i(\tau)$ and $v_i(\tau)$:

$$w_i^{(j)} = w_i(\tau_j), \qquad v_i^{(j)} = v_i(\tau_j). \tag{2.120}$$

Following the numerical method chosen, let us replace the integrals of convolution type, which enter into the system of equations (1.118), by the following finite sums:

$$\Phi * u\big|_{\tau=\tau_j} = \int_0^{\tau_j} \Phi(\tau_j - t)\, u(t)\, dt \approx \frac{d}{2}\, \Phi(0)\, u^{(j)} + \chi \bar{*} u,$$

$$\Phi \bar{*} u = d\left(\frac{1}{2}\, \Phi^{(j)} u^{(0)} + \sum_{l=1}^{j-1} \Phi^{(j-l)} u^{(l)}\right),$$

$$u^{(j)} = u(\tau_j) = u(jd), \qquad \Phi^{(j)} = \Phi(\tau_j) = \Phi(jd). \tag{2.121}$$

Then, the discrete analogy to the system of equations (2.118) will have the following form:

$$\mathbf{A}\,\mathbf{W}^{(j)} = \mathbf{B}^{(j)}, \qquad \mathbf{V}^{(j)} = \frac{2}{d}\,\mathbf{W}^{(j)} + \mathbf{C}^{(j)}; \tag{2.122}$$

here, A is the diagonal matrix of the dimensions $(n+1) \times (n+1)$, $\mathbf{W}^{(j)}$, $\mathbf{V}^{(j)}$, $\mathbf{B}^{(j)}$, and $\mathbf{C}^{(j)}$ are the columns of the dimensions $(n+1) \times 1$. Their elements can be determined as follows:

$$\mathbf{W}^{(j)} = \left(w_0^{(j)}, w_1^{(j)}, \ldots, w_n^{(j)}\right)^{\mathrm{T}}, \qquad \mathbf{V}^{(j)} = \left(v_0^{(j)}, v_1^{(j)}, \ldots, v_n^{(j)}\right)^{\mathrm{T}},$$

$$\mathbf{B}^{(j)} = \left(b_1^{(j)}, b_2^{(j)}, \ldots, b_{n+1}^{(j)}\right)^{\mathrm{T}}, \qquad \mathbf{C}^{(j)} = \left(c_1^{(j)}, c_2^{(j)}, \ldots, c_{n+1}^{(j)}\right)^{\mathrm{T}},$$

$$\mathbf{A} = (a_{kl}), \qquad a_{kl} = 0, \qquad k \neq l,$$

$$a_{i+1,\,i+1} = \frac{d}{2}\left[\Phi_{11}^{(i)}(0) - \Phi_{11}^{(i+1)}(0)\right] - \frac{\beta_i}{r_i}(1 - 2\kappa_i) + \frac{\beta_{i+1}}{r_i}(1 - 2\kappa_{i+1})$$

$$- \frac{2}{d}(\beta_i \gamma_i + \beta_{i+1}\gamma_{i+1}),$$

$$\Phi_{11}^{(i)}(0) = -\frac{\beta_i}{\gamma_i r_i^2}, \qquad \Phi_{00}^{(i+1)}(0) = -\frac{\beta_{i+1}}{\gamma_{i+1} r_i^2}, \qquad i = 0, 1, \ldots, n,$$

$$b_{i+1}^{(j)} = -\Phi_{01}^{(i)} \bar{*} w_{i-1}\Big|_{\tau=\tau_j} - \left(\Phi_{11}^{(i)} - \Phi_{00}^{(i+1)}\right) \bar{*} w_i\Big|_{\tau=\tau_j} + \Phi_{10}^{(i+1)} \bar{*} w_{i+1}\Big|_{\tau=\tau_j}$$

$$- \frac{2\beta_i}{r_i} \sum_{k=0}^{\infty}\left[\gamma_i r_{i-1} v_{i-1}(\tau_{1ik}) - \frac{\delta_i}{r_i}(2k+1)\, w_{i-1}(\tau_{1ik})\right]$$

$$- \frac{2\beta_i}{r_i} \sum_{k=1}^{\infty}\left[-\gamma_i r_i v_i(\tau_{0ik}) + 2k\frac{\delta_i}{r_{i-1}}\, w_i(\tau_{0ik})\right]$$

$$- \frac{2\beta_{i+1}}{r_i} \sum_{k=1}^{\infty}\left[-\gamma_{i+1} r_i v_i(\tau_{0,\,i+1,\,k}) + 2k\frac{\delta_{i+1}}{r_{i+1}}\, w_i(\tau_{0,\,i+1,\,k})\right]$$

$$-\frac{2\beta_{i+1}}{r_i}\sum_{k=0}^{\infty}\left[\gamma_{i+1}r_{i+1}v_{i+1}(\tau_{1,i+1,k})+(2k+1)\frac{\delta_{i+1}}{r_i}\,w_{i+1}(\tau_{1,i+1,k})\right]$$

$$-\frac{2}{d}\left(\beta_i\gamma_i+\beta_{i+1}\gamma_{i+1}\right)\left(1\bar{\mp}v_i\right)\Big|_{\tau=\tau_j}\,,\qquad i=1,2,\ldots,n-1,$$

$$b_1^{(j)}=-\left(\Phi_{11}^{(0)}-\Phi_{00}^{(1)}\right)\bar{*}w_0\Big|_{\tau=\tau_j}+\Phi_{10}^{(1)}\bar{*}w_1\Big|_{\tau=\tau_j}$$

$$-\frac{2\beta_1}{r_0}\sum_{k=1}^{\infty}\left[-\gamma_1r_0v_0(\tau_{01k})+2k\,\frac{\delta_1}{r_1}\,w_0(\tau_{01k})\right]$$

$$-\frac{2\beta_1}{r_0}\sum_{k=0}^{\infty}\left[\gamma_1r_0v_1(\tau_{11k})+(2k+1)\frac{\delta_1}{r_0}\,w_1(\tau_{11k})\right]$$

$$-\frac{2}{d}\left(\beta_0\gamma_0+\beta_1\gamma_1\right)\left(1\bar{\mp}v_i\right)\Big|_{\tau=\tau_j}\,,$$

$$b_{n+1}^{(j)}=-\Phi_{01}^{(n)}\bar{*}w_{n-1}\Big|_{\tau=\tau_j}-\left(\Phi_{11}^{(n)}-\Phi_{00}^{(n+1)}\right)\bar{*}w_n\Big|_{\tau=\tau_j}$$

$$-\frac{2\beta_n}{r_n}\sum_{k=0}^{\infty}\left[\gamma_nr_{n-1}v_{n-1}(\tau_{1nk})-(2k+1)\frac{\delta_n}{r_n}\,w_{n-1}(\tau_{1nk})\right]$$

$$-\frac{2\beta_n}{r_n}\sum_{k=1}^{\infty}\left[-\gamma_nr_nv_n(\tau_{0nk})-2k\,\frac{\delta_n}{r_{n-1}}\,w_n(\tau_{0nk})\right]$$

$$+\frac{\beta_{n+1}}{r_n}\sum_{k=1}^{\infty}(-1)^k\left[\gamma_{n+1}r_nv_n(\tau_{0,n+1,k})\right.$$

$$\left.+\,(2k-1+2\kappa_{n+1})\,w_n(\tau_{0,n+1,k})\right]$$

$$-\frac{2}{d}\left(\beta_n\gamma_n+\beta_{n+1}\gamma_{n+1}\right)\left(1\bar{\mp}v_n\right)\Big|_{\tau=\tau_j}$$

$$-\Phi_{00}^{(n+1)}*u_{\mathrm{s}}(r_n,\tau)\Big|_{\tau=\tau_j}+\sigma_{r\mathrm{s}}(r_n,\tau)$$

$$-\frac{\beta_{n+1}}{r_n}\sum_{k=0}^{\infty}(-1)^k\left[\gamma_{n+1}r_n\dot{u}_{\mathrm{s}}(r_n,\tau_{0,n+1,k})\right.$$

$$\left.+\,(2k-1+2\kappa_{n+1})\,u_{\mathrm{s}}(r_n,\tau_{0,n+1,k})\right].$$

The components of the stress–strain state in a converging wave from a point source corresponding to the potential (2.94) can be determined as follows:

$$u_{\mathrm{s}}(r,\tau)=-\frac{K}{r_n}\left[f(\tau)+r_n\gamma_{n+1}f'(\tau)\right]H(\tau),$$

$$v_{\mathrm{s}}(r,\tau)=-\frac{K}{r}\left[f'(\tau)+r_n\gamma_{n+1}f''(\tau)\right]H(\tau),$$

$$\sigma_{rs}(r, \tau) = K \frac{\beta_{n+1}}{r_n^2} \left[2 \left(1 - \kappa_{n+1} \right) f(\tau) \right.$$

$$\left. + 2r_n \gamma_{n+1} (1 - \kappa_{n+1}) f'(\tau) + r_n^2 \gamma_{n+1}^2 f''(\tau) \right] H(\tau). \tag{2.123}$$

We can determine the coefficient K in the formula for the potential $\varphi_s(r, \tau)$ from the following condition:

$$\lim_{\tau \to +0} \sigma_s(r_n, \tau) = -1, \qquad K = -\frac{1}{\beta_{n+1} \gamma_{n+1}^2 f''(0)}. \tag{2.124}$$

It is evident that the system of algebraic equations (2.124) allows us to arrange a relatively simple stepwise process of calculation of the columns $\mathbf{W}^{(j)}$ and $\mathbf{V}^{(j)}$. In order to start this process, it is only necessary to know $\mathbf{W}^{(0)}$ and $\mathbf{V}^{(0)}$. These columns can be obtained from the initial conditions in the problem (2.97) combining with the passage to the limit as $\tau \to 0$ in the system of equations (2.118). Then, we get

$$w_i(0) = v_i(0) = 0, \qquad i = 0, 1, \ldots, n - 1, \qquad w_n(0) = 0,$$

$$v_n(+0) = \frac{1}{\beta_n \gamma_n + \beta_{n+1} \gamma_{n+1}} \left\{ -\sigma_{rs}(r_n, +0) \right.$$

$$\left. + \frac{\beta_{n+1}}{r_n} \left[\gamma_{n+1} r_n \dot{u}_s(r_n, +0) + (2\kappa_{n+1} - 1) u_s(r_n, 0) \right] \right\}. \tag{2.125}$$

Let us note that, as it follows from the algorithm for calculation of the functions $\Phi_{mn}^{(i)}(\tau)$ (2.111)-(2.117), these functions can have the discontinuities of the first kind. For this reason, when replacing the integral of convolution type in the system of equations (2.118) by the finite sums of the function $\Phi_{mn}^{(i)}$ in the same time intervals $[\tau_j, \tau_{j+1}]$, where the discontinuity point τ_* is located, it is necessary to apply the trapezoid formula for each of the intervals $[\tau_j, \tau_*]$ and $[\tau_*, \tau_{j+1}]$.

Convergence of the scheme similar to that applied here for approximate solution of the integral equations of convolution type was demonstrated in Kubenko (1975c).

In the numerical examples presented below, we assume that the function, which specifies the law of variation in time of the stress–strain state at a front of disturbing wave, is of the form $f(\tau) = -\tau^2/2$.

The results of calculations based on the algorithm derived are presented in Figs. 2.33 and 2.34. A three-layer shell was studied; the outer layer ($i = 1$; $\delta_1 = 0.1$) and the inner layer ($i = 3$; $\delta_3 = 0.1$) were made from steel and the intermediate one ($i = 2$; $\delta_2 = 0.3$) was made from concrete. The shell was filled with water ($i = 4$; $\delta_4 = r_3 = 0.5$) and immersed into an infinite concrete medium ($i = 0$). The following dimensionless parameters were accepted: $\gamma_0 = 1$; $\gamma_1 = \gamma_3 = 0.694$; $\gamma_2 = 1.06$; $\gamma_4 = 2.76$; $\beta_0 = 1$; $\beta_1 = \beta_3 = 5.96$; $\beta_2 = 0.648$; $\beta_4 = 0.0514$; $\kappa_0 = \kappa_2 = 0.143$; $\kappa_1 = \kappa_3 = 0.393$; $\kappa_4 = 1$.

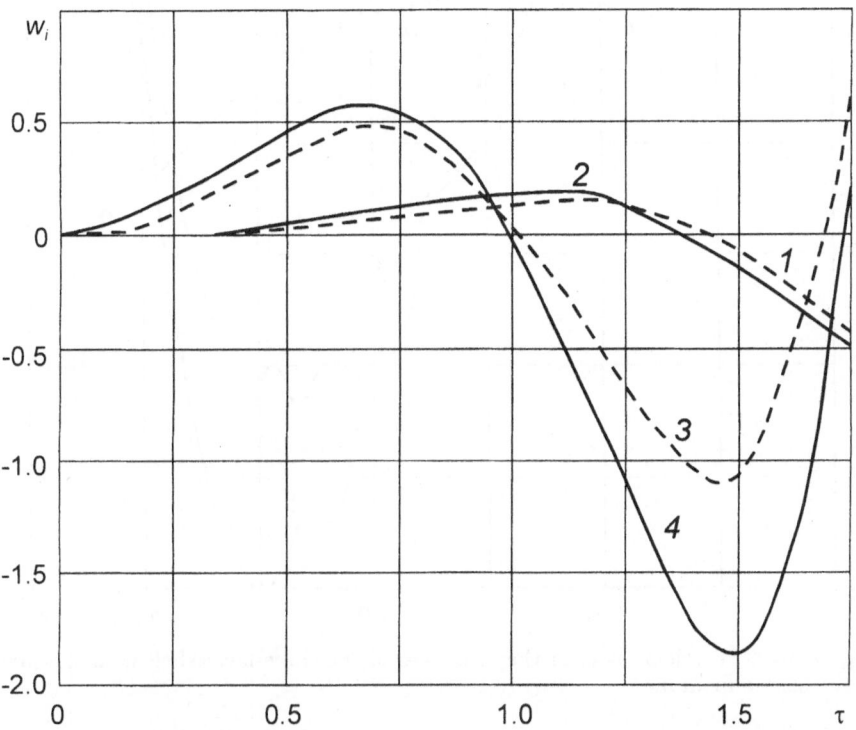

Fig. 2.33. The displacements w_i at the interfaces of the three-layer shell with the filler in the elastic medium

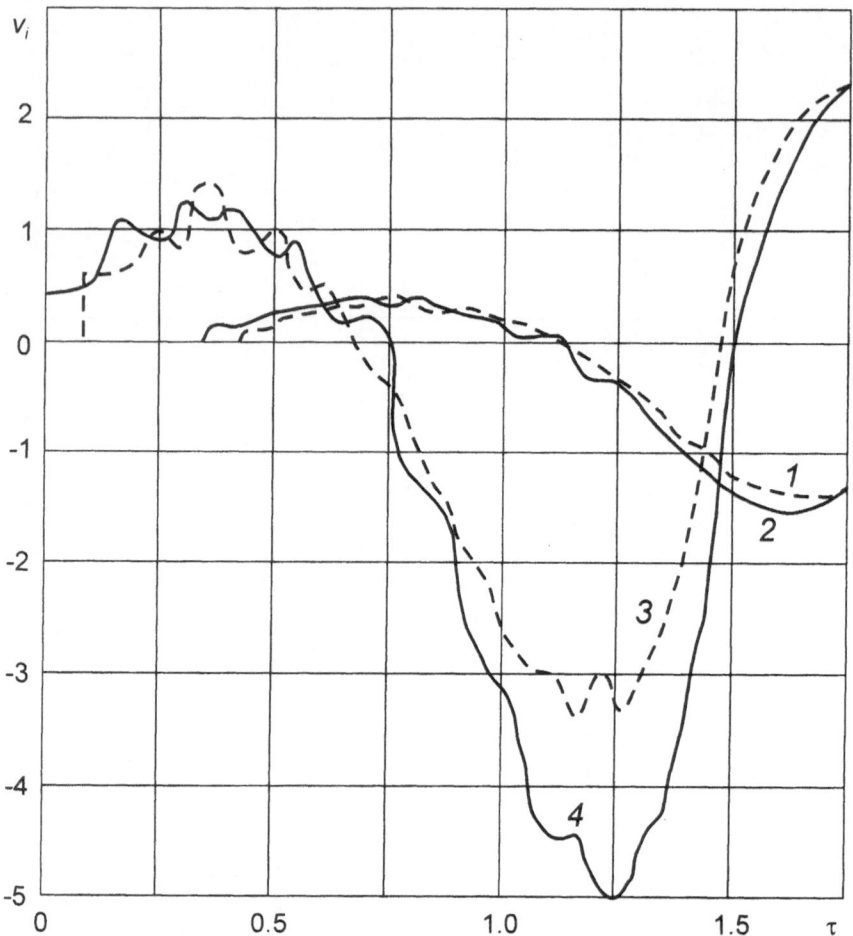

Fig. 2.34. The velocities v_i of the interfaces of the three-layer shell with the filler in the elastic medium

In Fig. 2.33, we plot the displacements w_i versus time, and in Fig. 2.34 we plot the speeds v_i versus time; the displacements w_i and speeds v_i, $i = 1, 2, 3, 4$, correspond to the interfaces of the media designated by $'i-1'$ and $'i'$, respectively. The dashed curves correspond to the external surfaces and the solid ones correspond to the internal surfaces of the steel thick-walled shells. Let us note that the calculations were performed using different values of time step $d = 0.05, 0.025, 0.0125, 0.00625, 0.003125$. The maximum discrepancy in displacements detected for the two last steps was $\approx 3\%$.

3. Diffraction of Waves by Elastic Spherical Bodies

When analyzing the problems of interaction of elastic or acoustic waves with deformable spherical obstacles, we run into some essential difficulties as compared to the problems considered in Chap. 2. The main peculiarity of the problems to be considered in this chapter is in the symmetry with respect to the axis that coincides with the direction of wave propagation. First, we get an increase in the problem dimension (the number of independent variables) and, second, we get an increase in the number of unknown functions (because the shear waves appear in addition to the compression waves).

The first stage and the first approximation of the solution of the problems of interaction ignore the additional loads that appear as a result of the interaction. In this case, we can assume that the load at the surface of elastic body is predetermined. The simplest problem of this kind is the problem of transient vibrations of a solid elastic sphere under the axially symmetric boundary conditions. A formal solution was derived for a solid sphere in Henneberg (1878–1879) by the method of complete separation of variables in the absence of shear waves. A similar solution for a hollow sphere was presented in Tsai (1973). In Usami (1962), the method of separation of variables was used to derive a general solution of the equations of motion of a homogeneous isotropic elastic sphere in the spherical coordinates. In Jaerisch (1880) directly followed after Clebsch (1862) and Henneberg (1878–1879), the equations for determination of the eigenfrequencies of a hollow sphere in the general case were derived, and that was a necessary component of the method of separation of variables. In Fridman (1976a), an expansion in terms of the eigenfunctions for arbitrary orthogonal coordinates and, as a particular case, for a sphere, was made. The author noted a weak convergence of the series at the wave front. Application of other methods presumed mainly a series expansion of the functions sought in terms of the spherical functions (Petrashen (1945), Petrashen (1949), Petrashen et al. (1953b)). This can be obviated by using the three-dimensional mesh method (Poverus, Myannil (1975), Bak (1978), Ryayamet (1975)).

The numerical method of characteristics was applied in Babichev (1969b) and Cheban, Sabodash (1972) for the cases when the loads were predetermined at the external surface of the sphere, and in Cherednichenko (1975), where the problem of impact of a thick-walled sphere against a stiff obstacle was analyzed. When solving such problems, the Fourier transform and the Laplace transform with respect to time followed by inversion by means of residues became very popular (Gray (1955), Sato et al. (1962), Sato (1963), Viswanathan, Biswas (1970) and Usami, Sato (1964), Jingu, Nezu (1985a), Petrashen (1950), Petrashen (1953), Saakyan (1973), respectively). At that, in the works mentioned, except Viswanathan, Biswas (1970), where the Watson transform was used for summation of the series by $P_n(\cos\theta)$, a homogeneous elastic sphere was analyzed. In Petrashen (1953), it was assumed that the medium was of acoustic type and an asymptotic solution was derived, which corresponded to the front's region. In Petrashen (1950), in order to solve the wave equation after separation of the angle, the method of sources followed by application of the Laplace transform was used. In Rakhmatulin et al. (1967a), the ordinary differential equations were derived for arbitrary functions corresponding to the converging and diverging waves for $n = 1$. The authors proposed to solve the general problem (when the load was predetermined at the internal surface) by means of superposition of the elementary waves.

The problems for the elastic bodies having spherical boundaries under given dynamical loads turned out to be more complicated. Here, accounting for the dependence of load on the angular coordinate brings us again to the series expansion of the functions sought in terms of the Legendre polynomials. In Buldyrev, Yanson (1965), the axially symmetric vibrations of a system: 'thick-walled sphere – elastic solid sphere' under a given surface load were considered. The method applied was the Laplace transform; the inversion was made by means of the theory of residues. An asymptotic analysis of the roots of corresponding transcendental equation was made. In Sato, Usami (1964), a sphere occupied by an acoustic medium was analyzed, and the problem was solved using the Fourier transform. In Sato et al. (1963), propagation of shear waves (S-waves) in a hollow sphere occupied by an acoustic medium was analyzed in the case when a point source acts in the sphere's cavity.

A solution of the problem of pulse application of a point force to the surface of contact of an infinite acoustic medium with a solid elastic sphere, a hollow elastic sphere, and a thick-walled sphere occupied by an acoustic medium was derived in Onishchuk (1980), Grilitsky, Onischuk (1980c), and Grilitsky, Onischuk (1980b), respectively. The method of solution used was the Fourier transform followed by an analysis of the solution at the front of the wave irradiated into the external medium. In Kamen (1984), an elastic sphere placed into an elastic medium in the case when the displacements at the interface were predetermined in the form of a finite initial segment of an infinite series (a truncated series) in terms of the spherical functions was considered. The authors derived a governing system of ordinary differential equations valid until the moment of arrival of a first reflected wave at the interface.

The problems of diffraction of elastic and acoustic waves by obstacles, whose motion is described by the equations of the linear theory of elasticity, have not been well studied. A statement of such problems and some solutions related to the echo-signals are presented in Nigul (1976) and Nigul et al. (1974). It should be pointed out that because of the axial symmetry, the method of incomplete separation of variables (the series expansion in terms of the Legendre polynomials $P_n(\cos\theta)$) was used in all of the works devoted to solution of these problems.

In the papers Makarov, Petrashen (1953), Petrashen et al. (1953a), and Buldyrev, Molotkov (1958), the solutions for the acoustic and elastic media were derived using the Laplace transform with respect to time followed by an asymptotic analysis of the images for the purpose of separating out the rapidly varying terms. In Hickling, Means (1968), the Fourier transform was used in order to solve the problem of diffraction of a plain acoustic pulse by a thick-walled hollow elastic sphere. A similar problem was studied in Berger (1969). The author performed an inversion of the Laplace transform numerically using a series expansion in terms of the ultraspherical polynomials, and obtained the numerical results for the radial vibrations only (when the load was a converging spherical wave). In Veksler (1975) and Veksler et al. (1970), a system 'obstacle – hollow elastic sphere' was studied; at that, the hollow elastic sphere was either without a filler or occupied by an acoustic medium or an absolutely rigid insert. In Veksler et al. (1970), the integral Fourier transform was employed and summation of the series by the Legendre polynomials was performed using the Watson transform followed by an asymptotic analysis. In Veksler (1975), an asymptotic analysis for the initial instants of time was made and a jump in the stresses at the wave front at a sphere was studied. The author marked out a qualitative difference of the behavior of thin-walled and thick-walled obstacles.

In Grigolyuk et al. (1976), Grigolyuk et al. (1977b), Tarlakovsky (1977), Gorshkov et al. (1979a), Grigolyuk et al. (1979), and Tarlakovsky (1981), the problems of interaction of plane and spherical elastic (acoustic) waves with a thick-walled elastic sphere, which either has a free internal surface or contains an elastic inclusion, were studied. The method of solution was based on derivation of a general solution of the hyperbolic equation obtained from the wave equation after separation of the angular coordinate (Grigolyuk et al. (1976)). (Let us note that a similar form of solution was also derived in Akkas (1977), Dikasov (1979), and Dikasov (1982).) Then, the functions sought were represented in the form of the sum of the generalized spherical waves, and this was a generalization of the method of summation of elementary spherical waves in the case of a central symmetry. For the Laplace images, the recursive relationships, which provided an opportunity to calculate the exact values of the originals, were derived. In Gorshkov et al. (1983), it was demonstrated that an analytical result can be obtained by means of direct expansion of the image of the function sought into the multiple series in terms of exponential functions.

In Podstrigach et al. (1980), the interaction of a spherical acoustic wave with a solid elastic spherical inclusion was studied. The solution was derived using the Fourier expansion with respect to time followed by numerical inversion. The numerical results were obtained mainly for the vicinity of the reflected wave front. An interaction of a plane acoustic wave with a thick-walled sphere occupied by an acoustic media and with a two-layered acoustic sphere was studied in Podstrigach et al. (1979) and Poddubnyak (1980), respectively. Besides the Fourier transform, the authors applied the Sommerfeld–Watson transform. The numerical results were obtained for the vicinity of the waves' fronts. Similar methods were used in Poddubnyak (1979), Poddubnyak et al. (1978), Poddubnyak (1984), and Podstrigach, Poddubnyak (1986). In Galazyuk, Gorechko (1983), the Laguerre transform with respect to time was used to solve the problem of diffraction of a spherical wave by a thick-walled shell surrounded by an acoustic medium. In Kubenko (1983), the Laplace transform by means of numerical solution of the integral equation was used.

The publications devoted to solution of the internal transient problems when a point source of spherical waves is located inside a sphere can be divided into two categories. The first one covers the problems of propagation of the seismic waves in the Earth. Among the works of this category, we should, first, mark out the cycle of papers Alterman, Kornfeld (1963), Alterman, Kornfeld (1964), Alterman, Abramovici (1965), Alterman, Kornfeld (1965a), Alterman, Kornfeld (1965b), Alterman, Abramovici (1966), and Alterman, Abramovici (1967). Besides the series expansion in terms of the Legendre polynomials, the authors used the Laplace transform with respect to time followed by summation of the infinite series in residues. In Alterman, Kornfeld (1963), Alterman, Kornfeld (1964), and Alterman, Kornfeld (1965a), it was assumed that a sphere was occupied by an acoustic medium. In Alterman, Abramovici (1965) and Alterman, Abramovici (1966), when analyzing an elastic sphere, it was assumed that the source generates the P-waves only. Besides the method mentioned above, a series expansion of the images in terms of the elementary waves was applied. In order to do this, the so-called coefficients of reflection were introduced ($P - P$, $P - S$, $S - P$, and $S - S$). These coefficients described a formation of the new P-waves and S-waves (the compression waves and the shear waves, respectively) when P-waves and S-waves encountered the boundaries. When analyzing the solution, the Watson transform was used. In Alterman, Kornfeld (1965b), a source, which created the purely shear waves, was considered, and in Alterman, Abramovici (1967), it was assumed that the source is a force applied at a point ($SV - P$-source).

In Gelchinsky (1958), Burridge (1963), Maiti (1969), and Tanyi (1966), propagation of a spherical P-wave from a point source in a solid elastic isotropic sphere was analyzed. At that, in Gelchinsky (1958), a formal solution in a space of the Laplace transform with respect to time was derived and the Rayleigh waves at the sphere's surface were studied. In Tanyi (1966), an estimation of convergence of the series in terms of $P_n(\cos\theta)$ was made and a comparison with the solution derived by means of the method of the geometrical optics was performed. In Maiti (1969), a point source was characterized by a velocity potential, and an integral transform with respect to the angle was used. In Burridge (1963), a solution was derived by means of the method of the geometrical optics using the Kirchhoff formula and Huygens principle.

In Jeffreys, Lapwood (1957), the Watson transform for a fluid sphere was used in assumption of the asymptotic representations by Debye for the cylindrical functions of semi-integer index. Let us note that application of the methods mentioned above, besides the Laplace transform, with the inversion by residues practically does not allow us to obtain the results for arbitrary instants of time.

The second category of the internal problems covers the problems of propagation of spherical waves in a fluid that occupies some spherical reservoir. Determination of the loads at the walls of an absolutely stiff spherical reservoir in the case of axial symmetry was presented in Babaev et al. (1974) and Gordienko, Kubenko (1983). The similar problems for the spherical reservoirs filled with and surrounded by the elastic media were studied in Gorshkov et al. (1978) and Tarlakovsky (1981). Similarly to the external problems of interaction, the method of summation of generalized spherical waves was applied when the Laplace transform followed by its exact inversion was used. Let us note that the Laplace transform in this case is an auxiliary body of mathematics. Basically, its application can be replaced by a successive integration of a recursive chain of systems of ordinary differential equations.

The problem of action of an asymmetrically located internal point source on a multilayer elastic spherical packet surrounded by and filled with the acoustic media was studied in Babaev (1985). Motion of each layer was described by the equations of the linear theory of elasticity. The problem was reduced to a system of integral equations in the coefficients of the functions sought.

In Slepyan (1963), when generalizing the results of Novozhilov (1959), the authors investigated the limiting values, as $t \to \infty$, of the displacements of a deformable body located in a continuum and subjected to an arbitrary pressure wave.

The dynamical problems for the thick-walled obstacles and an acoustic wave in accounting for different nonlinear effects were discussed in Galiev (1977) and Galiev (1981).

The general problems related to the analysis of the mixed initial and boundary value problems for the homogeneous and piecewise homogeneous elastic media having spherical boundaries were discussed in Natrashvili (1980)

and Natrashvili, Dzhagmaidze (1977). The integral formulas for the problems
of reflection of elastic waves based on the Huygens principle were presented
in Pao, Varatharajulu (1976).

In Israilov (1981), Israilov (1982), and Israilov (1983)), the existence theorems for a solution of the transient problems of diffraction of the elastic
waves by the rigid obstacles surrounded by the Lyapunov surface were proven
and a method of solution of these problems based on the method of small
parameter was presented.

In this chapter, the main attention will be concentrated on the problem
of diffraction of the elastic waves by a spherical deformable body, whose
motion is described by the equations of the theory of elasticity. Since the
solutions of the problems of transient vibrations of spherical elastic bodies
under predetermined dynamical loads can be obtained as a particular case of
this more general problem, we do not consider them separately.

The problems of diffraction of waves by a cavity and an absolutely stiff
sphere and the problems in which the elastic media interfaces are considered
as thin-walled shell are studied in Chapters 4 and 5.

3.1 Statement of the Problem

Let us now consider a problem of nonstationary interaction of the elastic
waves with a deformable spherical obstacle applying the results presented in
Tarlakovsky (1975), Grigolyuk et al. (1976), Grigolyuk et al. (1977a), Grigolyuk et al. (1977b), Tarlakovsky (1977), Gorshkov et al. (1978), Gorshkov et
al. (1979), and Tarlakovsky (1981).

We shall assume that a thick-walled elastic spherical shell of internal radius $R = a$ and external radius $R = b$ is placed into an infinite elastic medium
and this shell is occupied by an elastic medium (Gorshkov et al. (1978),
Gorshkov et al. (1979a)). Let us enumerate the media mentioned by the indexes 1, 0, and 2, respectively. Let the materials of all three media be homogeneous, isotropic, and linearly elastic. We shall analyze the media motion
in the spherical coordinate system (R, θ, ϑ) with the origin at the sphere's
center (the angle ϑ is assumed to be measured from the plane perpendicular
to the x axis (see Fig. 3.1)).

At the time $t = 0$, the sphere's surface comes in contact with an elastic
compression wave (P-wave). It can be a spherical wave generated by a point
source located inside the shell at the x axis at the distance D_2 ($D_2 < a$)
from the sphere's center (the internal problem). It can also be a plane wave
or a spherical wave propagating along the x axis in the external medium (the
external problem). In the last case, when considering a spherical wave, we
assume that the source is located at the x axis in the external medium at the
distance D_0 ($D_0 > b$) from the coordinate origin. We also assume that at the
initial instant of time, the system is at rest (the initial conditions are zeros).

Similarly to (1.11), let us introduce the dimensionless variables as follows:

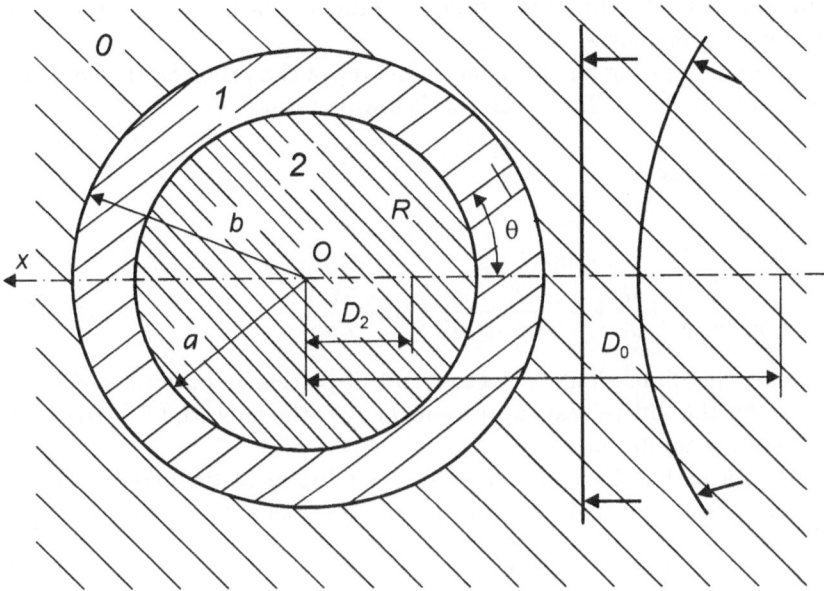

Fig. 3.1. The problem of interaction of a plane (spherical) wave with a thick-walled spherical shell

$$r = \frac{R}{b}, \qquad \tau = \frac{c_1^{(0)} t}{b}, \qquad \delta = \frac{b - a}{b}, \qquad r_1 = 1 - \delta,$$

$$\gamma_i = \frac{c_1^{(0)}}{c_1^{(i)}}, \qquad \eta_i = \frac{c_1^{(0)}}{c_2^{(i)}}, \qquad \kappa_i = \frac{\lambda_i}{\lambda_i + 2\mu_i},$$

$$\tilde{u}_i = \frac{u_i}{b}, \qquad \tilde{v}_i = \frac{v_i}{b}, \qquad \tilde{\varphi}_i = \frac{\varphi_i}{b^2}, \qquad \tilde{\psi}_i = \frac{\psi_i}{b^2},$$

$$\beta_i = \frac{\lambda_i + 2\mu_i}{\lambda_1 + 2\mu_1} = \frac{\rho_i c_1^{(i)\,2}}{\rho_1 c_1^{(1)\,2}}, \qquad \tilde{\sigma}_{\alpha\beta}^{(i)} = \frac{\sigma_{\alpha\beta}^{(i)}}{\lambda_1 + 2\mu_1}, \qquad i = 0, 1, 2; \quad (3.1)$$

the speeds

$$c_1^{(i)} = \sqrt{\frac{\lambda_i + 2\mu_i}{\rho_i}}, \qquad c_2^{(i)} = \sqrt{\frac{\mu_i}{\rho_i}}, \qquad i = 0, 1, 2,$$

are the speeds of propagation of the elastic compression waves (P-wave) and of the shear waves (S-wave), respectively. Here, λ_i and μ_i are the Lamé coefficients, ρ_i is the medium density, u_i and v_i are the radial and tangential displacements, $\sigma_{\alpha\beta}^{(i)}$, α, $\beta = r, \theta, \vartheta$, are the components of the stress tensor, φ_i and ψ_i are the scalar and nonzero components of the vector potential of elastic waves, which correspond to the undisturbed motion, and i is the medium number. We shall omit tildes in dimensionless values without any supplementing references.

Then, in accounting for the axial symmetry of the problem, the equations of motion of the ith medium take the following form:

$$\gamma_i^2 \frac{\partial^2 \varphi_i}{\partial \tau^2} = \Delta\varphi,$$

$$\eta_i^2 \frac{\partial^2 \psi_i}{\partial \tau^2} = \Delta\psi_i - \frac{\psi_i}{r^2 \sin\theta}, \qquad r \geq 0,$$

$$\Delta = \frac{1}{r^2} \frac{\partial}{\partial r}\left(r^2 \frac{\partial}{\partial r}\right) + \frac{1}{r^2 \sin\theta} \frac{\partial}{\partial \theta}\left(\sin\theta \frac{\partial}{\partial \theta}\right). \qquad (3.2)$$

At that, the dimensionless stresses and displacements are related to the potentials by the differential expressions (1.13) (the subscript i corresponds to the medium number).

In accounting for (1.69), the boundary conditions at the external surface $r = 1$ and the internal surface $r = r_1$ can be represented as follows ($\tau \geq 0$):

$$\sigma_{rr}^{(1)}\Big|_{r=1} = \sigma_{rr}^{(0)}\Big|_{r=1} + \sigma_{rr\,s}^{(0)}\Big|_{r=1},$$

$$u_1\big|_{r=1} = u_0\big|_{r=1} + u_{0\,s}\big|_{r=1},$$

$$\sigma_{r\theta}^{(1)}\Big|_{r=1} = \sigma_{r\theta}^{(0)}\Big|_{r=1} + \sigma_{r\theta\,s}^{(0)}\Big|_{r=1} = k_{10}(v_0 + v_{0\,s} - v_1)\big|_{r=1},$$

$$\sigma_{rr}^{(1)}\Big|_{r=r_1} = \sigma_{rr}^{(2)}\Big|_{r=r_1} + \sigma_{rr\,s}^{(2)}\Big|_{r=r_1},$$

$$u_1\big|_{r=r_1} = u_2\big|_{r=r_1} + u_{2\,s}\big|_{r=r_1},$$

$$\sigma_{r\theta}^{(1)}\Big|_{r=r_1} = \sigma_{r\theta}^{(2)}\Big|_{r=r_1} + \sigma_{r\theta\,s}^{(2)}\Big|_{r=r_1} = k_{12}(v_1 - v_2 - v_{2\,s})\big|_{r=r_1}; \qquad (3.3)$$

here, $u_{i\,s}$, $v_{i\,s}$, $\sigma_{rr\,s}^{(i)}$, and $\sigma_{r\theta\,s}^{(i)}$, $i = 0, 2$, are the components of the stress–strain state produced by the incoming wave. For the external problem, we have

$$u_{2\,s} = v_{2\,s} = \sigma_{rr\,s}^{(2)} = \sigma_{r\theta\,s}^{(2)} \equiv 0,$$

and for the internal one, we have

$$u_{0\,s} = v_{0\,s} = \sigma_{rr\,s}^{(0)} = \sigma_{r\theta\,s}^{(0)} \equiv 0.$$

We have introduced the coefficients k_{10} and k_{12} formally; it allows us to consider not only the intermediate cases but two extreme cases of the boundary conditions, as well; for $k_{10} = 0$ and $k_{12} = 0$, we have a free sliding of the boundary surfaces, and for $k_{10} = \infty$ and $k_{12} = \infty$, we have a stiff cohesion of the elastic media. Let us note that we do not consider the media lamination, which can take place in the first case.

The boundary conditions presumes the absence of disturbances at infinity (as $r \to \infty$) and a boundedness of the functions sought at the coordinate origin (as $r \to 0$), that is,

$$\varphi_2(r, \theta, \tau) = O(1), \qquad \psi_2(r, \theta, \tau) = O(1), \qquad r \to 0. \tag{3.4}$$

In order to state the boundary value problem (3.2)–(3.4), let us add the following initial conditions ($r \geq 0$):

$$\varphi_i|_{\tau=0} = \psi_i|_{\tau=0} = \left.\frac{\partial \varphi_i}{\partial \tau}\right|_{\tau=0} = \left.\frac{\partial \psi_i}{\partial \tau}\right|_{\tau=0} = 0, \qquad i = 0, 1, 2. \tag{3.5}$$

In order to derive an analytical solution of the problem (3.2)–(3.5), we should represent the functions sought in the form of the series:

$$\varphi_i = \sum_{n=0}^{\infty} \varphi_{in} P_n(\cos\theta), \qquad \psi_i = -\sin\theta \sum_{n=1}^{\infty} \psi_{in} C_{n-1}^{3/2}(\cos\theta),$$

$$u_i = \sum_{n=0}^{\infty} u_{in} P_n(\cos\theta), \qquad v_i = -\sin\theta \sum_{n=1}^{\infty} v_{in} C_{n-1}^{3/2}(\cos\theta),$$

$$\sigma_{rr}^{(i)} = \sum_{n=0}^{\infty} \sigma_{rrn}^{(i)} P_n(\cos\theta), \qquad \sigma_{r\theta}^{(i)} = \sin\theta \sum_{n=1}^{\infty} \sigma_{r\theta n}^{(i)} C_{n-1}^{3/2}(\cos\theta),$$

$$\sigma_{\theta\theta}^{(i)} = \sum_{n=0}^{\infty} \sigma_{\theta\theta n}^{(i)} P_n(\cos\theta) - \frac{1-\kappa_i}{r} \cos\theta \sum_{n=1}^{\infty} v_{in} C_{n-1}^{3/2}(\cos\theta),$$

$$\sigma_{\vartheta\vartheta}^{(i)} = \sum_{n=0}^{\infty} \sigma_{\vartheta\vartheta n}^{(i)} P_n(\cos\theta) + \frac{1-\kappa_i}{r} \cos\theta \sum_{n=1}^{\infty} v_{in} C_{n-1}^{3/2}(\cos\theta),$$

$$i = 0, 1, 2, \quad (3.6)$$

where $P_n(x)$ and $C_{n-1}^{(3/2)}(x)$ are the Legendre and Gegenbauer polynomials (Gradshtein, Ryzhik (1971), Abramowitz, Stegun (1965)).

The coefficients of the series (3.6) are interrelated as follows:

$$u_{in} = \frac{\partial \varphi_{in}}{\partial r} - \frac{n(n+1)}{r} \psi_{in},$$

$$\sigma_{rrn}^{(i)} = \beta_i \left[\frac{\partial u_{in}}{\partial r} - n(n+1) \frac{\kappa_i}{r} v_{in} + \frac{2\kappa_i}{r} u_{in} \right],$$

$$v_{in} = \frac{\varphi_{in} - \psi_{in}}{r} - \frac{\partial \psi_{in}}{\partial r},$$

$$\sigma_{r\theta n}^{(i)} = -\frac{1-\kappa_i}{2} \beta_i \left(\frac{\partial v_{in}}{\partial r} + \frac{u_{in} - v_{in}}{r} \right),$$

$$\sigma_{\theta\theta n}^{(i)} = \kappa_i \sigma_{rrn}^{(i)} + (1-\kappa_i)(1+2\kappa_i) \frac{u_{in}}{r} - n(n+1) \kappa_i(1-\kappa_i) \frac{v_{in}}{r},$$

$$\sigma_{\vartheta\vartheta n}^{(i)} = \sigma_{\theta\theta n}^{(i)} - n(n+1)(1-\kappa_i) \frac{v_{in}}{r}. \tag{3.7}$$

Hence, for an arbitrary number $n \geq 0$, we arrive at the following boundary value problem:

$$\gamma_i^2 \frac{\partial^2 \varphi_{in}}{\partial \tau^2} = \Delta_n \varphi_{in},$$

$$\eta_i^2 \frac{\partial^2 \psi_{in}}{\partial \tau^2} = \Delta_n \psi_{in},$$

$$\Delta_n = \frac{1}{r^2} \frac{\partial}{\partial r} \left(r^2 \frac{\partial}{\partial r} \right) - \frac{n(n+1)}{r^2},$$

$$\varphi_{in}|_{\tau=0} = \psi_{in}|_{\tau=0} = \frac{\partial \varphi_{in}}{\partial \tau}\bigg|_{\tau=0} = \frac{\partial \psi_{in}}{\partial \tau}\bigg|_{\tau=0} = 0, \qquad r > 0,$$

$$i = 0, 1, 2,$$

$$\sigma_{rrn}^{(1)}\bigg|_{r=1} = \sigma_{rrn}^{(0)}\bigg|_{r=1} + \sigma_{rrn\,s}^{(0)}\bigg|_{r=1},$$

$$u_{1n}|_{r=1} = u_{0n}|_{r=1} + u_{0n\,s}|_{r=1},$$

$$\sigma_{r\theta n}^{(1)}\bigg|_{r=1} = \sigma_{r\theta n}^{(0)}\bigg|_{r=1} + \sigma_{r\theta n\,s}^{(0)}\bigg|_{r=1} = k_{10}(v_{0n} + v_{0n\,s} - v_{1n})|_{r=1},$$

$$\sigma_{rrn}^{(1)}\bigg|_{r=r_1} = \sigma_{rrn}^{(2)}\bigg|_{r=r_1} + \sigma_{rrn\,s}^{(2)}\bigg|_{r=r_1},$$

$$u_{1n}|_{r=r_1} = u_{2n}|_{r=r_1} + u_{2n\,s}|_{r=r_1},$$

$$\sigma_{r\theta n}^{(1)}\bigg|_{r=r_1} = \sigma_{r\theta n}^{(2)}\bigg|_{r=r_1} + \sigma_{r\theta n\,s}^{(2)}\bigg|_{r=r_1} = k_{12}(v_{1n} - v_{2n} - v_{2n\,s})|_{r=r_1}.$$

$$(3.8)$$

The boundary conditions for the functions φ_{in} and ψ_{in}, $i = 0, 2$, as $r \to \infty$ and $r \to 0$, are similar to (3.4).

Here, $u_{in\,s}$, $v_{in\,s}$, $\sigma_{rrn\,s}^{(i)}$, and $\sigma_{r\theta n\,s}^{(i)}$ are the coefficients of the series expansion in terms of $P_n(\cos\theta)$ and $P_{n-1}^{3/2}(\cos\theta)$ of the components of the stress–strain state which corresponds to the incoming wave. For the external problem, we have

$$u_{2n\,s} = v_{2n\,s} = \sigma_{rrn\,s}^{(2)} = \sigma_{r\theta n\,s}^{(2)} \equiv 0,$$

and for the internal one, we have

$$u_{0n\,s} = v_{0n\,s} = \sigma_{rrn\,s}^{(0)} = \sigma_{r\theta n\,s}^{(0)} \equiv 0.$$

Let us assume that in the case of external problem, the disturbing load is produced by the P-wave in the external medium of the potential $\varphi_{0s}(r, \theta, \tau)$; for the plane wave, we obtain (Fig. 3.1)

$$\varphi_{0s}(r, \theta, \tau) = f(\tau + r\cos\theta - 1)\, H(\tau + r\cos\theta - 1), \qquad (3.9)$$

and for the spherical wave, we obtain

$$\varphi_{0s}(r,\,\theta,\,\tau) = \frac{d_0 - 1}{l}\, f(\tau + d_0 - 1 - l)\, H(\tau + d_0 - 1 - l)\,, \qquad (3.10)$$

where $f(x)$ is an arbitrary function, which determines a law of variation of the potential in time; here, $H(x)$ is the Heaviside unit function and

$$l = \sqrt{r^2 + d_0^2 - 2rd_0\,\cos\theta}\,, \qquad d_0 = \frac{D_0}{b}\,.$$

The parameter $(d_0 - 1)$ in (3.10) is used to make a passage to the limit when switching from the spherical wave to the plane one (as $d_0 \to \infty$ and $\varphi_{0s}(1,\,0,\,\tau) = f(\tau)$).

Since we shall apply the Laplace transform with respect to time τ, let us derive the expressions for the transforms of the function φ_{0ns}. Making the Laplace transform in (3.9) and (3.10) and using the addition formulas for the Bessel functions (Watson (1945), Gradshtein, Ryzhik (1971), Grigolyuk, Gorshkov (1976)), we obtain the relationships for the images of the coefficients of the series expansion of the potential φ_{0s} in terms of the Legendre polynomials (the superscript L designates the transform and s is the transform parameter); for a plane wave, we get

$$\varphi_{0ns}^{L}(r,\,s) = f^{L}(s)\,e^{-s}(2n+1)\,\sqrt{\frac{\pi}{2rs}}\,I_{n+1/2}(rs)$$

$$= f^{L}(s)\,\frac{(-1)^n(2n+1)}{2\,(rs)^{n+1}}\,e^{(r-1)\,s}\left[R_{n0}(-rs) - e^{-2rs}R_{n0}(rs)\right] \qquad (3.11)$$

and for a spherical wave, we get

$$\varphi_{0ns}^{L}(r,\,s) = f^{L}(s)\,\frac{d_0 - 1}{\sqrt{d_0}}\,e^{(d_0-1)\,s}(2n+1)\,I_{n+1/2}(rs)\,K_{n+1/2}(d_0 s)$$

$$= f^{L}(s)\,\frac{(-1)^n(d_0 - 1)(2n+1)\,R_{n0}(d_0 s)}{2s^{2n+1}(d_0 r)^{n+1}}\,e^{(r-1)\,s}$$

$$\times \left[R_{n0}(-rs) - e^{-2sr}R_{n0}(rs)\right]\,. \qquad (3.12)$$

When deriving (3.11) and (3.12), we have used the relationship between the Bessel functions of a semi-integer index and the elementary functions (1.126).

Considering the internal problem, we shall assume that a source of the spherical P-wave is located in the internal medium $i = 2$ at the distance D_2 from the sphere's center (Fig. 3.1). We shall predetermine the dimensionless potential of this wave by the formula

$$\varphi_{2s}(r,\,\theta,\,\tau) = \frac{r_1 - d_2}{l}\, f[\tau + \gamma_2(r_1 - l - d_2)]\, H[\tau + \gamma_2(r_1 - l - d_2)]\,, \qquad (3.13)$$

where $d_2 = D_2/b$; in this formula, l has the same meaning as in (3.9) and (3.10), if to replace d_0 by d_2. Introduction of the multiplier into the numerator of (3.13) allows us to satisfy the condition $\varphi_{2s}(r_1, 0, \tau) = f(\tau)$.

Similarly to the external problem, in the space of the Laplace transform, we obtain

$$\varphi_{2s}^L(r, s)$$

$$= f^L(s)(r_1 - d_2) e^{\gamma_2(r_1 - d_2)} \frac{2n+1}{\sqrt{rd_2}} I_{n+1/2}(\gamma_2 s d_2) K_{n+1/2}(\gamma_2 s r)$$

$$= f^L(s)(r_1 - d_2) \frac{(-1)^n (2n+1) R_{n0}(\gamma_2 s r)}{2(\gamma_2 s)^{2n+1}(d_2 r)^{n+1}} e^{\gamma_2(r_1 - r)s}$$

$$\times \left[R_{n0}(-\gamma_2 d_2 s) - e^{-2\gamma_2 d_2 s} R_{n0}(\gamma_2 d_2 s) \right] . \quad (3.14)$$

Various limiting cases can be analyzed using the statement of the problem presented above.

For example, according to (1.39), in order to study vibrations of the system 'acoustic medium $(i = 0)$ – thick-walled elastic sphere $(i = 1)$ – acoustic medium $(i = 2)$', it is sufficient to assume that there is no resistance to shear in the external and internal media (i.e., $\mu_i = 0$ for $i = 0, 2$). In other words, in accounting for (3.1), we must set

$$c_2^{(i)} = 0, \qquad \eta_i = \infty, \qquad \kappa_i = 1, \qquad \psi_{in} \equiv 0. \qquad (3.15)$$

Similarly, it is possible to study vibrations of the system in the case when only one of the media $i = 0$ or $i = 2$ is of acoustic type.

Let us consider two more limiting cases (Sect. 1.4).

(1) An absolutely stiff ring is placed at the boundary $r = 1$ or $r = r_1$:

$$u_{1n}|_{r=1} = v_{1n}|_{r=1} = 0 \qquad \text{or} \qquad u_{1n}|_{r=r_1} = v_{1n}|_{r=r_1} = 0.$$

(2) An external surface $r = 1$ or an internal surface $r = r_1$ is free:

$$\sigma_{rrn}^{(1)}\Big|_{r=1} = \sigma_{r\theta n}^{(1)}\Big|_{r=1} = 0 \qquad \text{or} \qquad \sigma_{rrn}^{(1)}\Big|_{r=r_1} = \sigma_{r\theta n}^{(1)}\Big|_{r=r_1} = 0.$$

It is possible to demonstrate that in the first case, it is necessary to set

$$c_1^{(i)} = c_2^{(i)} = \infty, \qquad \beta_i = \infty, \qquad \eta_i = \gamma_i = 0, \qquad i = 0, 2, \qquad (3.16)$$

and, in the second one,

$$c_1^{(i)} = c_2^{(i)} = 0, \qquad \eta_0 = \gamma_0 = \beta_i = 0,$$
$$\eta_2 = \gamma_2 = \infty, \qquad i = 0, 2. \qquad (3.17)$$

Thus, a solution of the problem in different limiting cases can be derived from a solution of the general problem (3.8); to obtain solutions for the particular cases, it is necessary to satisfy the conditions (3.15), (3.16), or (3.17), respectively.

In order to obtain a solution of the problem for a solid homogeneous sphere inserted into an infinite elastic medium, it is sufficient to set

$$\gamma_0 = \gamma_1, \qquad \eta_0 = \eta_1, \qquad \kappa_0 = \kappa_1, \qquad \beta_0 = \beta_1 = 1 \qquad (3.18)$$

in the solution of the general problem.

In order to get a solution of the problem in the case when the waves propagate in a homogeneous infinite medium, it is sufficient to set

$$\gamma_0 = \gamma_1 = \gamma_2, \qquad \eta_0 = \eta_1 = \eta_2,$$
$$\kappa_0 = \kappa_1 = \kappa_2, \qquad \beta_0 = \beta_1 = \beta_2 = 1 \qquad (3.19)$$

in the solution of the general problem.

3.2 The Laplace Transform

When solving the axially symmetric problems of the dynamics of a thick-walled sphere filled with and surrounded by the elastic or acoustic media, the Laplace transform with respect to time is usually applied. However, when following this approach, the problem of calculation of the original, which corresponds to the solution obtained in the space of images, makes difficulties. In Berger (1969), it was proposed to make inversion of the Laplace transform numerically expanding the images into the series in terms of the ultraspherical polynomials. An asymptotic solution valid for the initial instants of time of interaction of an acoustic wave with an obstacle was derived in Veksler (1975).

When deriving the original, it is desirable to employ the fact that in the class of problems under study, the images of the functions sought include the modified Bessel functions of a semi-integer index, which can be expressed via the elementary Bessel functions.

To calculate the originals, let us expand the images into the series in terms of the exponential functions. The formulas to be derived allow us to find an analytical solution of the problems and demonstrate an analogy with the solution obtained in Gorshkov et al. (1978) and Gorshkov et al. (1979a).

Applying the Laplace transform to the problem (3.8), we find out that the images of the potentials sought, which satisfy (3.8), can be represented in the following form:

$$\varphi_{in}^{L}(r, s) = \frac{1}{\sqrt{r}} \left[G_{in}^{(1)}(s) I_{n+1/2}(\gamma_i r s) + G_{in}^{(2)}(s) K_{n+1/2}(\gamma_i r s) \right],$$

$$\psi_{in}^{L}(r, s) = \frac{1}{\sqrt{r}} \left[G_{in}^{(3)}(s) I_{n+1/2}(\eta_i r s) + G_{in}^{(4)}(s) K_{n+1/2}(\eta_i r s) \right]; \qquad (3.20)$$

here, $G_{in}^{(j)}(s)$ are the constants of integration; $I_{n+1/2}(x)$ and $K_{n+1/2}(x)$ are the modified Bessel functions (Watson (1945)).

Taking into consideration the relationship between the series coefficients, which correspond to the potentials, the components of the vector of displacements, and the stress tensor (1.3), as well as the properties of the Bessel function (Watson (1945)), we arrive at the following expressions:

$$
u_{in}^L(r, s) = \frac{1}{r^{3/2}} \left\{ G_{in}^{(1)} \left[\gamma_i rs I_{n+3/2}(\gamma_i rs) + n I_{n+1/2}(\gamma_i rs) \right] \right.
$$

$$
+ G_{in}^{(2)} \left[-\gamma_i rs K_{n+3/2}(\gamma_i rs) + n K_{n+1/2}(\gamma_i rs) \right]
$$

$$
\left. - G_{in}^{(3)} \left[n(n+1) I_{n+1/2}(\eta_i rs) - G_{in}^{(4)} n(n+1) K_{n+1/2}(\eta_i rs) \right] \right\},
$$

$$
v_{in}^L(r, s) = \frac{1}{r^{3/2}} \left\{ G_{in}^{(1)} I_{n+1/2}(\gamma_i rs) + G_{in}^{(2)} K_{n+1/2}(\gamma_i rs) \right.
$$

$$
+ G_{in}^{(3)} \left[-\eta_i rs I_{n+3/2}(\eta_i rs) - (n+1) I_{n+1/2}(\eta_i rs) \right]
$$

$$
\left. + G_{in}^{(4)} \left[\eta_i rs K_{n+3/2}(\eta_i rs) - (n+1) K_{n+1/2}(\eta_i rs) \right] \right\},
$$

$$
\sigma_{rrn}^{(i)L}(r, s) = \frac{\beta_i}{r^{5/2}} \left\{ G_{in}^{(1)} \left[\gamma_i^2 r^2 s^2 I_{n+5/2}(\gamma_i rs) \right. \right.
$$

$$
+ \gamma_i rs(2n+1+2\kappa_i) I_{n+3/2}(\gamma_i rs) + n(n-1)(1-\kappa_i) I_{n+1/2}(\gamma_i rs) \Big]
$$

$$
+ G_{in}^{(2)} \left[\gamma_i^2 r^2 s^2 K_{n+5/2}(\gamma_i rs) - \gamma_i rs(2n+1+2\kappa_i) K_{n+3/2}(\gamma_i rs) \right.
$$

$$
\left. + n(n-1)(1-\kappa_i) I_{n+1/2}(\gamma_i rs) \right]
$$

$$
+ G_{in}^{(3)} n(n+1)(1-\kappa_i) \left[-\eta_i rs I_{n+3/2}(\eta_i rs) - (n-1) I_{n+1/2}(\eta_i rs) \right]
$$

$$
\left. + G_{in}^{(4)} n(n+1)(1-\kappa_i) \left[\eta_i rs K_{n+3/2}(\eta_i rs) - (n-1) K_{n+1/2}(\eta_i rs) \right] \right\},
$$

$$
\sigma_{r\theta n}^{(i)L}(r, s) = -\frac{\beta_i(1-\kappa_i)}{2r^{5/2}} \left\{ 2G_{in}^{(1)} \left[\gamma_i rs I_{n+3/2}(\gamma_i rs) \right. \right.
$$

$$
\left. + (n-1) I_{n+1/2}(\gamma_i rs) \right]
$$

$$
+ 2G_{in}^{(2)} \left[-\gamma_i rs K_{n+3/2}(\gamma_i rs) + (n-1) K_{n+1/2}(\gamma_i rs) \right]
$$

$$
+ G_{in}^{(3)} \left[-\gamma_i^2 r^2 s^2 I_{n+5/2}(\eta_i rs) - (2n+1) \eta_i rs I_{n+3/2}(\eta_i rs) \right.
$$

$$
\left. - 2(n^2-1) I_{n+1/2}(\eta_i rs) \right] + G_{in}^{(4)} \left[-\gamma_i^2 r^2 s^2 K_{n+5/2}(\eta_i rs) \right.
$$

$$
\left. \left. + (2n+1) \eta_i rs K_{n+3/2}(\eta_i rs) - 2(n^2-1) K_{n+1/2}(\eta_i rs) \right] \right\}. \quad (3.21)
$$

Considering a boundedness of the potential images, as $r \to 0$ and $r \to \infty$, and taking into account an unboundedness of $I_{n+1/2}(x)$ as $x \to \infty$ and of $K_{n+1/2}(x)$ as $x \to 0$ (Watson (1945)), we obtain

$$
G_{0n}^{(1)} = G_{0n}^{(3)} = G_{2n}^{(2)} = G_{2n}^{(4)} \equiv 0.
$$

Expressing the Bessel functions via the elementary functions (1.126), we can write (3.21) as follows:

$$u_{in}^{L}(r,\,s) = -\frac{1}{r^{n+2}} \left[C_{in}^{(1)} R_{n1}(-\gamma_i rs)\,e^{\gamma_i rs} + C_{in}^{(2)} R_{n1}(\gamma_i rs)\,e^{-\gamma_i rs} \right.$$
$$\left. +\, n(n+1)\, C_{in}^{(3)} R_{n0}(-\eta_i rs)\,e^{\eta_i rs} + C_{in}^{(4)} n(n+1)\, R_{n0}(\eta_i rs)\,e^{-\eta_i rs} \right],$$

$$v_{in}^{L}(r,\,s) = \frac{1}{r^{n+2}} \left[C_{in}^{(1)} R_{n0}(-\gamma_i rs)\,e^{\gamma_i rs} + C_{in}^{(2)} R_{n0}(\gamma_i rs)\,e^{-\gamma_i rs} \right.$$
$$\left. +\, C_{in}^{(3)} R_{n3}(-\eta_i rs)\,e^{\eta_i rs} + C_{in}^{(4)} R_{n3}(\eta_i rs)\,e^{-\eta_i rs} \right],$$

$$\sigma_{rrn}^{(i)\,L}(r,\,s) = \frac{\beta_i}{r^{n+3}} \left[C_{in}^{(1)} Q_{n1}^{(i)}(-\gamma_i rs)\,e^{\gamma_i rs} + C_{in}^{(2)} Q_{n1}^{(i)}(\gamma_i rs)\,e^{-\gamma_i rs} \right.$$
$$\left. +\, C_{in}^{(3)} n(n+1) Q_{n2}^{(i)}(-\eta_i rs)\,e^{\eta_i rs} + C_{in}^{(4)} n(n+1) Q_{n2}^{(i)}(\eta_i rs)\,e^{-\eta_i rs} \right],$$

$$\sigma_{r\theta n}^{(i)\,L}(r,\,s) = \frac{\beta_i}{r^{n+3}} \left[C_{in}^{(1)} Q_{n2}^{(i)}(-\gamma_i rs)\,e^{\gamma_i rs} + C_{in}^{(2)} Q_{n2}^{(i)}(\gamma_i rs)\,e^{-\gamma_i rs} \right.$$
$$\left. +\, C_{in}^{(3)} Q_{n3}^{(i)}(-\eta_i rs)\,e^{\eta_i rs} + C_{in}^{(4)} Q_{n3}^{(i)}(\eta_i rs)\,e^{-\eta_i rs} \right],$$

$$C_{in}^{(1)}(s) = \frac{(-1)^n}{\sqrt{2\pi}\,(\gamma_i s)^{n+1/2}}\, G_{in}^{(1)}(s),$$

$$C_{in}^{(2)}(s) = -\frac{(-1)^n}{\sqrt{2\pi}\,(\gamma_i s)^{n+1/2}}\, G_{in}^{(1)} + \frac{\sqrt{\pi/2}}{(\gamma_i s)^{n+1/2}}\, G_{in}^{(2)},$$

$$C_{in}^{(3)}(s) = \frac{(-1)^n}{\sqrt{2\pi}\,(\eta_i s)^{n+1/2}}\, G_{in}^{(3)}(s),$$

$$C_{in}^{(4)}(s) = -\frac{(-1)^n}{\sqrt{2\pi}\,(\eta_i s)^{n+1/2}}\, G_{in}^{(3)} + \frac{\sqrt{\pi/2}}{(\eta_i s)^{n+1/2}}\, G_{in}^{(4)},$$

$$C_{0n}^{(1)} = C_{0n}^{(3)} \equiv 0, \qquad C_{2n}^{(2)} = C_{2n}^{(1)}, \qquad C_{2n}^{(4)} = -C_{2n}^{(3)},$$

$$R_{n1}(s) = R_{n+1,0}(s) - n R_{n0}(s),$$

$$R_{n2}(s) = R_{n+2,0}(s) - (2n+1)\, R_{n+1,0}(s) + n(n-1)\, R_{n0}(s),$$

$$R_{n3}(s) = R_{n+1,0}(s) - (n+1)\, R_{n0}(s),$$

$$Q_{n1}^{(i)}(s) = R_{n2}(s) - 2\kappa_i R_{n1}(s) - n(n+1)\,\kappa_i R_{n0}(s),$$

$$Q_{n2}^{(i)}(s) = (1-\kappa_i)\, [R_{n1}(s) + R_{n0}(s)],$$

$$Q_{n3}^{(i)}(s) = \frac{1-\kappa_i}{2}\, [R_{n2}(s) + (n+2)(n-1)\, R_{n0}(s)]. \tag{3.22}$$

Substituting (3.22) into the boundary conditions (3.8), we arrive at the system of linear algebraic equations with respect to $C_{in}^{(j)}(s)$, $i = 0, 1, 2$, $j = 1, 2, 3, 4$,

$$\mathbf{AC} = \mathbf{B}, \qquad \mathbf{A}_{8\times 8} = \begin{pmatrix} \mathbf{A}_1 \\ \mathbf{A}_2 \end{pmatrix}. \tag{3.23}$$

Here, the following notation is used:

$$(\mathbf{A}_m)_{4\times 8} = \left(\mathbf{a}_1^{(m)}\,e^{-\gamma_0 r_m s},\; \mathbf{a}_2^{(m)}\,e^{-\eta_0 r_m s},\; \mathbf{a}_3^{(m)}\,e^{\gamma_1 r_m s},\; \mathbf{a}_4^{(m)}\,e^{-\gamma_1 r_m s}, \right.$$

$$\mathbf{a}_5^{(m)}\,e^{\eta_1 r_m s},\ \mathbf{a}_6^{(m)}\,e^{-\eta_1 r_m s},\ \mathbf{a}_7^{(m)\prime}\,e^{\gamma_2 r_m s}+\mathbf{a}_7^{(m)\prime\prime}\,e^{-\gamma_2 r_m s},$$

$$\mathbf{a}_8^{(m)\prime}\,e^{\eta_2 r_m s}+\mathbf{a}_8^{(m)\prime\prime}\,e^{-\eta_2 r_m s}\Big),\qquad m=0,1\,,$$

$$\mathbf{a}_k^{(m)}=\big(a_{1+4m,\,k},\,a_{2+4m,\,k},\,a_{3+4m,\,k},\,a_{4+4m,\,k}\big)^{\mathrm{T}}\,,$$

$$\mathbf{C}=\Big(C_{0n}^{(2)},\,C_{0n}^{(4)},\,C_{1n}^{(1)},\,C_{1n}^{(2)},\,C_{1n}^{(3)},\,C_{1n}^{(4)},\,C_{2n}^{(1)},\,C_{2n}^{(3)}\Big)^{\mathrm{T}}\,,$$

$$\mathbf{B}=\big(b_1,\,b_2,\,b_3,\,b_4,\,b_5,\,b_6,\,b_7,\,b_8\big)^{\mathrm{T}}\,,$$

$$b_1=\sigma_{rrns}^{(0)\mathrm{L}}(1,\,s)\,,\qquad b_2=u_{0ns}^{\mathrm{L}}(1,\,s)\,,$$

$$b_3=\sigma_{r\theta ns}^{(0)\mathrm{L}}(1,\,s)\,,\qquad b_4=k_{10}v_{0ns}^{\mathrm{L}}(1,\,s)\,,$$

$$b_5=\eta_1^3\sigma_{rrns}^{(2)\mathrm{L}}(r_1,\,s)\,,\qquad b_6=r_1^2 u_{2ns}^{\mathrm{L}}(r_1,\,s)\,,$$

$$b_7=r_1^3\sigma_{r\theta ns}^{(2)\mathrm{L}}(r_1,\,s)\,,\qquad b_8=k_{12}r_1^3 v_{2ns}(r_1,\,s)\,,$$

$$a_{l1}=a_{l2}=0\,,\qquad l=5,\,6,\,7,\,8\,,$$

$$a_{l7}'=a_{l7}''=a_{l8}'=a_{l8}''=0\,,\qquad l=1,\,2,\,3,\,4\,,$$

$$a_{11}(s)=-\beta_0 Q_{n1}^{(0)}(\gamma_0 s)\,,\qquad a_{12}(s)=-\beta_0 n(n+1)\,Q_{n2}^{(0)}(\eta_0 s)\,,$$

$$a_{21}(s)=R_{n1}(\gamma_0 s)\,,\qquad a_{22}(s)=n(n+1)\,R_{n0}(\eta_0 s)\,,$$

$$a_{31}(s)=-\beta_0 Q_{n2}^{(0)}(\gamma_0 s)\,,\qquad a_{32}(s)=-\beta_0 Q_{n3}^{(0)}(\eta_0 s)\,,$$

$$a_{41}(s)=-k_{10}R_{n0}(\gamma_0 s)\,,\qquad a_{42}(s)=-k_{10}R_{n3}(\eta_0 s)\,,$$

$$a_{1+4m,\,4}(s)=a_{1+4m,\,3}(-s)=\beta_1 Q_{n1}^{(1)}(\gamma_1 r_m s)\,,$$

$$a_{2+4m,\,4}(s)=a_{2+4m,\,3}(-s)=-R_{n1}(\gamma_1 r_m s)\,,$$

$$a_{3+4m,\,4}(s)=a_{3+4m,\,3}(-s)=\beta_1 Q_{n2}^{(1)}(\gamma_1 r_m s)\,,$$

$$a_{4+4m,\,4}(s)=a_{4+4m,\,3}(-s)$$
$$\qquad=\beta_1 Q_{n2}^{(1)}(\gamma_1 r_m s)+(-1)^m k_{1,\,2m}r_m R_{n0}(\gamma_1 r_m s)\,,$$

$$a_{1+4m,\,6}(s)=a_{1+4m,\,5}(-s)=\beta_1 n(n+1)\,Q_{n2}^{(1)}(\eta_1 r_m s)\,,$$

$$a_{2+4m,\,6}(s)=a_{2+4m,\,5}(-s)=-n(n+1)\,R_{n0}(\eta_1 r_m s)\,,$$

$$a_{3+4m,\,6}(s)=a_{3+4m,\,5}(-s)=\beta_1 Q_{n3}^{(1)}(\eta_1 r_m s)\,,$$

$$a_{4+4m,\,6}(s)=a_{4+4m,\,5}(-s)$$
$$\qquad=\beta_1 Q_{n3}^{(1)}(\eta_1 r_m s)+(-1)^m k_{1,\,2m}r_m R_{n3}(\eta_1 r_m s)\,,$$

$$a_{57}''(s)=-a_{57}'(-s)=\beta_2 Q_{n1}^{(2)}(\gamma_2 r_1 s)\,,$$

$$a_{67}''(s)=-a_{67}'(-s)=-R_{n1}(\gamma_2 r_1 s)\,,$$

$$a_{77}''(s)=-a_{77}'(-s)=-\beta_2 Q_{n2}^{(2)}(\gamma_2 r_1 s)\,,$$

$$a_{87}''(s)=-a_{87}'(-s)=-k_{12}r_1 R_{n0}(\gamma_2 r_1 s)\,,$$

$$a_{58}''(s)=-a_{58}'(-s)=\beta_2 Q_{n2}^{(2)}(\eta_2 r_1 s)\,,$$

$$a_{68}''(s)=-a_{68}'(-s)=-n(n+1)\,R_{n0}(\eta_2 r_1 s)\,,$$

$$a_{78}''(s)=-a_{78}'(-s)=-\beta_2 Q_{n3}^{(2)}(\eta_2 r_1 s)\,,$$

$$a''_{88}(s) = -a'_{88}(-s) = -k_{12}r_1 R_{n3}(\eta_2 r_1 s). \tag{3.24}$$

Here, $\mathbf{a}_k^{(m)}$ multiplied by the exponential function is the kth column of the matrix $\mathbf{A}_m$; at that, the seventh and eighth columns are the sums of two columns. All the coefficients $a_{ij}(s)$ are the polynomials. The vector $\mathbf{C}$ is the column of unknowns and the vector $\mathbf{B}$ is the column of arbitrary parameters, whose elements are the images of the corresponding stresses and displacements in the incoming waves.

Using the representations (3.11), (3.12), and (3.14), as well as the relationship of the potentials with the displacements and stresses, we obtain the following expressions for the elements of the column $\mathbf{B}$:

$$u^{L}_{0ns}(r, s) = -E_0(s) \frac{e^{\gamma_0 rs}}{r^{n+2}} \left[R_{n1}(-\gamma_0 rs) - e^{-2\gamma_0 rs} R_{n1}(\gamma_0 rs) \right],$$

$$v^{L}_{0ns}(r, s) = E_0(s) \frac{e^{\gamma_0 rs}}{r^{n+2}} \left[R_{n0}(-\gamma_0 rs) - e^{-2\gamma_0 rs} R_{n0}(\gamma_0 rs) \right],$$

$$\sigma^{(0)L}_{rrns}(r, s) = E_0(s) \beta_0 \frac{e^{\gamma_0 rs}}{r^{n+3}} \left[Q^{(0)}_{n1}(-\gamma_0 rs) - e^{-2\gamma_0 rs} Q^{(0)}_{n1}(\gamma_0 rs) \right],$$

$$\sigma^{(0)L}_{r\theta ns}(r, s) = E_0(s) \beta_0 \frac{e^{\gamma_0 rs}}{r^{n+3}} \left[Q^{(0)}_{n2}(-\gamma_0 rs) - e^{-2\gamma_0 rs} Q^{(0)}_{n2}(\gamma_0 rs) \right],$$

$$u^{L}_{2ns}(r, s) = -E_2(s) \frac{R_{n1}(\gamma_2 rs)}{(\gamma_2 s)^{n+1} r^{n+2}} e^{-\gamma_2 rs},$$

$$v^{L}_{2ns}(r, s) = E_2(s) \frac{R_{n0}(\gamma_2 rs)}{(\gamma_2 s)^{n+1} r^{n+2}} e^{-\gamma_2 rs},$$

$$\sigma^{(2)L}_{rrns}(r, s) = E_2(s) \beta_2 \frac{Q^{(2)}_{n1}(\gamma_2 rs)}{(\gamma_2 s)^{n+2} r^{n+3}} e^{-\gamma_2 rs},$$

$$\sigma^{(2)L}_{r\theta ns}(r, s) = E_2(s) \beta_2 \frac{Q^{(2)}_{n2}(\gamma_2 rs)}{(\gamma_2 s)^{n+2} r^{n+3}} e^{-\gamma_2 rs},$$

$$E_0(s) = f^{L}(s) e^{-\gamma_0 s} (-1)^n (n + 1/2) \Gamma_\xi, \qquad \xi = 1, 2,$$

$$\Gamma_1 = (\gamma_0 s)^{-n-1}, \qquad \Gamma_2 = \frac{R_{n0}(\gamma_0 d_0 s)}{(\gamma_0 s)^{2n+1} d_0^n},$$

$$E_2(s) = (r_1 - d_2) f^{L}(s) e^{\gamma_2 r_1 s} \frac{(-1)^n (2n + 1)}{2 (\gamma_2 s)^n d_2^{n+1}} \left[R_{n0}(-\gamma_2 d_2 s) \right.$$
$$\left. - e^{-2\gamma_2 s d_2} R_{n0}(\gamma_2 d_2 s) \right]; \tag{3.25}$$

here, $f^{L}(s)$ is the image of the function, which defines a law of variation in time τ of the potential in the incoming wave, Γ_1 corresponds to the plane wave, and Γ_2 corresponds to the spherical one.

The elements of the matrix $\mathbf{A}$ turn out to be simpler in the case when the coefficients k_{10} and k_{12} are of the limiting values. Thus, in the case of absolutely stiff cohesion of the media, as $k_{10} \to \infty$ and $k_{12} \to \infty$, we get

$$a_{41}(s) = -R_{n0}(\gamma_0 s), \qquad a_{42}(s) = -R_{n3}(\eta_0 s),$$

$$a_{4+4m,4}(s) = a_{4+4m,3}(-s) = (-1)^m r_m R_{n0}(\gamma_1 r_m s),$$

$$a_{4+4m,6}(s) = a_{4+4m,5}(-s) = (-1)^m r_m R_{n3}(\eta_1 r_m s),$$

$$a_{87}''(s) = -a_{87}''(-s) = r_1 R_{n0}(\gamma_2 r_1 s),$$

$$a_{88}''(s) = -a_{88}''(-s) = -r_1 R_{n3}(\eta_2 r_1 s),$$

$$b_4 = v_{0ns}^L(1, s), \qquad b_8 = -r_1^3 v_{2ns}^L(r, s), \tag{3.26}$$

and in the case of free sliding, when $k_{10} = k_{12} = 0$, we get

$$a_{41}(s) = a_{42}(s) = a_{87}''(s) = a_{87}'(s) = a_{88}''(s) = a_{88}'(s) \equiv 0,$$

$$a_{4+4m,4}(s) = a_{4+4m,3}(-s) = a_{3+4m,4}(s),$$

$$a_{4+4m,6}(s) = a_{4+4m,5}(-s) = a_{3+4m,6}(s),$$

$$b_4 = b_8 \equiv 0. \tag{3.27}$$

The solution of the system of equations (3.23) is as follows:

$$C_{0n}^{(2)} = \frac{\Delta_1}{\Delta_0}, \qquad C_{0n}^{(4)} = \frac{\Delta_2}{\Delta_0},$$

$$C_{2n}^{(1)} = \frac{\Delta_7}{\Delta_0}, \qquad C_{2n}^{(3)} = \frac{\Delta_8}{\Delta_0},$$

$$C_{1n}^{(j)} = \frac{\Delta_{j+2}}{\Delta_0}, \qquad j = 1, 2, 3, 4, \qquad \Delta_0 = \det A; \tag{3.28}$$

this solution and the formulas (3.22) give us a complete solution of the boundary value problem (3.8) in the space of the Laplace images. In (3.28), Δ_j is the determinant derived from the determinant of the matrix A by replacement of the jth column by the free column.

Let us now consider calculation of the originals of the functions sought. As it follows from the structure of the matrix A, the determinants Δ_j, $j = 0, 1, \ldots, 8$, are the sums of the exponential functions multiplied by the polynomials with respect to the argument s. Since the determinants are similar in form, let us evaluate the determinant Δ_0. In order to separate the exponential functions, we shall apply the Laplace theorem on the determinant and represent Δ_0 in the following form (at that, we take into account the presence of zero elements in the matrix A):

$$\Delta_0 = M_{1234}^{1234} M_{5678}^{5678} - M_{1235}^{1234} M_{4678}^{5678} + M_{1236}^{1234} M_{4578}^{5678}$$

$$+ M_{1245}^{1234} M_{3678}^{5678} - M_{1246}^{1234} M_{3578}^{5678} + M_{1256}^{1234} M_{3478}^{5678}. \tag{3.29}$$

Here, $M_{j_1 j_2 j_3 j_4}^{i_1 i_2 i_3 i_4}$ is the minor of the fourth order of the matrix A, that is, the determinant of the matrix of the fourth order composed from the entries of the matrix A located at the intersections of the rows i_1, i_2, i_3, and i_4, and the columns j_1, j_2, j_3, and j_4, respectively.

Taking into account the properties of the determinants, we can write (3.29) as follows:

$$\Delta_0 = e^{\tau_{02}s} \sum_{l=0}^{4} \sum_{m=0}^{3} X_{lm}(s) e^{-\nu_{lm}s},$$

$$\tau_{02} = -(\gamma_0 + \eta_0) + (\gamma_1 + \eta_1)\delta + (\gamma_2 + \eta_2) r_1,$$

$$\nu_{1m} = 2\gamma_1\delta + \alpha_m, \qquad \nu_{2m} = (\gamma_1 + \eta_1)\delta + \alpha_m,$$

$$\nu_{3m} = 2\eta_1\delta + \alpha_m, \qquad \nu_{4m} = 2(\gamma_1 + \eta_1)\delta + \alpha_m, \qquad \nu_{0m} = \alpha_m,$$

$$m = 0, 1, 2, 3,$$

$$\alpha_0 = 0, \qquad \alpha_1 = 2\gamma_2 r_1, \qquad \alpha_2 = 2\eta_2 r_1, \qquad \alpha_3 = 2(\gamma_2 + \eta_2) r_1,$$

$$X_{0m}(s) = -Z_{35}(s) Y_{46}^{(m)}(s), \qquad X_{1m}(s) = Z_{45}(s) Y_{36}^{(m)}(s),$$

$$X_{2m}(s) = Z_{34}(s) Y_{46}^{(m)}(s) + Z_{56}(s) Y_{34}^{(m)}(s),$$

$$X_{3m}(s) = Z_{36}(s) Y_{45}^{(m)}(s), \qquad X_{4m}(s) = -Z_{46}(s) Y_{35}^{(m)}(s). \tag{3.30}$$

The polynomials $Z_{i_1 i_2}(s)$ and $Y_{j_1 j_2}^{(m)}(s)$ can be determined as follows: $Z_{i_1 i_2}(s) = M_{12 i_1 i_2}^{1234}$ for $\gamma_i = \eta_i \equiv 0$, $i = 0, 1, 2$, and $Y_{j_1 j_2}^{(m)}(s) = M_{j_1 j_2 78}^{5678}$ under the same assumptions; at that, for $m = 0$, in the last two columns there are the elements with one prime; for $m = 1$, in the third column there are the elements with two primes, and in the fourth one there are the elements with one prime; for $m = 2$, in the third column there are the elements with one prime, and in the third one there are the elements with two primes; for $m = 3$, in the last two columns there are the elements with two primes.

When deriving the originals, the main difficulty is to obtain an inverse of the expression $\Delta_0^{-1}(s)$. Let us note that, in accounting for the asymptotic properties of the exponential function ($\nu_{lm} > 0$), there exists a value of s_0 such that

$$\left| \sum_{l=0}^{4} \sum_{\substack{m=0 \\ l+m \neq 0}}^{3} \frac{X_{lm}(s)}{X_{00}(s)} e^{-\nu_{lm}s} \right| < 1, \qquad \mathrm{Re}\, s > \mathrm{Re}\, s_0. \tag{3.31}$$

Hence, using the known series expansion of the sum of geometric progression, in the half-plane $\mathrm{Re}\, s > \mathrm{Re}\, s_0$ we have

$$\Delta_0^{-1}(s) = \frac{e^{\tau_{02}s}}{X_{00}(s)} \sum_{p=0}^{\infty} (-1)^p \left(\sum_{l=0}^{4} \sum_{\substack{m=0 \\ l+m \neq 0}}^{3} \frac{X_{lm}(s)}{X_{00}(s)} e^{-\nu_{lm}s} \right)^p. \tag{3.32}$$

Raising the series coefficients to the power p, we transform the expression (3.32) into the following form:

$$\Delta_0^{-1}(s) = e^{\tau_{02}s} \sum_{p=0}^{N} (-1)^p X_{00}^{-p-1}(s) V_p(s) e^{-\tau_p s}, \qquad N = E\left(\frac{\tau}{\nu_{10}} \right),$$

$$V_p(s) = \sum_{|\alpha|=p} (p;\alpha) \prod_{l=0}^{4} \prod_{\substack{m=0 \\ l+m \neq 0}}^{3} X_{lm}^{p_{lm}}(s),$$

$$\tau_p = \sum_{l=0}^{4} \sum_{\substack{m=0 \\ l+m \neq 0}}^{3} \nu_{lm} p_{lm},$$

$$(p;\alpha) = \frac{p!}{p_{01}! \, p_{02}! \, \cdots \, p_{43}!},$$

$$|\alpha| = \sum_{l=0}^{4} \sum_{\substack{m=0 \\ l+m \neq 0}}^{3} p_{lm}. \tag{3.33}$$

Here, $\alpha = (p_{01}, p_{02}, \ldots p_{43})$ is the multi-index, $|\alpha|$ is the length of the multi-index, $(p;\alpha)$ is the multinomial coefficient, and $E(x)$ is the integer part of x.

We replaced the upper limit of summation in (3.33) by the finite one since, when inverting this expression for the finite instant τ, the originals of all the successive terms become zero because of the presence of the exponential functions with retarded arguments. At that, it is possible to prove that for the fixed number p, the value τ_p reaches the maximum at $p_{01} = p$ and $p_{lm} = 0$, where $l \neq 0$ and $m \neq 1$.

The structure of (3.33) shows that it represents a sum of the products of rational functions and exponential functions. Hence, the representations of the function transforms, which characterize the stress–strain state of the system, will be of a similar form. Exact computation of the originals of such functions is not difficult and can be performed, for example, by means of the algorithm presented in Chap. 2.

Thus, the expressions (3.22), (3.28), and (3.33) in combination with the algorithm of inversion of the Laplace transform for the functions of this class give us the exact solutions for the boundary value problem (3.8). Performing the corresponding summation of series by the Legendre polynomials, we can obtain the solutions of the internal and external dynamical problems for thick-walled spheres contacting elastic media.

Let us assume that the external medium $i = 0$ and the internal medium $i = 2$ are of acoustic type; this assumption corresponds to the absence of resistance to shear. Making a passage to the limit, as $\eta_0 \to \infty$ and $\eta_2 \to \infty$ (i.e., as $\nu_{12} \to \infty$ and $\nu_{13} \to \infty$), we obtain the simplified relationships (see (3.33))

$$V_p(s) = \sum_{|\alpha|=p} (p;\alpha) \prod_{l=0}^{4} \prod_{\substack{m=0 \\ l+m \neq 0}}^{1} X_{lm}^{p_{lm}}(s),$$

$$\tau_p = \sum_{l=0}^{4} \sum_{\substack{m=0 \\ l+m \neq 0}}^{1} \nu_{lm} p_{lm},$$

$$\psi_{in}(r,\tau) = \sum_{j,k,l,m} \left[\psi_{0jklm}^{(i)} H\left(\tau - S_{0jklm}^{(i)}\right) + \psi_{1jklm}^{(i)} H\left(\tau - S_{1jklm}^{(i)}\right) \right],$$

$$\varphi_{0jklm}^{(i)} = \frac{(-1)^n}{r^{n+1}} \sum_{p=0}^{n} (-\gamma_i r)^{n-p} A_{np} \frac{d^{n-p}}{d\tau^{n-p}} f_{0jklm}^{(i)}\left(\tau - T_{0jklm}^{(i)}\right),$$

$$\psi_{0jklm}^{(i)} = \frac{(-1)^n}{r^{n+1}} \sum_{p=0}^{n} (-\eta_i r)^{n-p} A_{np} \frac{d^{n-p}}{d\tau^{n-p}} g_{0jklm}^{(i)}\left(\tau - S_{0jklm}^{(i)}\right),$$

$$\varphi_{1jklm}^{(i)} = \frac{1}{r^{n+1}} \sum_{p=0}^{n} (\gamma_i r)^{n-p} A_{np} \frac{d^{n-p}}{d\tau^{n-p}} f_{1jklm}^{(i)}\left(\tau - T_{1jklm}^{(i)}\right),$$

$$\psi_{1jklm}^{(i)} = \frac{1}{r^{n+1}} \sum_{p=0}^{n} (\eta_i r)^{n-p} A_{np} \frac{d^{n-p}}{d\tau^{n-p}} g_{1jklm}^{(i)}\left(\tau - S_{1jklm}^{(i)}\right),$$

$$T_{0jklm}^{(i)} = \gamma_i(r_{0i} - r) + \tau_{jklm}, \qquad T_{1jklm}^{(i)} = \gamma_i(r - r_{1i}) + \tau_{jklm},$$

$$S_{0jklm}^{(i)} = \eta_i(r_{0i} - r) + \tau_{jklm}, \qquad S_{1jklm}^{(i)} = \eta_i(r - r_{1i}) + \tau_{jklm},$$

$$\tau_{jklm} = (\gamma_2 k + \eta_2 j) r_1 + (\gamma_1 m + \eta_1 l) \delta,$$

$$r_{02} = r_{11} = r_1, \qquad r_{10} = r_{01} = 1, \qquad r_{12} = 0. \tag{3.36}$$

Here, $f_{0jklm}^{(i)}$ and $f_{1jklm}^{(i)}$ are arbitrary functions corresponding to the converging and diverging generalized spherical P-waves; $g_{0jklm}^{(i)}$ and $g_{1jklm}^{(i)}$ are arbitrary functions corresponding to the similar S-waves.

The functions φ_{in} and ψ_{in} defined by (3.36) convert the equations (3.8) into the identities and satisfy the initial conditions if to set

$$\frac{d^q f_{0jklm}^{(i)}}{d\tau^q}(0) = \frac{d^q f_{1jklm}^{(i)}}{d\tau^q}(0) = \frac{d^q g_{0jklm}}{d\tau^q}(0)$$

$$= \frac{d^q g_{1jklm}}{d\tau^q}(0) = 0, \qquad q = 0, 1, \ldots, n+1. \tag{3.37}$$

Making the Laplace transform of (3.36), we obtain

$$\varphi_{in}^L(r,s) = \frac{1}{r^{n+1}} \sum_{j,k,l,m=0}^{\infty} \left[(-1)^n R_{n0}(-\gamma_i rs) f_{0jklm}^{(i)L}(s) e^{-T_{0jklm}^{(i)} s} \right.$$

$$\left. + R_{n0}(\gamma_i rs) f_{1jklm}^{(i)L}(s) e^{-T_{1jklm}^{(i)} s} \right],$$

$$\psi_{in}^L(r,s) = \frac{1}{r^{n+1}} \sum_{j,k,l,m=0}^{\infty} \left[(-1)^n R_{n0}(-\eta_i rs) g_{0jklm}^{(i)L}(s) e^{-S_{0jklm}^{(i)} s} \right.$$

$$\left. + R_{n0}(\eta_i rs) g_{1jklm}^{(i)L}(s) e^{-S_{1jklm}^{(i)} s} \right]. \tag{3.38}$$

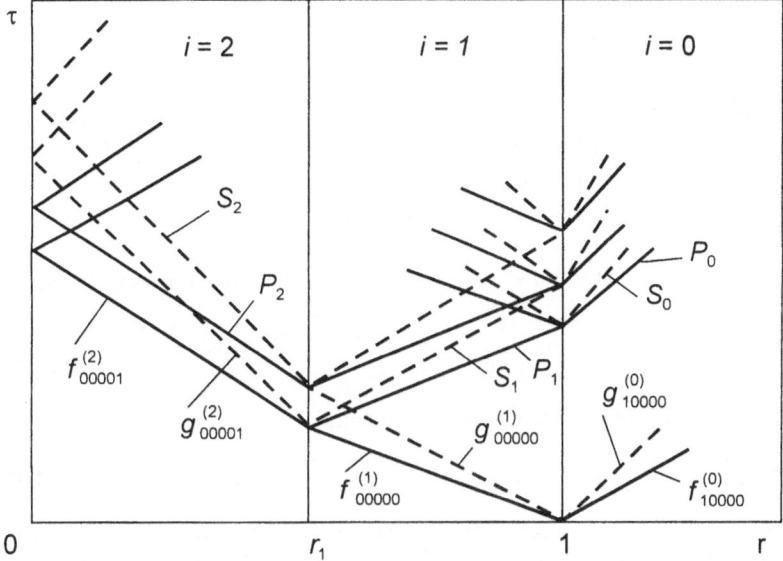

Fig. 3.2. Formation of the P-waves (solid lines) and the S-waves (dashed lines). The external problem

The boundary condition as $r \to \infty$ is equivalent to the absence of the converging elastic waves in the external medium $i = 0$. It follows that $f^{(0)}_{0jklm} = g^{(0)}_{0jklm} = 0$.

In order to study the boundary condition as $r \to 0$ for the internal medium $i = 2$, let us take into account the relationship between the polynomials $R_{n0}(s)$ and the modified Bessel functions $I_{n+1/2}(s)$ (1.126) and the following expansion of the modified Bessel functions into the power series (Watson (1945), Gradshtein, Ryzhik (1971), Kuznetsov (1965)):

$$
I_{n+1/2}(s) = \frac{2^{n+1/2}}{\sqrt{\pi}} \, s^{n+1/2} \sum_{k=0}^{\infty} \frac{(n+k)!}{k!(2n+2k+1)!} \, s^{2k}
$$

$$
= \frac{2^{n+1/2}}{\sqrt{\pi}} \, s^{n+1/2} \frac{n!}{(2n+1)!} + O(s^{n+5/2}), \qquad s \to 0. \tag{3.39}
$$

From (1.126) and (3.39) it immediately follows that

$$
e^s R_{n0}(-s) - e^{-s} R_{n0}(s)
$$
$$
= (-1)^n I_{n+1/2}(s) \, s^{n+1/2} \sqrt{2\pi} = O(s^{2n+1}), \qquad s \to 0. \tag{3.40}
$$

From (3.38), we obtain

$$\lim_{r \to 0} \varphi_{2n}^{L}(r, s)$$

$$= \sum_{j, k, l, m=0}^{\infty} \lim_{r \to 0} \frac{1}{r^{n+1}} \left\{ R_{n0}(\gamma_2 rs) e^{-\gamma_2 rs} \left[(-1)^n f_{0j, k-1, lm}^{(2)L} + f_{1jklm}^{(2)L} \right] \right.$$

$$\left. + O\left[(\gamma_2 rs)^{2n+1} \right] \right\} e^{-\tau_{jklm} s}. \quad (3.41)$$

The expression (3.41) shows that, as $r \to 0$, the functions $\varphi_{2n}^{L}(r, s)$ are bounded if and only if

$$(-1)^n f_{0j, k-1, lm}^{(2)} + f_{1jklm}^{(2)} = 0. \quad (3.42)$$

Similarly, for the elastic S-waves, we obtain

$$(-1)^n g_{0, j-1, klm}^{(2)} + g_{1jklm}^{(2)} = 0. \quad (3.43)$$

We can prove that the conditions (3.42) and (3.43) provide also a boundedness of the displacements and stresses at the sphere's center $r = 0$.

From (3.42), (3.43), and the analysis of the pattern of propagation of elementary generalized spherical waves (Fig. 3.2), it follows that $f_{1jklm}^{(2)} \equiv 0$ when j is odd and k is even, $g_{1jklm}^{(2)} \equiv 0$ when j is even and k is odd, and the functions $f_{1jklm}^{(i)}$, $f_{0jklm}^{(i)}$, $g_{1jklm}^{(i)}$, and $g_{0jklm}^{(i)}$, $i = 1, 2$, differ from zero only when j and k are even.

Let us designate

$$f_{jklm}^{(2)} = f_{1, 2j, 2k+1, lm}^{(2)} = (-1)^{n+1} f_{0, 2j, 2k, lm}^{(2)},$$

$$g_{jklm}^{(2)} = g_{1, 2j+1, 2k, lm}^{(2)} = (-1)^{n+1} g_{0, 2j, 2k, lm}^{(2)},$$

$$f_{jklm}^{(0)} = f_{1, 2j, 2k, lm}^{(0)},$$

$$g_{jklm}^{(0)} = g_{1, 2j, 2k, lm}^{(0)}, \qquad j, k = 0, 1, \ldots. \quad (3.44)$$

Let us also assume that for the other arbitrary functions, the indexes j and k take only even values.

Taking into account the considerations presented above and the formulas (1.13) and (3.38), we obtain the following expressions for the transforms of displacements and stresses:

$$u_{0n}^{L}(r, s) = -\frac{1}{r^{n+2}} \sum_{j, k, l, m} \left[R_{n1}(rs) f_{jklm}^{(0)L}(s) e^{-(r-1)s} \right.$$

$$\left. + n(n + 1) R_{n0}(\eta_0 rs) g_{jklm}^{(0)L}(s) e^{-\eta_0(r-1)s} \right] e^{-\tau_{jklm} s},$$

$$v_{0n}^{L}(r, s) = \frac{1}{r^{n+2}} \sum_{j, k, l, m} \left[R_{n0}(rs) f_{jklm}^{(0)L}(s) e^{-(r-1)s} \right.$$

$$+ R_{n3}(\eta_0 rs) \, g_{jklm}^{(0)\mathrm{L}}(s) \, \mathrm{e}^{-\eta_0(r-1)s} \Big] \, \mathrm{e}^{-\tau_{jklm}s} \,,$$

$$u_{1n}^{\mathrm{L}}(r,s) = -\frac{1}{r^{n+2}} \sum_{j,k,l,m} \Big[(-1)^n R_{n1}(-\gamma_1 rs) \, f_{0jklm}^{(1)\mathrm{L}}(s) \, \mathrm{e}^{-\gamma_1(1-r)s}$$

$$+ R_{n1}(\gamma_1 rs) \, f_{1jklm}^{(1)\mathrm{L}}(s) \, \mathrm{e}^{-\gamma_1(r-r_1)s}$$

$$+ (-1)^n n(n+1) R_{n0}(-\eta_1 rs) \, g_{0jklm}^{(1)\mathrm{L}}(s) \, \mathrm{e}^{-\eta_1(1-r)s}$$

$$+ n(n+1) R_{n0}(\eta_1 rs) \, g_{1jklm}^{(1)\mathrm{L}}(s) \, \mathrm{e}^{-\eta_1(r-r_1)s} \Big] \, \mathrm{e}^{-\tau_{jklm}s} \,,$$

$$v_{1n}^{\mathrm{L}}(r,s) = \frac{1}{r^{n+2}} \sum_{j,k,l,m} \Big[(-1)^n R_{n0}(-\gamma_1 rs) \, f_{0jklm}^{(1)\mathrm{L}}(s) \, \mathrm{e}^{-\gamma_1(1-r)s}$$

$$+ R_{n0}(\gamma_1 rs) \, f_{1jklm}^{(1)\mathrm{L}}(s) \, \mathrm{e}^{-\gamma_1(r-r_1)s}$$

$$+ (-1)^n R_{n3}(-\eta_1 rs) \, g_{0jklm}^{(1)\mathrm{L}}(s) \, \mathrm{e}^{-\eta_1(1-r)s}$$

$$+ R_{n3}(\eta_1 rs) \, g_{1jklm}^{(1)\mathrm{L}}(s) \, \mathrm{e}^{-\eta_1(r-r_1)s} \Big] \, \mathrm{e}^{-\tau_{jklm}s} \,,$$

$$u_{2n}^{\mathrm{L}}(r,s) = -\frac{1}{r^{n+2}} \sum_{j,k,l,m} \Big\{ \Big[-R_{n1}(-\gamma_2 rs) \, \mathrm{e}^{-\gamma_2(r_1-r)s}$$

$$+ R_{n1}(\gamma_2 rs) \, \mathrm{e}^{-\gamma_2(r+r_1)s} \Big] \, f_{jklm}^{(2)\mathrm{L}}(s)$$

$$+ n(n+1) \Big[-R_{n0}(-\eta_2 rs) \, \mathrm{e}^{-\eta_2(r_1-r)s}$$

$$+ R_{n0}(\eta_2 rs) \, \mathrm{e}^{-\eta_2(r+r_1)s} \Big] \, g_{jklm}^{(2)\mathrm{L}}(s) \Big\} \, \mathrm{e}^{-\tau_{jklm}s} \,,$$

$$v_{2n}^{\mathrm{L}}(r,s) = \frac{1}{r^{n+2}} \sum_{j,k,l,m} \Big\{ \Big[-R_{n0}(-\gamma_2 rs) \, \mathrm{e}^{-\gamma_2(r_1-r)s}$$

$$+ R_{n0}(\gamma_2 rs) \, \mathrm{e}^{-\gamma_2(r+r_1)s} \Big] \, f_{jklm}^{(2)\mathrm{L}}(s)$$

$$+ \Big[-R_{n3}(-\eta_2 rs) \, \mathrm{e}^{-\eta_2(r_1-r)s}$$

$$+ R_{n3}(\eta_2 rs) \, \mathrm{e}^{-\eta_2(r+r_1)s} \Big] \, g_{jklm}^{(2)\mathrm{L}}(s) \Big\} \, \mathrm{e}^{-\tau_{jklm}s} \,,$$

$$\sigma_{rrn}^{(0)\,\mathrm{L}}(r,s) = \frac{\beta_0}{r^{n+3}} \sum_{j,k,l,m} \Big[Q_{n1}^{(0)}(rs) \, f_{jklm}^{(0)\,\mathrm{L}}(s) \, \mathrm{e}^{-(r-1)s}$$

$$+ n(n+1) Q_{n2}^{(0)}(\eta_0 rs) \, g_{jklm}^{(0)\mathrm{L}}(s) \, \mathrm{e}^{-\eta_0(r-1)s} \Big] \, \mathrm{e}^{-\tau_{jklm}s} \,,$$

$$\sigma_{r\theta n}^{(0)\,\mathrm{L}}(r,s) = \frac{\beta_0}{r^{n+3}} \sum_{j,k,l,m} \Big[Q_{n2}^{(0)}(rs) \, f_{jklm}^{(0)\,\mathrm{L}}(s) \, \mathrm{e}^{-(r-1)s}$$

$$+ Q_{n3}^{(0)}(\eta_0 rs) \, g_{jklm}^{(0)\mathrm{L}}(s) \, \mathrm{e}^{-\eta_0(r-1)s} \Big] \, \mathrm{e}^{-\tau_{jklm}s} \,,$$

$$\sigma_{rrn}^{(1)\,\mathrm{L}}(r,s) = \frac{1}{r^{n+3}} \sum_{j,k,l,m} \Big[(-1)^n Q_{n1}^{(1)}(-\gamma_1 rs) \, f_{0jklm}^{(1)\mathrm{L}}(s) \, \mathrm{e}^{-\gamma_1(1-r)s}$$

$$+ Q_{n1}^{(1)} (\gamma_1 r s)\, f_{1jklm}^{(1)\text{L}} (s)\, e^{-\gamma_1 (r-r_1)\, s}$$

$$+ (-1)^n n(n+1)\, Q_{n2}^{(1)} (-\eta_1 r s)\, g_{0jklm}^{(1)\text{L}} (s)\, e^{-\eta_1 (1-r)\, s}$$

$$\left. + n(n+1)\, Q_{n2}^{(1)} (\eta_1 r s)\, g_{1jklm}^{(1)\text{L}} (s)\, e^{-\eta_1 (r-r_1)\, s} \right] e^{-\tau_{jklm} s},$$

$$\sigma_{r\theta n}^{(1)\text{L}} (r,\, s) = \frac{1}{r^{n+3}} \sum_{j,\, k,\, l,\, m} \left[(-1)^n Q_{n2}^{(1)} (-\gamma_1 r s)\, f_{0jklm}^{(1)\text{L}} (s)\, e^{-\gamma_1 (1-r)\, s} \right.$$

$$+ Q_{n2}^{(1)} (\gamma_1 r s)\, f_{1jklm}^{(1)\text{L}} (s)\, e^{-\gamma_1 (r-r_1)\, s}$$

$$+ (-1)^n Q_{n3}^{(1)} (-\eta_1 r s)\, g_{0jklm}^{(1)\text{L}} (s)\, e^{-\eta_1 (1-r)\, s}$$

$$\left. + Q_{n3}^{(1)} (\eta_1 r s)\, g_{1jklm}^{(1)\text{L}} (s)\, e^{-\eta_1 (r-r_1)\, s} \right] e^{-\tau_{jklm} s},$$

$$\sigma_{rrn}^{(2)\text{L}} (r,\, s) = \frac{\beta_2}{r^{n+3}} \sum_{j,\, k,\, l,\, m} \left\{ \left[-Q_{n1}^{(2)} (-\gamma_2 r s)\, e^{-\gamma_2 (r_1-r)\, s} \right. \right.$$

$$\left. + Q_{n1}^{(2)} (\gamma_2 r s)\, e^{-\gamma_2 (r+r_1)\, s} \right] f_{jklm}^{(2)\text{L}} (s)$$

$$+ n(n+1) \left[-Q_{n2}^{(2)} (-\eta_2 r s)\, e^{-\eta_2 (r_1-r)\, s} \right.$$

$$\left. \left. + Q_{n2}^{(2)} (\eta_2 r s)\, e^{-\eta_2 (r+r_1)\, s} \right] g_{jklm}^{(2)\text{L}} (s) \right\} e^{-\tau_{jklm} s},$$

$$\sigma_{r\theta n}^{(2)\text{L}} (r,\, s) = \frac{\beta_2}{r^{n+3}} \sum_{j,\, k,\, l,\, m} \left\{ \left[-Q_{n2}^{(2)} (-\gamma_2 r s)\, e^{-\gamma_2 (r_1-r)\, s} \right. \right.$$

$$\left. + Q_{n2}^{(2)} (\gamma_2 r s)\, e^{-\gamma_2 (r+r_1)\, s} \right] f_{jklm}^{(2)\text{L}} (s)$$

$$+ \left[-Q_{n3}^{(2)} (-\eta_2 r s)\, e^{-\eta_2 (r_1-r)\, s} \right.$$

$$\left. \left. + Q_{n3}^{(2)} (\eta_2 r s)\, e^{-\eta_2 (r+r_1)\, s} \right] g_{jklm}^{(2)\text{L}} (s) \right\} e^{-\tau_{jklm} s}. \quad (3.45)$$

The polynomials $Q_{n1}^{(i)} (s)$, $Q_{n2}^{(i)} (s)$, $Q_{n3}^{(i)} (s)$, $R_{n0} (s)$, $R_{n1} (s)$, and $R_{n2} (s)$ are defined by (1.126) and (3.22); at that, the last two can also be defined by the relationships

$$R_{n1} (s) = \sum_{p=0}^{n+1} B_{np} s^{n+1-p}, \qquad R_{n2} (s) = \sum_{p=0}^{n+2} C_{np} s^{n+2-p},$$

$$B_{np} = A_{np} + p A_{n,\, p-1}, \qquad C_{np} = B_{np} + p B_{n,\, p-1},$$

$$\left[R_{n0} (s)\, s^{-n-1} e^{-s} \right]' = -R_{n1} (s)\, s^{-n-2} e^{-s},$$

$$\left[R_{n1} (s)\, s^{-n-2} e^{-s} \right]' = -R_{n2} (s)\, s^{-n-3} e^{-s}. \quad (3.46)$$

The coefficients τ_{jklm} in (3.45) are as follows:

$$\tau_{jklm} = 2\, (j\eta_2 + k\gamma_2)\, r_1 + (l\eta_1 + m\gamma_1)\, \delta. \quad (3.47)$$

Substituting (3.45) into the boundary conditions (3.8) for $r = 1$ and $r = r_1$ and equating the expressions corresponding to the equal powers of the exponential functions, we obtain the recursive relationships, which allow us to determine arbitrary functions for $j + k + m + l > 0$:

$$
\begin{pmatrix} f_{0jklm}^{(1)L} \\ g_{0jklm}^{(1)L} \\ f_{jklm}^{(0)L} \\ g_{jklm}^{(0)L} \end{pmatrix} = \frac{Y_n}{X_{n0}} \begin{pmatrix} f_{1jkl,\,m-1}^{(1)L} \\ g_{1jk,\,l-1,\,m}^{(1)L} \\ 0 \\ 0 \end{pmatrix} ,
$$

$$
\begin{pmatrix} f_{1jklm}^{(1)L} \\ g_{1jklm}^{(1)L} \\ f_{jklm}^{(2)L} \\ g_{jklm}^{(2)L} \end{pmatrix} = \frac{Z_n}{X_{n1}} \begin{pmatrix} f_{0jkl,\,m-1}^{(1)L} \\ g_{0jk,\,l-1,\,m}^{(0)L} \\ f_{j,\,k-1,\,lm}^{(2)L} \\ g_{j-1,\,klm}^{(2)L} \end{pmatrix} ; \tag{3.48}
$$

here, $X_{n0}(s)$ and $X_{n1}(s)$ are the polynomials in s of the degree $4n + 7$; Z_n and Y_n are the matrices related to the matrices $M_n = (M_{nij})_{4\times 4}$, $N_n = (N_{nij})_{4\times 4}$, $K_n = (K_{nij})_{4\times 4}$, and $L_n = (L_{nij})_{4\times 4}$ as follows:

$$
Y_n = -X_{n0}M_n^{-1}N_n , \qquad Z_n = -X_{n1}K_n^{-1}L_n ,
$$
$$
X_{n0} = \det M_n , \qquad X_{n1} = \det K_n . \tag{3.49}
$$

All the matrices considered are of polynomial type and the powers of their elements are smaller than or equal to $4n + 7$; the entries of the matrices M_n, N_n, K_n, and L_n are as follows:

$$
M_{n11} = (-1)^n Q_{n1}^{(1)}(-\gamma_1 s) , \qquad M_{n12} = (-1)^n n(n + 1) Q_{n2}^{(1)}(-\eta_1 s) ,
$$
$$
M_{n13} = -\beta_0 Q_{n1}^{(1)}(s) , \qquad M_{n14} = -\beta_0 n(n + 1) Q_{n2}^{(0)}(\eta_0 s) ,
$$
$$
M_{n21} = (-1)^n R_{n1}(-\gamma_1 s) , \qquad M_{n22} = (-1)^n n(n + 1) R_{n0}(-\eta_1 s) ,
$$
$$
M_{n23} = -R_{n1}(s) , \qquad M_{n24} = -n(n + 1) R_{n0}(\eta_0 s) ,
$$
$$
M_{n31} = (-1)^n Q_{n2}^{(1)}(-\gamma_1 s) , \qquad M_{n32} = (-1)^n Q_{n3}^{(1)}(-\eta_1 s) ,
$$
$$
M_{n33} = -\beta_0 Q_{n2}^{(0)}(s) , \qquad M_{n34} = -\beta_0 Q_{n3}^{(0)}(\eta_0 s) ,
$$
$$
M_{n41} = (-1)^n \left[Q_{n2}^{(1)}(-\gamma_1 s) + k_{10} R_{n0}(-\gamma_1 s) \right] ,
$$
$$
M_{n42} = (-1)^n \left[Q_{n3}^{(1)}(-\eta_1 s) + k_{10} R_{n3}(-\eta_1 s) \right] ,
$$
$$
M_{n43} = -k_{10} R_{n0}(s) , \qquad M_{n44} = -k_{10} R_{n3}(\eta_0 s) ,
$$
$$
N_{n11} = Q_{n1}^{(1)}(\gamma_1 s) , \qquad N_{n12} = n(n + 1) Q_{n2}^{(1)}(\eta_1 s) ,
$$
$$
N_{n21} = R_{n1}(\gamma_1 s) , \qquad N_{n22} = n(n + 1) R_{n0}(\eta_1 s) ,
$$
$$
N_{n31} = Q_{n2}^{(1)}(\gamma_1 s) , \qquad N_{n32} = Q_{n3}^{(1)}(\eta_1 s) ,
$$
$$
N_{n41} = Q_{n2}^{(1)}(\gamma_1 s) + k_{10} R_{n0}(\gamma_1 s) , \qquad N_{n42} = Q_{n3}^{(1)}(\eta_1 s) + k_{10} R_{n3}(\eta_1 s) ,
$$
$$
N_{n13} = N_{n14} = N_{n23} = N_{n24} = N_{n33} = N_{n34} = N_{n43} = N_{n44} = 0 ,
$$

$$K_{n11} = Q_{n1}^{(1)}(\gamma_1 r_1 s), \qquad K_{n12} = n(n+1) Q_{n2}^{(1)}(\eta_1 r_1 s),$$

$$K_{n13} = \beta_2 Q_{n1}^{(2)}(-\gamma_1 r_1 s), \qquad K_{n14} = \beta_2 n(n+1) Q_{n2}^{(2)}(-\eta_2 r_1 s),$$

$$K_{n21} = R_{n1}(\gamma_1 r_1 s), \qquad K_{n22} = n(n+1) R_{n0}(\eta_1 r_1 s),$$

$$K_{n23} = R_{n1}(-\gamma_2 r_1 s), \qquad K_{n24} = n(n+1) R_{n0}(-\eta_2 r_1 s),$$

$$K_{n31} = Q_{n2}^{(1)}(\gamma_1 r_1 s), \qquad K_{n32} = Q_{n3}^{(1)}(\eta_1 r_1 s),$$

$$K_{n33} = \beta_2 Q_{n2}^{(2)}(-\gamma_2 r_1 s), \qquad K_{n34} = \beta_2 Q_{n3}^{(2)}(-\eta_2 r_1 s),$$

$$K_{n41} = Q_{n2}^{(1)}(\gamma_1 r_1 s) - k_{12} r_1 R_{n0}(\gamma_1 r_1 s),$$

$$K_{n42} = Q_{n3}^{(1)}(\eta_1 r_1 s) - k_{12} r_1 R_{n3}(\eta_1 r_1 s),$$

$$K_{n43} = -k_{12} r_1 R_{n0}(-\gamma_2 r_1 s), \qquad K_{n44} = -k_{12} r_1 R_{n3}(-\eta_2 r_1 s),$$

$$L_{n11} = (-1)^n Q_{n1}^{(1)}(-\gamma_1 r_1 s), \qquad L_{n12} = (-1)^n n(n+1) Q_{n2}^{(1)}(-\eta_1 r_1 s),$$

$$L_{n13} = -\beta_2 Q_{n2}^{(2)}(\gamma_2 r_1 s), \qquad L_{n14} = -\beta_2 n(n+1) Q_{n2}^{(2)}(\eta_2 r_1 s),$$

$$L_{n21} = (-1)^n R_{n1}(-\gamma_1 r_1 s), \qquad L_{n22} = (-1)^n n(n+1) R_{n0}(-\eta_1 r_1 s),$$

$$L_{n23} = -R_{n1}(\gamma_2 r_1 s), \qquad L_{n24} = -n(n+1) R_{n0}(\eta_2 r_1 s),$$

$$L_{n31} = (-1)^n Q_{n2}^{(1)}(-\gamma_1 r_1 s), \qquad L_{n32} = (-1)^n Q_{n3}^{(1)}(-\eta_1 r_1 s),$$

$$L_{n33} = -\beta_2 Q_{n2}^{(2)}(\gamma_2 r_1 s), \qquad L_{n34} = -\beta_2 Q_{n3}^{(2)}(\eta_2 r_1 s),$$

$$L_{n41} = (-1)^n \left[Q_{n2}^{(1)}(-\gamma_1 r_1 s) - k_{12} r_1 R_{n0}(-\gamma_1 r_1 s) \right],$$

$$L_{n42} = (-1)^n \left[Q_{n3}^{(1)}(-\eta_1 r_1 s) - k_{12} r_1 R_{n3}(-\eta_1 r_1 s) \right],$$

$$L_{n43} = k_{12} r_1 R_{n0}(\gamma_2 r_1 s), \qquad L_{n44} = k_{12} r_1 R_{n3}(\eta_2 r_1 s). \tag{3.50}$$

Let us notice that when using the recursive relationships (3.48), even though one of the indexes of the arbitrary functions will be smaller than zero, the arbitrary functions will be identically equal to zero. From the pattern of wave formation (Fig. 3.2)) it follows that when $m + l$ is even, we get

$$f_{jklm}^{(2)} = g_{jklm}^{(2)} = f_{1jklm}^{(1)} = g_{1jklm}^{(1)} \equiv 0,$$

and when $m + l$ is odd, we get

$$f_{jklm}^{(0)} = g_{jklm}^{(0)} = f_{0jklm}^{(1)} = g_{0jklm}^{(1)} \equiv 0.$$

We should supplement the recursive formulas (3.50) by the initial conditions for $j = k = l = m = 0$.

Taking into account the expressions for the transforms of components of the stress–strain state in the incoming wave (3.25) and considering the formulas (3.45) and the boundary conditions (3.8) for $r = 1$, we arrive at

Table 3.1. The orders of poles of the images. An axially symmetric problem

Function	s_{0j}	s_{1j}	s_α
External problem			
$f_{jklm}^{(0)L}, g_{jklm}^{(0)L}, f_{0jklm}^{(1)L}, g_{0jklm}^{(1)L}$	$1+(m+l)/2$	$j+k+(m+l)/2$	1
$f_{1jklm}^{(1)L}, g_{1jklm}^{(1)L}, f_{jklm}^{(2)L}, g_{jklm}^{(2)L}$	$(m+l+1)/2$	$j+k+(m+l+1)/2$	1
Internal problem			
$f_{jklm}^{(0)L}, g_{jklm}^{(0)L}, f_{0jklm}^{(1)L}, g_{0jklm}^{(1)L}$	$(m+l+1)/2$	$j+k+(m+l+1)/2$	1
$f_{1jklm}^{(1)L}, g_{1jklm}^{(1)L}, f_{jklm}^{(2)L}, g_{jklm}^{(2)L}$	$(m+l)/2$	$j+k+1+(m+l)/2$	1

$$
\begin{pmatrix}
f_{00000}^{(1)L} \\
g_{00000}^{(1)L} \\
f_{0000}^{(0)L} \\
g_{0000}^{(0)L}
\end{pmatrix}
= \frac{E_0(s)}{X_{n0}(s)} \left[\mathbf{F}_0(s) - e^{-2s} \mathbf{F}_1(s) \right] ,
\tag{3.51}
$$

where $\mathbf{F}_0$ and $\mathbf{F}_1$ are the polynomial columns of dimension 4×1, which can be determined by the following manner:

$$
\mathbf{F}_0(s) = \left(F_{01}(s), F_{02}(s), F_{03}(s), F_{04}(s) \right)^{\mathrm{T}}
$$
$$
= -X_{n0} M_n^{-1} \left(M_{n13}(-s), M_{n23}(-s), M_{n33}(-s), M_{n43}(-s) \right)^{\mathrm{T}} ,
$$
$$
\mathbf{F}_1(s) = -X_{n0} M_n^{-1} \left(M_{n13}(s), M_{n23}(s), M_{n33}(s), M_{n43}(s) \right)^{\mathrm{T}}
$$
$$
= \left(0, 0, -X_{n0}(s), 0 \right)^{\mathrm{T}} .
\tag{3.52}
$$

Thus, the recursive relationships (3.48) with the initial conditions (3.51) allow us to determine all the arbitrary functions in the Laplace transform space. At that, their images are the rational functions or the rational functions multiplied by an exponential function. Basically, it is possible to derive a solution of the recursive system of equations (3.48) and (3.51). However, the formulas derived by such a manner for arbitrary values of j, k, l, and m will be very cumbersome and not useful for calculations. For this reason, it is more convenient to search for the originals using the original recursive system of equations directly. At that, it is necessary to know the orders of the transform poles, which can be obtained by means of analysis of the relationships (3.48) and (3.51). The orders of poles for the external and internal problems are presented in Table 3.1.

Let us notice that since $E_0(s)$ in (3.25) has the pole $s = 0$ of the order $n + 1$ for a plane wave, and of the order $2n + 1$ for a spherical one, the transforms of all the arbitrary functions will have the same peculiarity.

A solution of the internal problem can be obtained similarly. In this case, the source of the spherical P-wave is located in the internal medium at the

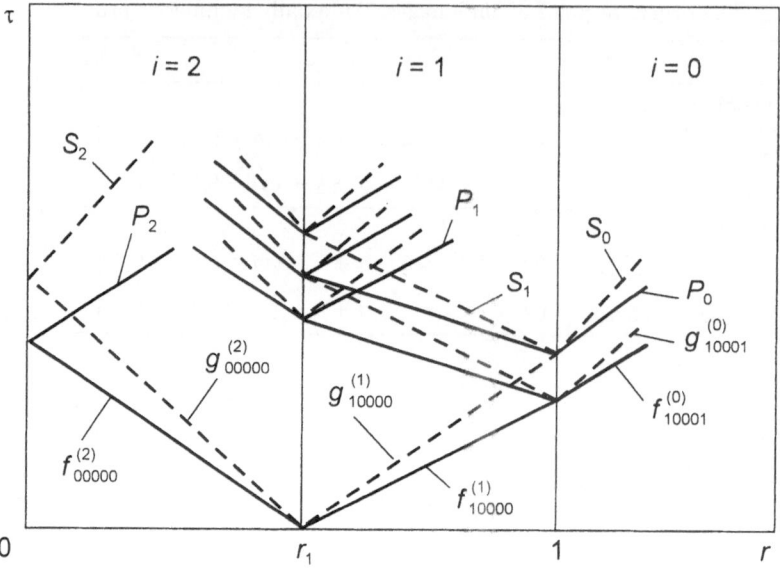

Fig. 3.3. Formation of the P-waves (solid lines) and the S-waves (dashed lines). The internal problem)

distance D_2 from the sphere's center (Fig. 3.1). The dimensionless potential of this wave is defined by (3.13).

The boundary value problem for determination of the potentials of the P-waves and the S-waves, as before, is stated by (3.8). Using a phase plane $r - \tau$, we illustrate in Fig. 3.3 a general pattern of propagation and formation of the generalized spherical waves for the fixed n. We should search for the potentials $\varphi_{in}(r, \tau)$ and $\psi_{in}(r, \tau)$ in the form of superposition of converging and diverging waves (3.36). All the argumentation presented when considering the external problem and the formulas (3.36)–(3.50), inclusive, remain valid for the case of internal problem, as well. At that, when $m + l$ is odd, we get

$$f_{jklm}^{(2)} = g_{jklm}^{(2)} = f_{1jklm}^{(1)} = g_{1jklm}^{(1)} \equiv 0,$$

and when $m + l$ is even, we get

$$f_{jklm}^{(0)} = g_{jklm}^{(0)} = f_{0jklm}^{(1)} = g_{0jklm}^{(1)} \equiv 0.$$

We should also change the initial conditions for the recursive system of equations (3.48) for the case $j = k = l = m = 0$.

Substituting (3.25) into the boundary conditions (3.8) for $r = r_1$, we get

$$
\begin{pmatrix}
f^{(1)L}_{10000} \\
g^{(1)L}_{10000} \\
f^{(2)L}_{0000} \\
g^{(2)L}_{0000}
\end{pmatrix}
=
\frac{E_2(s)}{X_{n1}(s)}
\begin{pmatrix}
Z_{n13}(s) \\
Z_{n23}(s) \\
Z_{n33}(s) \\
Z_{n43}(s)
\end{pmatrix}.
\tag{3.53}
$$

The transforms of the arbitrary functions, similarly to the case of the external problem, are the rational functions or the rational functions multiplied by an exponential function. The orders of poles of the corresponding images for the case of the internal problem are presented in Table 3.1. According to (3.25), the order of the pole $s = 0$ is equal to $2n + 1$.

Let us mark out two particular cases related to the conditions of cohesion of the surfaces in contact.

The general conclusions and the recursive relationships (3.48) with the initial conditions (3.51) or (3.53) remain valid in these cases, as well. Only the expressions for some elements of the matrices M_n, N_n, K_n, and L_n are simplified.

(1) When sliding takes place on the surface of contact $r = 1$ (i.e., $k_{10} = 0$), we get

$$
\begin{aligned}
& M_{n41} = (-1)^n Q^{(1)}_{n2}(-\gamma_1 s), \qquad M_{n42} = (-1)^n Q^{(1)}_{n3}(-\eta_1 s), \\
& M_{n43} = M_{n44} \equiv 0, \\
& N_{n41} = Q^{(1)}_{n2}(\gamma_1 s), \qquad N_{n42} = Q^{(1)}_{n3}(\eta_1 s).
\end{aligned}
\tag{3.54}
$$

(2) When stiff cohesion of the media $i = 0$ and $i = 1$ takes place (i.e., $k_{10} = \infty$), we get

$$
\begin{aligned}
& M_{n41} = (-1)^n R_{n0}(-\gamma_1 s), \qquad M_{n42} = (-1)^n Q^{(1)}_{n3}(-\eta_1 s), \\
& M_{n43} = -R_{n0}(s), \qquad M_{n44} = -R_{n3}(\eta_0 s), \\
& N_{n41} = R_{n0}(\gamma_1 s), \qquad N_{n42} = R_{n3}(\eta_1 s).
\end{aligned}
\tag{3.55}
$$

(3) When sliding takes place on the surface of contact $r = r_1$ (i.e., $k_{12} = 0$), we get

$$
\begin{aligned}
& K_{n41} = Q^{(2)}_{n2}(\gamma_1 r_1 s), \qquad K_{n42} = Q^{(1)}_{n3}(\eta_1 r_1 s), \qquad K_{n43} = K_{n44} \equiv 0, \\
& L_{n41} = (-1)^n Q^{(1)}_{n2}(-\gamma_1 r_1 s), \qquad L_{n42} = (-1)^n Q^{(1)}_{n3}(-\eta_1 r_1 s), \\
& L_{n43} = L_{n44} \equiv 0.
\end{aligned}
\tag{3.56}
$$

(4) When stiff cohesion of the media $i = 1$ and $i = 2$ takes place (i.e., $k_{12} = \infty$), we get

$$K_{n41} = -R_{n0}(\gamma_1 r s), \qquad K_{n42} = -R_{n3}(\eta_1 r_1 s),$$
$$K_{n43} = -R_{n0}(-\gamma_2 r_1 s), \qquad K_{n44} = -R_{n3}(-\eta_2 r_1 s),$$
$$L_{n41} = (-1)^{n+1} R_{n0}(-\gamma_1 r_1 s), \qquad L_{n42} = (-1)^{n+1} R_{n3}(-\eta_1 r_1 s),$$
$$L_{n43} = R_{n0}(\gamma_2 r_1 s), \qquad L_{n44} = R_{n3}(\eta_2 r_1 s). \tag{3.57}$$

It is interesting to compare the recursive relationships obtained here with the results of Sect. 3.2. The comparison shows that the formulas obtained in Sect. 3.2, like (3.33), represent the solution of the recursive relationships (3.48). At that, the polynomial $X_{00}(s)$, which presents in the denominator of all the rational functions in (3.33), is related to the denominator of the recursive relationships (3.48) as follows: $X_{00} = -X_{n1}X_{n0}$. The difference in signs can be explained by the similar difference in signs of the numerators.

3.4 Diffraction Problems for an Acoustic Medium

Based on the results obtained in Sect. 3.3, let us derive a solution of the important particular problem, when the external medium $i = 0$ and the internal medium $i = 2$ are of acoustic type (Grigolyuk et al. (1976), Grigolyuk et al. (1977b)). According to the notice made in Sect. 1.2 (see (1.39)), in order to make a passage to the limit, it is sufficient to set $\kappa_0 = \kappa_2 = 0$, $\eta_0 = \eta_2 = \infty$, and $k_{10} = k_{12} = \infty$.

Before starting a derivation, let us obtain some relationships interrelating the polynomials $R_{n0}(s)$, $R_{n1}(s)$, and $R_{n2}(s)$.

Using the relationship of the polynomial $R_{n0}(s)$ with the Bessel functions of a semi-integer index (1.126) and taking into account the formula (Watson (1945), Gradshtein, Ryzhik (1971))

$$I_\nu(x) K_{\nu+1}(x) + I_{\nu+1}(x) K_\nu(x) = x^{-1}, \tag{3.58}$$

we obtain

$$R_{n+1,0}(s) R_{n0}(-s) - R_{n0}(s) R_{n+1,0}(s) = 2(-1)^n s^{2n+1} \tag{3.59}$$

or, taking into consideration the relationship between R_{n1} and R_{n0}, we obtain

$$R_{n1}(s) R_{n0}(-s) - R_{n0}(s) R_{n1}(-s) = 2(-1)^n s^{2n+1}. \tag{3.60}$$

Let us divide (3.60) by s^{2n+3}. Then, performing differentiation and taking into account (3.46), we arrive at

$$R_{n2}(s) R_{n0}(-s) - R_{n0}(s) R_{n2}(-s) = 4(-1)^n s^{2n+1}. \tag{3.61}$$

Dividing the equality for R_{n1} in (3.22) by s^{n+2} and making its differentiation, we obtain

$$R_{n2}(s) = R_{n+1,1}(s) - n[R_{n0}(s) + R_{n1}(s)]$$
$$= R_{n+1,1}(s) - nR_{n+1,0}(s) + n(n-1)R_{n0}(s). \tag{3.62}$$

From (3.61) and (3.62), we obtain

$$R_{n+1,1}(s)R_{n0}(-s) - R_{n0}(s)R_{n+1,1}(-s) = 2(-1)^n(n+2)s^{2n+1}. \tag{3.63}$$

Let us now consider (3.60) and (3.62) as a system of linear algebraic equations with respect to $R_{n0}(s)$ and $R_{n0}(-s)$. Then, following the Cramer rule, we get

$$R_{n0}(s) = \frac{\Delta_+}{\Delta},$$

$$\Delta = R_{n2}(s)R_{n1}(-s) - R_{n2}(-s)R_{n1}(s),$$

$$\Delta_+ = 2(-1)^n s^{2n+1}[2R_{n1}(s) - R_{n2}(s)]. \tag{3.64}$$

Separating the integer part of the rational fraction, we arrive at

$$\frac{R_{n2}(s) - 2R_{n1}(s)}{R_{n0}(s)} = \sum_{k=0}^{2} \alpha_k s^{2-k} + \frac{1}{R_{n0}(s)}\sum_{k=0}^{n-1} \beta_k s^{n-1-k}, \tag{3.65}$$

where α_k and β_k are unknown coefficients.

Multiplying the last equality by $R_{n0}(s)$ and equating the coefficients corresponding to the similar powers s, in accounting for the formulas for A_{nk}, B_{nk}, and C_{nk} (1.113) and (3.46), we obtain $\alpha_0 = 1$, $\alpha_1 = 0$, $\alpha_2 = n(n+1)$, and $\beta_k = 0$ ($k = 0, 1, \ldots, n-1$). It follows that

$$R_{n2}(s) - 2R_{n1}(s) = [s^2 + n(n+1)]R_{n0}(s). \tag{3.66}$$

From (3.64) and (3.66), we get

$$R_{n2}(s)R_{n1}(-s) - R_{n2}(-s)R_{n1}(s) = (-1)^{(n+1)}2s^{2n+1}[s^2 + n(n+1)]. \tag{3.67}$$

Taking into account the expressions for the polynomials $Q_{nj}^{(i)}(s)$, $i = 1, 2, 3$, and the formulas (3.59)–(3.67), we get the following relationships:

$$Q_{n1}^{(i)}(s) = 2(1 - \kappa_i)R_{n1}(s) + R_{n0}(s)[s^2 + (1 - \kappa_i)n(n+1)],$$

$$Q_{n1}^{(i)}(s)R_{n0}(-s) - Q_{n1}^{(i)}(-s)R_{n0}(s) = 4(-1)^n(1 - \kappa_i)s^{2n+1},$$

$$Q_{n1}^{(i)}(s)R_{n1}(-s) - Q_{n1}^{(i)}(-s)R_{n1}(s)$$
$$= 2(-1)^{n+1}s^{2n+1}[s^2 + (1 - \kappa_i)n(n+1)],$$

$$Q_{n2}^{(i)}(s)R_{n0}(-s) - Q_{n2}^{(i)}(-s)R_{n0}(s) = 2(-1)^n(1 - \kappa_i)s^{2n+1},$$

$$Q_{n2}^{(i)}(s)R_{n1}(-s) - Q_{n2}^{(i)}(-s)R_{n1}(s) = -2(1 - \kappa_i)(-1)^n s^{2n+1},$$

$$Q_{n3}^{(i)}(s) R_{n0}(-s) - Q_{n3}^{(i)}(-s) R_{n0}(s) = 2(1 - \kappa_i)(-1)^n s^{2n+1},$$

$$Q_{n1}^{(i)}(s) Q_{n2}^{(i)}(-s) - Q_{n1}^{(i)}(-s) Q_{n2}^{(i)}(s)$$
$$= -2(-1)^n s^{2n+1}(1 - \kappa_i)[s^2 + (1 - \kappa_i)(n+2)(n-1)],$$

$$Q_{n2}^{(i)}(s) Q_{n3}^{(i)}(-s) - Q_{n2}^{(i)}(-s) Q_{n3}^{(i)}(s)$$
$$= (1 - \kappa_i)^2 (-1)^n s^{2n+1}[s^2 + 2(n-1)(n+2)]. \tag{3.68}$$

In order to switch to acoustic media, we set $g_{jklm}^{(0)} = g_{jklm}^{(2)} \equiv 0$ (i.e., we assume the absence of shear in the media with the indexes $i = 0, 2$) and take into account that arbitrary functions do not depend on j. Saving, for a simplicity, all the notation of Sect. 3.3, and omitting the index j, we arrive at the following recursive system of equations for $k + m + l > 0$:

$$\begin{pmatrix} f_{0klm}^{(1)L} \\ g_{0klm}^{(1)L} \\ f_{klm}^{(0)L} \end{pmatrix} = \frac{Y_n}{X_{n0}} \begin{pmatrix} f_{1kl,m-1}^{(1)L} \\ g_{1k,l-1,m}^{(1)L} \\ 0 \end{pmatrix},$$

$$\begin{pmatrix} f_{1klm}^{(1)L} \\ g_{1klm}^{(1)L} \\ f_{klm}^{(2)L} \end{pmatrix} = \frac{Z_n}{X_{n1}} \begin{pmatrix} f_{0kl,m-1}^{(1)L} \\ g_{0k,l-1,m}^{(1)L} \\ f_{k-1,lm}^{(2)L} \end{pmatrix}. \tag{3.69}$$

Here, Y_n, Z_n, X_{n0}, and X_{n1} have the same meaning as in (3.49). However, the matrices M_n, N_n, K_n, and L_n should be considered as the matrices of the dimension 3×3 arranged from the elements of the first three rows and columns of the corresponding matrices (3.49). At that, some of the elements are simplified. Since $\kappa_0 = \kappa_2 = 1$, we have

$$Q_{n1}^{(i)}(s) = s^2 R_{n0}(s), \qquad Q_{n2}^{(i)}(s) = Q_{n3}^{(i)}(s) \equiv 0, \qquad i = 0, 2,$$
$$M_{n13} = -\beta_0 s^2 R_{n0}(s), \qquad K_{n13} = \beta_2(\gamma_2 r_1 s)^2 R_{n0}(-\gamma_2 r_1 s),$$
$$L_{n13} = -\beta_2(\gamma_2 r_1 s)^2 R_{n0}(\gamma_2 r_1 s), \qquad M_{n33} = K_{n33} = L_{n33} = 0. \tag{3.70}$$

The other coefficients have the same form as in (3.50).

Let us notice that we can also obtain (3.69) by a passage to the limit, as $\eta_i \to \infty$ and $\kappa_i \to 1$ $(i = 0, 2)$, if to take into account that in (3.48) we have cancelled out common factors of the form $\exp(-\eta_i s)$ and that

$$\lim_{x \to +\infty} P_n(x) e^{-x} = 0,$$

where $P_n(x)$ is the polynomial of arbitrary power n. Since the dimensions of the matrices here are smaller than these in Sect. 3.3, it is easier to derive the explicit expressions for the polynomials X_{n0} and X_{n1}, and the elements of the matrices Y_n and Z_n, as well. From (3.49), (3.50), and (3.60)–(3.70), we obtain

$$X_{n0}(s) = -E_{n0}(s, -\gamma_1 s, -\eta_1 s),$$

$$Y_{n11}(s) = (-1)^n E_{n0}(s, \gamma_1 s, -\eta_1 s),$$

$$Y_{n22}(s) = (-1)^n E_{n0}(s, -\gamma_1 s, \eta_1 s),$$

$$X_{n1}(s) = E_{n2}(-\gamma_2 r_1 s, \gamma_1 r_1 s, \eta_1 r_1 s),$$

$$Z_{n11}(s) = -(-1)^n E_{n2}(-\gamma_2 r_1 s, -\gamma_1 r_1 s, \eta_1 r_1 s),$$

$$Z_{n22}(s) = -(-1)^n E_{n2}(-\gamma_2 r_1 s, \gamma_1 r_1 s, -\eta_1 r_1 s),$$

$$Z_{n33}(s) = E_{n2}(\gamma_2 r_1 s, \gamma_1 r_1 s, \eta_1 r_1 s),$$

$$E_{ni}(x, y, z) = \beta_i x^2 R_{n0}(x) D_{n1}^{(1)}(y, z) - R_{n1}(x) D_{n2}^{(1)}(y, z),$$

$$D_{n1}^{(i)}(y, z) = R_{n1}(y) Q_{n3}^{(i)}(z) - n(n+1) Q_{n2}^{(i)}(y) R_{n0}(z),$$

$$D_{n2}^{(i)}(y, z) = Q_{n1}^{(i)}(y) Q_{n3}^{(i)}(z) - n(n+1) Q_{n2}^{(i)}(y) Q_{n2}^{(i)}(z),$$

$$Y_{n12} = -n(n+1)(1-\kappa_1)(\eta_1 s)^{2n+1}$$
$$\times \left\{ (1-\kappa_1) R_{n1}(s) \left[\eta_1^2 s^2 + 2(n-1)(n+2) \right] + 2\beta_0 s^2 R_{n0}(s) \right\},$$

$$Y_{n21} = -2(1-\kappa_1)(\gamma_1 s)^{2n+1}$$
$$\times \left\{ R_{n1}(s) \left[\gamma_1^2 s^2 + (1-\kappa_1)(n-1)(n+2) \right] + \beta_0 s^2 R_{n0}(s) \right\},$$

$$Y_{n31} = 2(-1)^n (\gamma_1 s)^{2n+1} \left\{ -Q_{n3}^{(1)}(-\eta_1 s) \left[\gamma_1^2 s^2 + (1-\kappa_1) n(n+1) \right] \right.$$
$$+ n(n+1)(1-\kappa_1) R_{n0}(-\eta_1 s) \left[\gamma_1^2 s^2 + (1-\kappa_1)(n-1)(n+2) \right]$$
$$\left. + n(n+1)(1-\kappa_1) Q_{n2}^{(1)}(-\eta_1 s) \right\},$$

$$Y_{n32} = (-1)^n n(n+1)(1-\kappa_1)(\eta_1 s)^{2n+1}$$
$$\times \left\{ (1-\kappa_1) R_{n1}(-\gamma_1 s) \left[\eta_1^2 s^2 + 2(n-1)(n+2) \right] \right.$$
$$\left. + 2Q_{n1}^{(1)}(-\gamma_1 s) - 2Q_{n2}^{(1)}(-\gamma_1 s) \right\},$$

$$Y_{n13} = Y_{n23} = Y_{n33} \equiv 0,$$

$$Z_{n12} = -n(n+1)(1-\kappa_1)(\eta_1 r_1 s)^{2n+1}$$
$$\times \left\{ (1-\kappa_1) \left[(\eta_1 r_1 s)^2 + 2(n-1)(n+2) \right] R_{n1}(-\gamma_2 r_1 s) \right.$$
$$\left. + 2\beta_2 (\gamma_2 r_1 s)^2 R_{n0}(-\gamma_2 r_1 s) \right\},$$

$$Z_{n13} = 2(-1)^n \beta_2 (\gamma_2 r_1 s)^{2n+3} Q_{n3}^{(1)}(\eta_1 r_1 s),$$

$$Z_{n21} = -2(1-\kappa_1)(\gamma_1 r_1 s)^{2n+1}$$
$$\times \left\{ R_{n1}(-\gamma_2 r_1 s) \left[(\gamma_1 r_1 s)^2 + (1-\kappa_1)(n-1)(n+2) \right] \right.$$
$$\left. + \beta_2 (\gamma_2 r_1 s)^2 R_{n0}(-\gamma_2 r_1 s) \right\},$$

$$Z_{n23} = 2\,(-1)^n \beta_2 (\gamma_2 r_1 s)^{2n+3} Q_{n2}^{(1)}(\gamma_1 r_1 s)\,,$$

$$Z_{n31} = -2\,(\gamma_1 r_1 s)^{2n+1} \left\{ Q_{n3}^{(1)}(\eta_1 r_1 s)\,\left[(\gamma_1 r_1 s)^2 + (1 - \kappa_1)\,n(n+1)\right]\right.$$

$$- (1 - \kappa_1)\,n(n+1)\,R_{n0}(\eta_1 r_1 s)\,\left[(\gamma_1 r_1 s)^2\right.$$

$$+ (1 - \kappa_1)\,(n-1)\,(n+2)] - (1 - \kappa_1)\,n(n+1)\,Q_{n2}^{(1)}(\eta_1 r_1 s)\Big\}\,,$$

$$Z_{n32} = -2\,(1 - \kappa_1)\,n(n+1)\,(\eta_1 r_1 s)^{2n+1}$$

$$\times \left\{ n(n+1)\,Q_{n2}^{(1)}(\gamma_1 r_1 s) - Q_{n1}^{(1)}(\gamma_1 r_1 s)\right.$$

$$+ (1 - \kappa_1)\,R_{n1}(\gamma_1 r_1 s)\,\left[(\gamma_1 r_1 s)^2 + 2\,(n-1)\,(n+2)\right]\Big\}\,. \quad (3.71)$$

The degrees of all the polynomials in the left-hand side of (3.71) are smaller or equal to $3n + 5$. Considering (3.51) and (3.53), we can obtain the conditions for the recursive system of equations (3.69). In the case of the external problem, we get

$$\begin{pmatrix} f_{0000}^{(1)\mathrm{L}} \\ g_{0000}^{(1)\mathrm{L}} \\ f_{0000}^{(0)\mathrm{L}} \end{pmatrix} = -E_0(s) \left[\begin{pmatrix} \mathbf{F}_0(s) \\ X_{n0}(s) \end{pmatrix} + e^{-2s} \begin{pmatrix} 0 \\ 0 \\ 1 \end{pmatrix} \right]; \qquad (3.72)$$

in the case of the internal problem, we get

$$\begin{pmatrix} f_{1000}^{(1)\mathrm{L}} \\ g_{1000}^{(1)\mathrm{L}} \\ f_{000}^{(2)\mathrm{L}} \end{pmatrix} = -\frac{E_2(s)}{X_{n1}(s)} \begin{pmatrix} Z_{n13}(s) \\ Z_{n23}(s) \\ Z_{n33}(s) \end{pmatrix}. \qquad (3.73)$$

Here, the functions $E_0(s)$ and $E_2(s)$ are determined by (3.25), and $\mathbf{F}_0(s)$ have the following form:

$$\mathbf{F}_0(s) = -X_{n0}(s)\mathrm{M}_n^{-1}\,\left(M_{n13}(-s),\ M_{n23}(-s),\ 0\right)^{\mathrm{T}}. \qquad (3.74)$$

In (3.72) and (3.73), the sign was changed to the opposite since it was supposed that compressive stress (pressure) in the acoustic media is positive.

The orders of the singularities of images can be obtained from Table 3.1 if $j = 0$. Let us also notice that for $n > 0$, the polynomials X_{n0}, X_{n1}, Y_{nij}, and Z_{nij} $(i, j = 1, 2, 3)$ are the homogeneous functions of the argument s in the second degree. For all of the polynomials, except X_{n0}, X_{n1}, Y_{nii}, and Z_{nii} $(i = 1, 2, 3)$, this statement is evident. Before starting to prove this statement for the last polynomials, let us notice that the coefficients at x^0 and x^1 of the polynomials $R_{nj}(x)$ and $Q_{n,j+1}^{(1)}(x)$, $j = 0, 1, 2$, as it follows from (1.113) and (3.46), are equal to

$$R_{n0}(0) = R'_{n0}(0) = A_{nn}, \qquad R_{n1}(0) = R'_{n0}(0) = B_{nn} = (n+1) A_{nn},$$

$$R_{n2}(0) = R'_{n2}(0) = C_{nn} = (n+1)(n+2) A_{nn},$$

$$Q^{(i)}_{n1}(0) = Q^{(i)}_{n1}{}'(0) = (n+1)(n+2)(1-\kappa_i) A_{nn},$$

$$Q^{(i)}_{n2}(0) = Q^{(i)}_{n2}{}'(0) = (n+2)(1-\kappa_i) A_{nn},$$

$$Q^{(i)}_{n3}(0) = Q^{(i)}_{n3}{}' = n(n+2)(1-\kappa_i) A_{nn}, \tag{3.75}$$

respectively.

From the definition of the polynomials in two variables $D^{(i)}_{n1}(y, z)$ and $D^{(i)}_{n2}(y, z)$ (3.71) it follows that

$$D^{(1)}_{n1}(0, 0) = R_{n1}(0) Q^{(1)}_{n3}(0) - n(n+1) Q^{(1)}_{n2}(0) R_{n0}(0) = 0,$$

$$D^{(1)}_{n1,\,y}{}'(0, 0) = R'_{n1}(0) Q^{(1)}_{n3}(0) - n(n+1) Q^{(1)}_{n2}{}'(0) R_{n0}(0) = 0,$$

$$D^{(1)}_{n1,\,z}{}'(0, 0) = R_{n1}(0) Q^{(1)}_{n3}{}'(0) - n(n+1) Q^{(1)}_{n2}(0) R'_{n0}(0) = 0 \tag{3.76}$$

and, similarly,

$$D^{(1)}_{n2}(0, 0) = D^{(1)}_{n2,\,y}{}'(0, 0) = D^{(1)}_{n2,\,z}{}'(0, 0) = 0.$$

This result jointly with the equalities (3.71) proves the homogeneity of the polynomials under consideration. Since the recursive relationships enter into these polynomials, it is possible to cancel out the common factors s^2 and, thus, to reduce the maximum degree of the polynomials to $3(n+1)$.

Let us also notice that the point $s = 0$ is still the pole of the order $2n+1$ for the internal problem and this point will not be a singularity for the external problem. In fact, in accounting for (3.50), (3.60), and (3.70), from (3.74) it follows that

$$\mathbf{F}_0 = \left(F_{01},\, F_{02},\, F_{03}\right)^{\mathrm{T}},$$

$$F_{01} = 2\beta_0 s^{2n+3} Q^{(1)}_{n3}(-\eta_1 s),$$

$$F_{02}(s) = -2\beta_0 s^{2n+3} Q^{(1)}_{n2}(-\gamma_1 s),$$

$$F_{03}(s) = \beta_0 s^2 R_{n0}(-s) D^{(1)}_{n1}(-\gamma_1 s,\, \eta_1 s) - R_{n1}(-s) D^{(1)}_{n2}(-\gamma_1 s,\, -\eta_1 s). \tag{3.77}$$

Thus, the multipliers s^{2n+3} enter into the numerators and denominators of the transforms of all of the arbitrary functions except $f^{(0)L}_{000}$. But the image $f^{(0)L}_{000}$ does not enter into the recursive relationships (3.69); it should be calculated via the functions $f^{(1)L}_{0000}$ and $g^{(1)L}_{00000}$. That is the reason for the statement made for the point $s = 0$. We can prove that a similar conclusion can be performed for the general external problem (3.48) and (3.51).

Let us now consider the following particular case of the external problem: the inner surface of a thick-walled sphere is free and there is an acoustic medium outside of the sphere (Grigolyuk et al. (1979), Tarlakovsky (1977)). Then, for the expressions obtained in this section, in accordance with Sect. 1.4, we should set $\beta_2 = 0$ and $\gamma_2 = \infty$. The last condition brings us to the identity $f_{klm}^{(2)} \equiv 0$, as well as to the fact of independence of the other arbitrary functions from the index k. Saving all the previous notation, from (3.69) we obtain the following recursive relationships:

$$
\begin{pmatrix} f_{0lm}^{(1)L} \\ g_{0lm}^{(1)L} \\ f_{lm}^{(0)L} \end{pmatrix} = \frac{Y_n}{X_{n0}} \begin{pmatrix} f_{1l,m-1}^{(1)L} \\ g_{1,l-1,m}^{(1)L} \\ 0 \end{pmatrix} ,
$$

$$
\begin{pmatrix} f_{1lm}^{(1)L} \\ g_{1lm}^{(1)L} \end{pmatrix} = \frac{Z_n}{X_{n1}} \begin{pmatrix} f_{0l,m-1}^{(1)L} \\ g_{0,l-1,m}^{(1)L} \end{pmatrix} .
\tag{3.78}
$$

Here, Y_n and X_{n0} remain unchanged and Z_n and X_{n1} have the same meaning as in (3.49). However, the matrices K_n and L_n should be regarded as the matrices of the dimension 2×2 arranged from the elements of the first two rows and columns of the corresponding matrices (3.49). Besides the simplifications in (3.69) mentioned, let us note that since $\beta_2 = 0$, as it follows from (3.71), the polynomials X_{n1} and Z_{nij} $(i, j = 1, 2)$ have the multiplier $R_{n1}(-\gamma_2 r_1 s)$. These polynomials are the numerators and denominators of the recursive relationships. Hence, making a passage to the limit, as $\gamma_2 \to \infty$, we can write

$$
\begin{aligned}
X_{n1}(s) &= -D_{n2}^{(1)}(\gamma_1 r_1 s, \eta_1 r_1 s) , \\
Z_{n11}(s) &= (-1)^n D_{n2}^{(1)}(-\gamma_1 r_1 s, \eta_1 r_1 s) , \\
Z_{n22}(s) &= (-1)^n D_{n2}^{(1)}(\gamma_1 r_1 s, -\eta_1 r_1 s) , \\
Z_{n12}(s) &= -n(n+1)(1-\kappa_1)^2(\eta_1 r_1 s)^{2n+1} \left[(\eta_1 r_1 s)^2 + 2(n-1)(n+2)\right] , \\
Z_{n21}(s) &= -2(1-\kappa_1)(\gamma_1 r_1 s)^{2n+1} \left[(\gamma_1 r_1 s)^2 + (n-1)(n+2)\right] .
\end{aligned}
\tag{3.79}
$$

As before, the initial conditions for the recursive equations (3.78) are of the form (3.72). The orders of the singularities (poles) of the images can be obtained from Table 3.1 if $j = k = 0$.

3.5 Numerical Examples

The recursive relationships derived in Sects. 3.3 and 3.4 for the transforms of arbitrary functions, as well as the formulas (3.6) and (3.45), give us an opportunity to obtain the Laplace transforms of all of the components of the stress–strain state for any medium $i = 0, 1, 2$. It is easy to calculate their originals by means of the algorithm described in Chap. 2 (see (2.35)–(2.38)).

Table 3.2. The partial sums S_n and the Cesaro means σ_n of the series in terms of the Legendre polynomials for a disturbing plane wave ($\theta = 0$, $r = 1$)

n	0	1	2	3	4	5	6	7	8	9	10
$\tau = 0.5$											
S_n	0.250	0.812	1.281	1.363	1.099	0.793	0.726	0.925	1.170	1.228	1.062
σ_n	0.250	0.531	0.781	0.926	0.961	0.933	0.903	0.906	0.935	0.965	0.973
$\tau = 1.0$											
S_n	0.500	1.250	1.250	0.812	0.812	1.156	1.156	0.863	0.863	1.123	1.123
σ_n	0.500	0.875	1.000	0.953	0.925	0.963	0.991	0.975	0.962	0.978	0.991
$\tau = 3.0$											
S_n	1.000	1.000	1.000	1.000	1.000	1.000	1.000	1.000	1.000	1.000	1.000
σ_n	1.000	1.000	1.000	1.000	1.000	1.000	1.000	1.000	1.000	1.000	1.000

The same method can be applied for obtaining the originals of the functions derived in Sect. 3.2.

We have discussed above the method of calculation of the coefficients of the series expansion of the functions sought in terms of the Legendre polynomials. For the solution obtained, it is necessary to perform summation of the series (3.6).

It is very difficult to test the validity of representation of the functions in the form of the series (3.6). Let us point to the works, which allow us to make some indirect judgment about a uniform convergence of these series on the sets of points, which do not include the vicinities of the discontinuity points. The general theorems on the series expansion in terms of the spherical functions are presented in Hobson (1955). Insofar as the elastic or acoustic media having one spherical interface are concerned, the uniform convergence was studied in Buldyrev, Molotkov (1958), Makarov, Petrashen (1953), and Petrashen (1953). The authors made their studies applying the inverse Laplace transform with respect to time and the method of decomposition of the series into the *slowly* and *rapidly* varying parts. In Tupholme (1967), the author marked out a uniform convergence of the series used for solution of the problems of propagation of an axially symmetric disturbance from a spherical cavity in an elastic medium. In Natrashvili (1980), the author has proven absolute and uniform convergence of the series in terms of the spherical functions for some hyperbolic operator (the equations of the linear theory of elasticity make a particular case) determined in the Euclid space E^3 with one spherical interface.

When working directly with the series (3.6), a question about their rate of convergence appears. One more difficulty is caused by the Gibbs effect in the vicinity of the waves' fronts. In order to improve the convergence, different methods of summation were proposed (Veksler (1973), Berger, Klein (1972)).

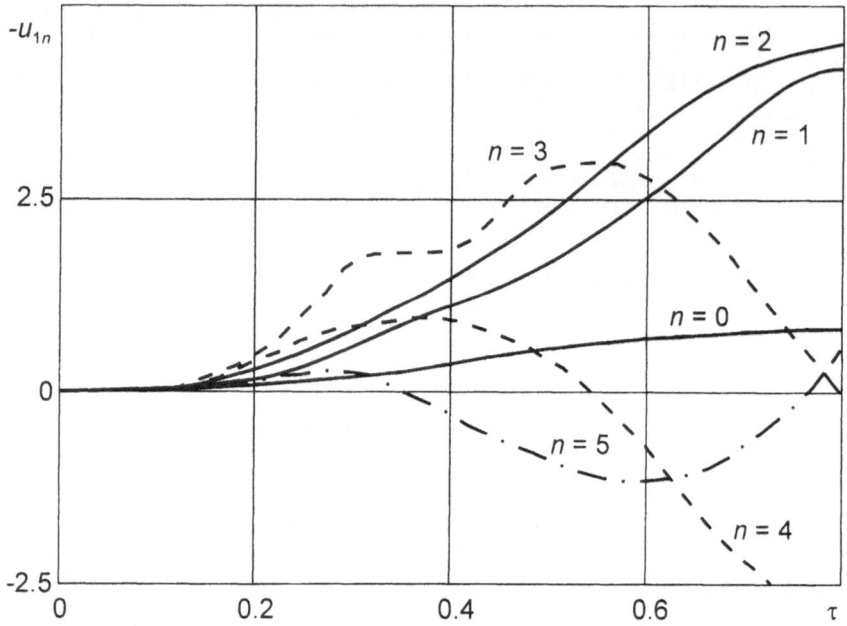

Fig. 3.4. The contribution of the terms of the series in terms of the Legendre polynomials into the sum of the series for a radial displacement (n is the term number, $r = 1$, $\theta = 0$)

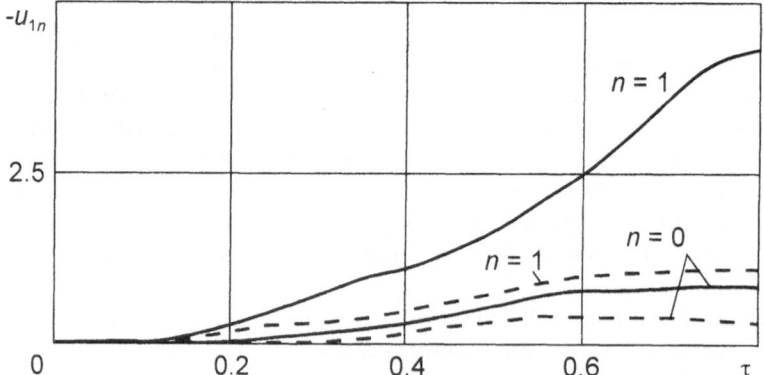

Fig. 3.5. The influence of thickness of a hollow shell on the radial displacement

Let us dwell on the method of summation of series by the Cesaro means of the first order (Alexits (1961), Hobson (1955)). The method is based on the statement that if the sequence of partial sums S_n of some series converges to S, then the sequence of their arithmetic means converges to S, as well (Alexits (1961)), that is, if

$$\lim_{n \to \infty} S_n = S,$$

then

$$\lim_{n \to \infty} \sigma_n = S, \qquad \sigma_n = \frac{1}{n+1} \sum_{i=0}^{n} S_i.$$

To make summation by this method, let us derive a recursive formula, which interrelates σ_n and σ_{n-1}. We get

$$\sigma_n - \sigma_{n-1} = \frac{1}{n+1} \sum_{i=0}^{n} S_i - \frac{1}{n} \sum_{i=0}^{n-1} S_i$$

$$= \frac{1}{n(n+1)} \sum_{i=0}^{n} [nS_i - (n+1)S_i] + \frac{Sn}{n} = \frac{S_n - \sigma_n}{n}.$$

It follows that

$$\sigma_n = \frac{1}{n+1} (n\sigma_{n-1} + S_n). \tag{3.80}$$

In order to estimate a rate of convergence of the series in terms of the Legendre polynomials $P_n(\cos \theta)$ and to make a comparison of the conventional sums with the Cesaro sums, let us make summation of the potential in the incoming wave (3.9) expanded into the series with the coefficients determined by (3.11). At that, let us assume that $f(\tau) = 1$ $(f^L(s) = s^{-1})$ and use the algorithm (2.35)–(2.38) for calculation of the originals. The results of calculations for the point $(r = 1, \theta = 0)$ are presented in Table 3.2.

The computations illustrate the Cesaro method advantages. Besides, we can make a conclusion that for $n = 5$ or $n = 6$, the Cesaro means give us a good approximation to the initial Heaviside unit function.

Let us consider, for example, a problem of interaction of a plane acoustic wave, which propagates in an external medium (water), with a hollow thick-walled shell. At that, let us assume that the pressure in the incoming wave varies following the law

$$p(r, \theta, \tau) = \Delta p \, e^{-\alpha(\tau + r\cos\theta - 1)} H(\tau + r\cos\theta - 1). \tag{3.81}$$

For $\alpha = 0$ and $\Delta p = 1$, the load has a form of the Heaviside unit function.

When performing calculations, we shall use the following values of the dimensionless parameters: $\alpha = 0$, $\Delta p = 1$, $\gamma_1 = 0.252$, $\beta_0 = 0.00862$, and $\kappa_1 = 0.393$.

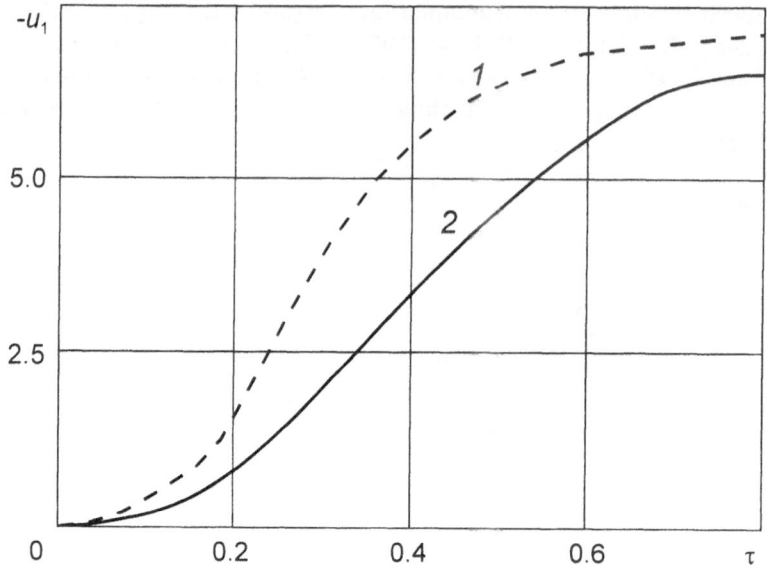

Fig. 3.6. The radial displacements of a thick-walled spherical shell at the point $r = 1$, $\theta = 0$ (the solid line corresponds to the Cesaro summation and the dashed one corresponds to the conventional summation)

Some results of calculations are illustrated in Figs. 3.4–3.7. All the curves correspond to the shell's point with the coordinates $r = 1$ and $\theta = 0$.

Let us note that, as it follows from the method of summation of elementary waves, the number of the elementary waves increases with a decrease in the relative thickness of the shell δ and an increase of time τ; this leads to a pronounced increase in computational effort.

In Fig. 3.4, we demonstrate a dependence of the forms of the radial displacement of the shell $u_{1n}(1, \tau)$ versus time τ for $\delta = 0.3$. The numbers at curves correspond to the numbers of terms of the series ($n = 0, 1, \ldots, 5$). The curves show that the contribution of all of the forms considered into the sum of the series is significant, but, it decreases with a decrease of the number n.

In Fig. 3.5, we demonstrate the influence of the obstacle thickness on the displacements (the solid lines correspond to $\delta = 0.3$, and the dashed ones correspond to $\delta = 0.5$). The curves were plotted for $n = 0$ and $n = 1$ (the numbers at curves correspond to n).

In Fig. 3.6, we demonstrate the curves which characterize a dependence of the radial displacement (the corresponding series was summed in accounting for the first six terms, $n = 0, 1, \ldots, 5$) on time τ for $\delta = 0.3$. The dashed line *1* corresponds to a conventional partial sum of the series and the solid line *2* corresponds to the Cesaro mean. It follows that, when τ increases, a relative error caused by the difference in the methods of summation decreases.

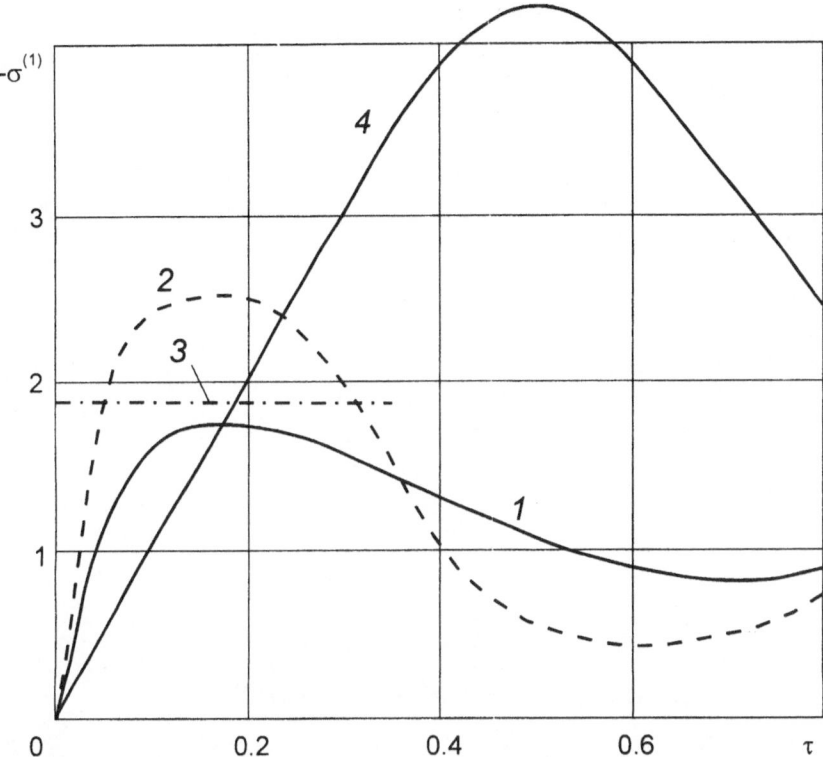

Fig. 3.7. The stresses at the frontal point of a shell (curves *1–3* correspond to $\sigma_{rr}^{(1)}$ and curve *4* corresponds to $\sigma_{\theta\theta}^{(1)}$)

In Fig. 3.7, we plot the stresses versus time for a shell of the relative thickness $\delta = 0.3$. Curve *4* corresponds to the stress $\sigma_{\theta\theta}^{(1)}$, and the others correspond to the radial stress $\sigma_{rr}^{(1)}$. Summation of the corresponding series was performed in accounting for the first six terms. The dashed curve *2* was plotted using the results of conventional summation and the solid ones were plotted using the results of summation of the Cesaro means.

Curve *3* corresponds to the asymptotic solution derived in Veksler (1975). It follows that the conventional sums give us an upper estimate of an exact solution and the Cesaro sums give us the lower one.

4. Axially Symmetric Vibrations of Elastic Media Having a Spherical Cavity or a Stiff Inclusion

The results to be presented in this chapter are closely related to those presented in Chapters 2 and 3 and, basically, can be derived by the corresponding passages to the limits in the solutions presented in Chap. 3.

At the same time, the problems considered in Chap. 2 make a particular case of the problems to be considered here. We consider them in a separate chapter because the problem of propagation of axially symmetric disturbances from a cavity and diffraction of the waves by an absolutely soft obstacle (a cavity) or an absolutely stiff obstacle (an immovable sphere) is of a great practical importance. On the other hand, the results of this study should demonstrate that the problems to be considered are similar to the more complicated ones, when the media boundaries are separated by the thin-walled shells (Chap. 5). Unification of the problems of determination of transition functions (Sect. 4.1), of propagation of disturbances from a cavity (Sect. 4.2), and of diffraction (Sect. 4.3) is caused, first of all, by the fact that, from a mathematical point of view, their solutions can be derived using similar methods.

The classical problems to be considered in this chapter were analyzed in many publications. The extensive literature was reviewed in Inouye (1937), Brune (1970), Babich, Molotkov (1977), Gorshkov (1979), Vestyak et al. (1983), and Vestyak et al. (1984). The mathematical aspects of the problems of diffraction of acoustic waves by the obstacles were considered in Friedlender (1958).

For the case of axial symmetry, the problem of propagation of disturbances from a spherical cavity in an acoustic medium was solved for the first time by Love (1905). A solution for the case of diverging waves was derived on the basis of representation of the general integral of the wave equations (1.5) in series form of the spherical harmonics (the series in terms of the Legendre polynomials $P_n(\cos\theta)$ and their derivatives); the coefficients of these series were expressed via arbitrary functions. A similar solution was presented in Lamb (1931). An asymptotic solution of the problem when the velocity was predetermined at the cavity's boundary was presented in Ansell, Tupholme (1972).

One of the first detailed studies of a similar problem for an elastic medium was presented in Sezawa, Kanai (1941–1942). However, the authors

considered only one term of the series expansion of the functions sought
in terms of the Legendre polynomials. In Azizov (1973) and Vaněk (1953),
this problem was solved by means of the Laplace transform with respect to
time. In Rakhmatulin et al. (1967b), a general solution of the wave equation was derived in accounting for two terms of the Legendre polynomials.
In Singh, Rosenmann (1973), a problem of propagation of the shear disturbances from a cavity was considered under a similar assumption. At that, the
authors used the Fourier transform with respect to time. An axially symmetric problem in the case of more general boundary conditions was studied in
Tupholme (1967), Saakyan (1973), Singh (1973), Moodie et al. (1983), and
Tupholme (1983). At that, the authors marked out the uniform convergence
of the series in terms of the Legendre polynomials used.

In Podilchuk, Rubtsov (1981) and Podilchuk, Rubtsov (1986), the ray
method was used for a similar problem. The authors presented the numerical
results obtained in accounting for sixteen terms of the ray series. A solution
of a three-dimensional problem of propagation of disturbances from a spherical cavity in acoustic and elastic media was derived in Smirnov (1937a) and
Smirnov (1937b). The Volterra integral equations were derived in order to
solve a corresponding boundary value problem with the boundary conditions
at the cavity's surface. A similar approach was used in Mindlin (1940a) and
Mindlin (1940b).

In Vaněk (1956), Eringen (1957), Chadwick, Trowbridge (1967a), and
Chadwick, Trowbridge (1967b), a three-dimensional problem of propagation
of the disturbances from a spherical cavity in an elastic medium under arbitrary boundary conditions was studied. The authors used the series expansion in terms of the spherical functions and the Fourier transform or
the Laplace transform with respect to time. Inversion of the transforms was
either not performed at all or performed for a particular case of radial vibrations. A particular case of distribution of pressure over a cavity's surface
was considered in the late 1930s in Japan (Nishimura (1937), Inouye (1938),
Nishimura, Takayama (1938)). In order to derive a solution, the authors used
the Fourier transform with respect to time followed by inversion and application of the residue theory. They made a detailed study of the effect of a
pulse shape on the stress–strain state of the medium. The numerical results
were presented for a spherical wave front. The special forms of boundary
conditions (the *Honda model* and the *Scholte model*), which approximate
the real conditions of an explosion in an elastic medium, were considered in
Honda (1960a), Honda (1960b), and Hirasawa (1964).

Propagation of waves in infinite space from the sources of different types
was studied in Ogurtsov et al. (1957). In Usami (1962), a general solution
of the equation of motion of an isotropic linearly elastic medium in the
spherical coordinates was presented; the authors applied the method of separation of variables. An axially symmetric problem of application of two
diametrically located transient forces to a spherical cavity was studied in

Jingu, Nezu (1985b). The specific boundary conditions in the form of a wave travelling at the boundary of a spherical cavity in an acoustic medium were studied in Bazhina (1957). An analysis accounting for variability of the radius of a cavity (a linear law of variation in time) in an elastic medium was presented in Yanyutin (1983).

Some problems related to diffraction of transient waves by the plane and absolutely stiff obstacles were analyzed, for example, in Pao, Mow (1973) and Guz et al. (1978a). The works devoted to these problems were reviewed in Guz et al. (1978b) and Gorshkov (1979).

The main results have been obtained for the cavities of cylindrical and spherical shapes. At that, the functions sought, as a rule, were expanded into the trigonometric Fourier series or series in terms of the Legendre polynomials. In Kubenko (1975b), the Laplace integral transform was applied; inversion of the transform was performed by means of the Volterra integral equations. For a spherical cavity, a similar technique was applied in Panasyuk (1978), Kubenko (1979), and Guz et al. (1980). At that, the authors of the second work analyzed a spherical wave and the author of the third one analyzed a plane wave. In the last study, it was remarked that in order to determine the stresses, it is sufficient to take into account seven terms of the Legendre polynomials.

In order to solve the problem of interaction of an elastic pressure wave with a spherical cavity, the method of inversion of the Laplace transform by means of residues was used in Bakarat (1960), Yen (1964), Cohen, Handelman (1965), Norwood, Miklowitz (1967), McLeary (1969), Huang, Wang (1972), and Heale, Raddy (1975). In Shaw, English (1972), the Kirchhoff formula was applied; as a result, the problem of diffraction of a plane wave or an acoustic wave by a spherical cavity was reduced to an integral equation. A numerical procedure was also presented. In Kovshov (1979), the detailed numerical results for the case of an elastic plane wave were obtained using finite differencing.

A problem of diffraction of an acoustic wave by a spherical cavity in application to the echo-signals was studied in Podstrigach, Poddubnyak (1986). The Fourier transform with respect to time in the case of diffraction of a plane elastic wave by a spherical cavity was used in Peralta et al. (1966). An asymptotic solution valid for the initial instants of time was obtained in Burdun, Sazonov (1985). In Akkas (1977), in order to derive a solution of a similar problem for an acoustic medium, a representation of the general integral in the form of converging waves was applied.

The complicated models of the conditions at the boundaries of a spherical cavity, when interacting with a plane wave, were studied in Baron (1957) and Geers (1975); in the latter work, a linear dependence of the pressure on the displacement was assumed, in the former one, it was assumed that the pressure was a nonlinear function of the displacement. The problem was reduced to an integral equation, which was solved numerically. The models of the boundary conditions presented allowed the authors to take into account approximately the obstacle elasticity without a simultaneous increase of the solution complexity as compared to a free cavity.

The problems of diffraction of the waves by the stiff immovable inclusions of a spherical or cylindrical shape were discussed in Ting, Lee (1969) and Sysoev, Shugaev (1979). Interaction of acoustic and spherical waves with a stiff sphere and a cylinder was studied in Kubenko, Panasyuk (1977).

The problems of diffraction of plane elastic or acoustic waves by an immovable sphere were solved in Kharkevich (1950), Zamyshlyaev, Yakovlev (1967), Slepyan (1972), Nigul et al. (1974), Grigolyuk, Gorshkov (1976), and Poruchikov (1986). The same problems were studied in Allen, Robinson (1966), Lazarenko (1966), and Auphan, Matthys (1978). In Poruchikov (1981), a series expansion in terms of the Legendre polynomials on a variable interval was applied. In Galazyuk, Gorechko (1980a), the Laguerre transform with respect to time was applied, and in Grilitsky, Poddubnyak (1980), the problem of diffraction of a plane shear wave of torsion (SH-wave) by an immovable sphere was solved.

In Babichev et al. (1976), the resultant force produced by a plane elastic compression wave acting on an immovable sphere was determined. In Veksler (1973), behavior of an immovable sphere in an acoustic medium was studied by applying a special method of summation in order to improve convergence of the series in terms of the Legendre polynomials. In Lauvstad (1965), the case of an arbitrary obstacle located in an ideal fluid was studied using the Fourier transform with respect to time and the Green function of the Helmholtz equation. For example, a spherical obstacle was considered. In Pekurovsky et al. (1983), the problem of interaction of a plane acoustic wave with an immovable sphere, which had a thin coating, was studied.

In the sections of this chapter followed, we present the solutions of the problems mentioned above based on the systematic usage of the concept of *generalized* spherical waves.

4.1 Interaction Forces

When solving the problems of interaction of deformable media, we do not know the stresses and displacements at the interfaces and should determine them as part of the solution process (it follows from the conditions of contact (1.69)). For this reason, it is very important to know the functions, using which and applying a convolution operator we can determine the stresses (the pressure) corresponding to the given displacements of the contact surface. In the fluid mechanics, these functions are usually termed the *transition functions*. They are of a great value when solving some special problems (Grigolyuk, Gorshkov (1976), Zamyshlyaev, Yakovlev (1967), Mnev, Pertsev (1970)).

Let us consider a problem of determining of the pressure at the sphere's external surface $r = 1$ for $k_{10} = \infty$. As it follows from the boundary conditions of the problem (3.8), the stresses $\sigma_{rrn}^{(1)}(1, \tau)$ and $\sigma_{r\theta n}^{(1)}(1, \tau)$ depend on the components of the stress tensor in the incoming wave and on the stresses of the diffracted and reflected elastic waves in the external medium $i = 0$. At that, the latter depend on the kinematic parameters of the sphere's surface motion. They can be determined by the following manner.

We shall assume that $\xi_{\varphi u}(r, \tau)$ and $\xi_{\psi u}(r, \tau)$ are the solutions of the problem

$$\frac{\partial^2 \xi_{\varphi u}}{\partial \tau^2} = \Delta_n \xi_{\varphi u},$$

$$\eta_0^2 \frac{\partial^2 \xi_{\psi u}}{\partial \tau^2} = \Delta_n \xi_{\psi u},$$

$$\xi_{\varphi u}|_{\tau=0} = \xi_{\psi u}|_{\tau=0} = \left.\frac{\partial \xi_{\varphi u}}{\partial \tau}\right|_{\tau=0} = \left.\frac{\partial \xi_{\psi u}}{\partial \tau}\right|_{\tau=0} = 0, \qquad r > 1,$$

$$u_{0n}|_{r=1} = H(\tau), \qquad v_{0n}|_{r=1} = 0,$$

$$\lim_{r \to \infty} \xi_{\varphi u} = \lim_{r \to \infty} \xi_{\psi u} = 0, \qquad \tau \geq 0, \tag{4.1}$$

and the functions $\xi_{\varphi v}(r, \tau)$ and $\xi_{\psi v}(r, \tau)$ can be determined by solution of the problem

$$\frac{\partial^2 \xi_{\varphi v}}{\partial \tau^2} = \Delta_n \xi_{\varphi v},$$

$$\eta_0^2 \frac{\partial^2 \xi_{\psi v}}{\partial \tau^2} = \Delta_n \xi_{\psi v},$$

$$\xi_{\varphi v}|_{\tau=0} = \xi_{\psi v}|_{\tau=0} = \left.\frac{\partial \xi_{\varphi v}}{\partial \tau}\right|_{\tau=0} = \left.\frac{\partial \xi_{\psi v}}{\partial \tau}\right|_{\tau=0} = 0, \qquad r > 1,$$

$$u_{0n}|_{r=1} = 0, \qquad v_{0n}|_{r=1} = H(\tau),$$

$$\lim_{r \to \infty} \xi_{\varphi v} = \lim_{r \to \infty} \xi_{\psi v} = 0, \qquad \tau \geq 0. \tag{4.2}$$

Here, $\xi_{\varphi u}$ and $\xi_{\psi u}$ correspond to the elastic potential of P-waves and $\xi_{\varphi v}$ and $\xi_{\psi v}$ correspond to the potential of S-waves. These functions are related to the displacements and stresses by the formulas (3.7)

According to the Duhamel method (Arsenin (1974)), the potentials φ_{0n} and ψ_{0n}, in accounting for the boundary conditions for $r = 1$, can be represented in the form of convolution as follows:

$$\varphi_{0n}(r, \tau) = \frac{\partial \xi_{\varphi u}}{\partial \tau} * (u_{1n} - u_{0ns})|_{r=1} + \frac{\partial \xi_{\varphi v}}{\partial \tau} * (v_{1n} - v_{0ns})|_{r=1},$$

$$\psi_{0n}(r, \tau) = \frac{\partial \xi_{\psi u}}{\partial \tau} * (u_{1n} - u_{0ns})|_{r=1} + \frac{\partial \xi_{\psi v}}{\partial \tau} * (v_{1n} - v_{0ns})|_{r=1}. \tag{4.3}$$

An asterisk designates the operation of convolution.

Then, the stresses $\sigma_{rrn}^{(0)}(1, \tau)$ and $\sigma_{r\theta n}^{(0)}(1, \tau)$ can be determined from (4.3) by means of the differential expressions (3.7).

By analogy with the fluid mechanics, the functions

$$-\frac{\partial \xi_{\varphi u}}{\partial \tau}\bigg|_{r=1}, \qquad -\frac{\partial \xi_{\varphi v}}{\partial \tau}\bigg|_{r=1}, \qquad -\frac{\partial \xi_{\psi u}}{\partial \tau}\bigg|_{r=1}, \qquad -\frac{\partial \xi_{\psi v}}{\partial \tau}\bigg|_{r=1}$$

can be named the *transition functions*. As it follows from (4.1) and (4.2), determination of these functions is related to solution of the problem of propagation of a nonstationary disturbance from a spherical cavity in an infinite elastic medium.

For the sake of simplicity, let us concentrate on determination of a transition function only for the case of acoustic medium (Grigolyuk et al. (1976)). Using (1.39), we clarify that the problem reduces to determination of one function $\xi_{\varphi u}(r, \tau)$. Let us designate

$$\xi_{\varphi u}(r, \tau) = \xi(r, \tau), \qquad \frac{\partial \xi}{\partial \tau}(1, \tau) = -\chi_n(\tau).$$

Then, the function $\xi(r, \tau)$ is the solution of the following boundary value problem:

$$\frac{\partial^2 \xi}{\partial \tau^2} = \Delta_n \xi,$$

$$\xi|_{\tau=0} = \frac{\partial \xi}{\partial \tau}\bigg|_{\tau=0} = 0, \qquad r > 1,$$

$$\frac{\partial \xi}{\partial r}\bigg|_{r=1} = H(\tau), \qquad \lim_{r \to \infty} \xi = 0. \tag{4.4}$$

In order to obtain $\xi(r, \tau)$, let us apply the representation of the wave equation solution in the form (1.114). Following the condition at infinity, we should search for the function $\xi(r, \tau)$ in the form of a diverging spherical wave (the first sum in (1.114)). Taking into consideration the boundary condition in (4.4) for $r = 1$, we obtain the ordinary differential equation with the constant coefficients for determination of the arbitrary function $f(\tau)$:

$$\sum_{k=0}^{n+1} B_{nk} f^{(n+1-k)}(\tau) = -1,$$

$$B_{nk} = k A_{n,k-1} + A_{nk}, \qquad f(\tau+1) = g_1(\tau). \tag{4.5}$$

If to set

$$f^{(i)}(0) = 0, \qquad i = 0, 1, \cdots, n, \tag{4.6}$$

then the initial conditions (4.4) will be satisfied.

Let us assume a particular solution of (4.5) in the form

$$f_*(\tau) = -\frac{1}{(n+1)\,A_{nn}}\,. \tag{4.7}$$

Then, satisfying the conditions (4.6), we can write the solution of the Cauchy initial value problem (4.5) and (4.6) as follows:

$$f(\tau) = \frac{1}{(n+1)\,A_{nn}} \left[\frac{1}{W(0)} \sum_{i=0}^{n+1} W_{1i}(0)\,y_i(\tau) - 1 \right]; \tag{4.8}$$

here, $y_1, \ldots, y_{n+1}$ is the fundamental system of solutions of the homogeneous equation corresponding to (4.5), $W(\tau)$ is the Wronskian of this system, and $W_{1i}(\tau)$ is the algebraic cofactor corresponding to the element of the first row and the ith column of the Wronskian.

Taking into account the relationship between the transition function and $\xi(1, \tau)$, as well as the expressions (1.114) and (4.8), we finally get

$$\chi_n(\tau) = -\frac{1}{(n+1)\,A_{nn}W(0)} \sum_{i=1}^{n+1} W_{1i}(0) \sum_{k=0}^{n} A_{nk}y_i^{(n+1-k)}(\tau). \tag{4.9}$$

Let us consider the case $n = 1$ as an example of application of (4.9). From (1.114), we obtain $A_{10} = A_{11} = 1$. The coefficients of (4.5) are equal to $B_{10} = 1$ and $B_{11} = B_{12} = 2$. Then, we obtain

$$y_1(\tau) = e^{-\tau} \cos \tau, \qquad y_2(\tau) = e^{-\tau} \sin \tau,$$
$$W(0) = W_{11}(0) = W_{12}(0) = 1.$$

Hence, $\chi_1(\tau)$ has the form

$$\chi_1(\tau) = e^{-\tau} \cos \tau;$$

this result corresponds to the solution presented in Kharkevich (1950) and Grigolyuk, Gorshkov (1976).

In order to obtain the transition function $\chi_n(\tau)$, we can apply the Laplace transform with respect to time τ directly to the problem (4.4). Then, for the image $\chi_n^{\mathsf{L}}(s)$, we get (Grigolyuk, Gorshkov (1976))

$$\chi_n^{\mathsf{L}}(s) = \frac{K_{n+1/2}(s)}{(n+1)\,K_{n+1/2}(s) + sK_{n-1/2}(s)}, \tag{4.10}$$

where $K_\nu(x)$ are the modified Bessel functions (the Macdonald functions) of the order ν. Taking into account the relationship of $K_\nu(x)$ with the elementary functions (1.126) and the formulas (3.22), we obtain

$$\chi_n^{\mathsf{L}}(s) = \frac{R_{n0}(s)}{R_{n1}(s)}\,. \tag{4.11}$$

The formula (4.11) can also be obtained if the Laplace transform is applied directly to the problem (4.5)–(4.6).

Let us now consider another form of representation of the transition function $\chi_n(\tau)$.

In order to solve the boundary value problem (4.4), we can apply the method of separation of variables (Grigolyuk et al. (1979)). Since r varies in the infinite interval $[1, \infty)$, we can expect that the eigenvalue spectrum of the problem (4.4) will not be discrete. Hence, in order to use this method, it is necessary to apply the general theory of ordinary differential operators (Levitan, Sargsyan (1970), Naimark (1969)).

Let us reduce the problem (4.4) to a problem with homogeneous boundary conditions. Separating the stationary part $\Phi_{0n}(r)$ in the solution, let us represent $\xi(r, \tau)$ in the following form:

$$\xi(r, \tau) = \Psi_{1n}(r, \tau) + \Phi_{0n}(r).$$

Then, Ψ_{1n} and Φ_{0n} should satisfy the following problems, respectively:

$$\frac{\partial^2 \Psi_{1n}}{\partial \tau^2} = \Delta_n \Psi_{1n},$$

$$\Psi_{1n}\big|_{\tau=0} = -\Phi_{0n}(r), \qquad \frac{\partial \Psi_{1n}}{\partial \tau}\bigg|_{\tau=0} = 0, \qquad r > 1,$$

$$\frac{\partial \Psi_{1n}}{\partial r}\bigg|_{r=1} = 0, \qquad \lim_{r \to \infty} \Psi_{1n} = 0, \qquad \tau \geq 0; \tag{4.12}$$

$$\Delta_n \Phi_{0n} = 0, \qquad \Phi'_{0n}(1) = 1, \qquad \lim_{r \to \infty} \Phi_{0n} = 0. \tag{4.13}$$

Searching for the particular solutions of (4.13) in the form $\Phi_{0n} = r^\alpha$ and satisfying the boundary conditions, we get

$$\Phi_{0n}(r) = -\frac{1}{(n + 1) r^{n+1}}. \tag{4.14}$$

In order to reduce the differential expression in the right-hand side of (4.12) to a self-adjoint form, let us make the following substitution:

$$\Psi_n(r, \tau) = r \Psi_{1n}(r, \tau). \tag{4.15}$$

Then, for the function Ψ_n, we obtain

$$\frac{\partial^2 \Psi_n}{\partial \tau^2} = \frac{\partial^2 \Psi_n}{\partial r^2} - \frac{n(n + 1)}{r^2} \Psi_n,$$

$$\Psi_n\big|_{\tau=0} = \frac{1}{(n + 1) r^n}, \qquad \frac{\partial \Psi_n}{\partial \tau}\bigg|_{\tau=0} = 0, \qquad r > 1,$$

$$\left(\frac{1}{r}\frac{\partial \Psi_n}{\partial r} - \frac{\Psi_n}{r^2}\right)\bigg|_{r=1} = 0, \qquad \lim_{r \to \infty} \Psi_n = 0. \tag{4.16}$$

Let us search for a particular solution of (4.16) in the form

$$\Psi_n(r, \tau) = R_n(r)\, T_n(\tau)\,. \tag{4.17}$$

Designating

$$l_n(y) = -y'' + \frac{n(n+1)}{r^2}\, y$$

and substituting (4.17) into (4.16), we arrive at the following equations for determination of the functions $R_n(r)$ and $T_n(\tau)$:

$$l_n(R_n) = \lambda R_n\,, \qquad T_n''(\tau) + \lambda T_n(\tau) = 0\,; \tag{4.18}$$

here, λ is the eigenvalue of the boundary value problem (4.16).

Let us seek the eigenfunction $R_n(r)$ in the space of the square integrable functions $L_2(1, \infty)$ with the norm

$$\|R_n\|^2 = \int_1^\infty R_n^2(r)\, \mathrm{d}r\,.$$

At that, the boundary condition (4.12), as $r \to \infty$, is satisfied. For an arbitrary number n, we have

$$\Psi_n(0, r) \in L_2(1, \infty)\,.$$

Thus, for the function $R_n(r)$, we get

$$l_n(R_n) = \lambda R_n\,, \qquad R_n'(1) = R_n(1)\,, \qquad R_n(r) \in L_2(1, \infty)\,. \tag{4.19}$$

It is known that the eigenvalue spectrum of the problem (4.19) is continuous (Naimark (1969)).

Let us then assume that $U_n(r, \lambda)$ and $V_n(r, \lambda)$ satisfy the following problems, respectively:

$$l_n(U_n) = \lambda U_n\,, \qquad U_n(1, \lambda) = 1\,, \qquad U_n'(1, \lambda) = U_n(1, \lambda)\,; \tag{4.20}$$
$$l_n(V_n) = \lambda V_n\,, \qquad V_n(1, \lambda) = 0\,, \qquad V_n'(1, \lambda) = -1\,. \tag{4.21}$$

Then, according to Naimark (1969), an arbitrary function $f(r) \in L_2(1, \infty)$ can be represented in the form

$$f(r) = \int_{-\infty}^\infty \varphi(\lambda)\, U_n(r, \lambda)\, \sigma_n'(\lambda)\, \mathrm{d}\lambda\,,$$
$$\varphi(\lambda) = \int_1^\infty f(r)\, U_n(r, \lambda)\, \mathrm{d}r\,, \tag{4.22}$$

where $\sigma_n(\lambda)$ is the spectral function of an operator corresponding to (4.19).

For the self-adjoint differential operators of the second order, we have (Naimark (1969))

$$\sigma_n(\lambda) = \text{const} + \lim_{\epsilon \to +0} \frac{1}{\pi} \int_0^1 \text{Im}\, M_n(\lambda + i\epsilon)\, d\lambda, \qquad (4.23)$$

where $M_n(\lambda)$ is the characteristic function, which can be determined by the following condition:

$$V_n(r, \lambda) + M_n(\lambda)\, U_n(r, \lambda) \in L_2(1, \infty), \qquad \text{Im}\, \lambda > 0. \qquad (4.24)$$

Thus, in order to derive a solution of the problem (4.16), it is sufficient to obtain $M_n(\lambda)$ and $\sigma_n'(\lambda)$. By solving (4.20) and (4.21), we get

$$U_n(r, \lambda) = \frac{\pi\sqrt{r}}{8i} \left\{ \left[H^{(2)}_{n+1/2}(\sqrt{\lambda}) - 2\sqrt{\lambda}\, H^{(2)'}_{n+1/2}(\sqrt{\lambda}) \right] H^{(1)}_{n+1/2}(\sqrt{\lambda}\, r) \right.$$
$$\left. - \left[H^{(1)}_{n+1/2}(\sqrt{\lambda}) - 2\sqrt{\lambda}\, H^{(1)'}_{n+1/2}(\sqrt{\lambda}) \right] H^{(2)}_{n+1/2}(\sqrt{\lambda}\, r) \right\},$$

$$V_n(r, \lambda) = \frac{\pi\sqrt{r}}{4i} \left[-H^{(2)}_{n+1/2}(\sqrt{\lambda})\, H^{(1)}_{n+1/2}(\sqrt{\lambda}\, r) \right.$$
$$\left. + H^{(1)}_{n+1/2}(\sqrt{\lambda})\, H^{(2)}_{n+1/2}(\sqrt{\lambda}\, r) \right], \qquad (4.25)$$

where $H^{(1)}_\nu(x)$ and $H^{(2)}_\nu(x)$ are the Hankel functions of the order ν.

Substituting (4.25) into (4.24) and using the relationships

$$H^{(1)}_{n+1/2}(x) \in L_2(1, \infty), \qquad H^{(2)}_{n+1/2}(x) \notin L_2(1, \infty),$$

(which are valid for $\text{Im}\, x > 0$ (Watson (1945), Kuznetsov (1965))), we obtain

$$M_n(\lambda) = \frac{2 H^{(1)}_{n+1/2}(\sqrt{\lambda})}{H^{(1)}_{n+1/2}(\sqrt{\lambda}) - 2\sqrt{\lambda}\, H^{(1)'}_{n+1/2}(\sqrt{\lambda})}. \qquad (4.26)$$

Taking into account (4.26), from (4.23) we obtain the derivative of the spectral function $\sigma_n'(\lambda)$. As a result, for $\lambda \leq 0$, we have $\sigma_n'(\lambda) \equiv 0$ (the eigenvalue spectrum is positive), and for $\lambda > 0$, we have

$$\sigma_n'(\lambda) = \frac{2}{\pi^2} \left\{ \left[\sqrt{\lambda}\, J_{n-1/2}(\sqrt{\lambda}) - (n+1)\, J_{n+1/2}(\sqrt{\lambda}) \right]^2 \right.$$
$$\left. + \left[\sqrt{\lambda}\, N_{n-1/2}(\sqrt{\lambda}) - (n+1)\, N_{n+1/2}(\sqrt{\lambda}) \right]^2 \right\}^{-1}. \qquad (4.27)$$

When deriving (4.27), we applied the following relationship between the Hankel functions, the Bessel functions $J_{n+1/2}(x)$, and the Neumann functions $N_{n+1/2}(x)$ (Watson (1945), Kuznetsov (1965)) (a bar designates complex conjugation):

$$H^{(1)}_{n+1/2}(x) = J_{n+1/2}(x) + i\, N_{n+1/2}(x),$$
$$H^{(2)}_{n+1/2}(x) = \overline{H^{(1)}_{n+1/2}(x)}. \qquad (4.28)$$

Let us express $U_n(r, \lambda)$ via the real-valued Bessel functions; in accounting for (4.28) and (4.25), we get

$$U_n(r, \lambda) = \frac{\pi \sqrt{r}}{4} \left\{ \left[J_{n+1/2}(\lambda) - 2\sqrt{\lambda}\, J'_{n+1/2}(\sqrt{\lambda}) \right] N_{n+1/2}(\sqrt{\lambda}\, r) \right.$$
$$\left. - \left[N_{n+1/2}(\sqrt{\lambda}) - 2\sqrt{\lambda}\, N'_{n+1/2}(\sqrt{\lambda}) \right] J_{n+1/2}(\sqrt{\lambda}\, r) \right\} . \qquad (4.29)$$

Let us now revert to the problem (4.16). According to (4.22), the solution $\Psi_n(r, \tau)$ can be represented in the following form:

$$\Psi_n(r, \tau) = \int_0^\infty T_n(\tau, \lambda)\, U_n(r, \lambda)\, \sigma'_n(\lambda)\, d\lambda . \qquad (4.30)$$

We can represent the value of the function $\Psi_n(r, \tau)$ for $\tau = 0$ in the form of the integral

$$\Psi_n(r, 0) = \frac{1}{(n+1)\, r^n} \int_0^\infty A_n(\lambda)\, U_n(r, \lambda)\, \sigma'_n(\lambda)\, d\lambda . \qquad (4.31)$$

Then, according to (4.22), we get

$$A_n(\lambda) = \int_1^\infty \Psi_n(r, 0)\, U_n(r, \lambda)\, dr .$$

Calculating the last integral in accounting for (4.29), we obtain $A_n(\lambda) = 1/\lambda$ (Watson (1945)).

As it follows from (4.26) and (4.28), the function $T_n(\tau, \lambda)$ can be determined from the initial value problem

$$T''_n + \lambda T_n = 0 , \qquad T_n(0, \lambda) = \frac{1}{\lambda} , \qquad T'_n(0, \lambda) = 0 . \qquad (4.32)$$

In this case, the solution has the following form:

$$T_n(\tau, \lambda) = \frac{\cos(\sqrt{\lambda}\, \tau)}{\lambda} .$$

Finally, considering the representation of $\xi(r, \tau)$ in the form of a sum of $\Psi_{1n}(r, \tau)$ and $\Phi_{0n}(r)$, from (4.14), (4.15), and (4.30), we obtain

$$\xi(r, \tau) = \frac{1}{r} \int_0^\infty \frac{\cos(\sqrt{\lambda}\, \tau)}{\lambda}\, U_n(r, \lambda)\, \sigma'_n(\lambda)\, d\lambda - \frac{1}{(n+1)\, r^{n+1}} . \qquad (4.33)$$

Taking into consideration a relationship between the transition function and $\xi(r, \tau)$, making the substitution $\sqrt{\lambda} = x$ in (4.33), and taking into account (4.27) and (4.29), we arrive at the following integral representation for $\chi_n(\tau)$:

$$\chi_n(\tau) = -\frac{\partial \xi}{\partial \tau}(1,\,\tau) = \frac{4}{\pi^2} \int_0^\infty \frac{\sin(x\tau)\,\mathrm{d}x}{J_{1n}^2(x) + N_{1n}^2(x)}\,,$$

$$J_{1n}(x) = (n+1)\,J_{n+1/2}(x) - xJ_{n-1/2}(x)\,,$$

$$N_{1n}(x) = (n+1)\,N_{n+1/2}(x) - xN_{n-1/2}(x)\,. \tag{4.34}$$

Let us compare the formula (4.34) with the Fourier transform of the function $\chi_n(\tau)$ with respect to time τ. We can obtain the last one from its Laplace transform (4.10). In order to do this, we should consider that all the positive poles of the image $\chi_n^{(L)}(s)$ are located to the left of the imaginary axis in the complex plane. Then, the function $\chi_n(\tau)$ can be represented in the following form (Krylov, Skoblya (1974), Lavrentev, Shabat (1987)):

$$\chi_n(\tau) = \frac{1}{\pi} \int_0^\infty g(x)\,\cos(x\tau)\,\mathrm{d}x + \frac{i}{\pi} \int_0^\infty h(x)\,\sin(x\tau)\,\mathrm{d}x\,,$$

$$g(x) = \frac{1}{2}\left[\chi_n^{L}(ix) + \chi_n^{L}(-ix)\right]\,,$$

$$h(x) = \frac{1}{2}\left[\chi_n^{L}(ix) - \chi_n^{L}(-ix)\right]\,. \tag{4.35}$$

Making the substitutions $s = ix$ and $s = -ix$ and taking into account the relationships for the Bessel functions (Watson (1945), Kuznetsov (1965))

$$K_{n+1/2}(ix) = -\frac{\pi}{2}\,i^{-n+1}\,e^{-\pi i/4}H_{n+1/2}^{(2)}(x)\,,$$

$$K_{n+1/2}(-ix) = \frac{\pi}{2}\,i^{n+1}\,e^{\pi i/4}H_{n+1/2}^{(1)}(x)\,,$$

we obtain

$$g(x) = \mathrm{Re}\chi_n^{L}(ix)\,, \qquad h(x) = i\,\mathrm{Im}\chi_n^{L}(ix)\,.$$

It follows that

$$\chi_n(\tau) = \frac{2}{\pi^2} \int_0^\infty \frac{\sin(x\tau)\,\mathrm{d}x}{J_{1n}^2(x) + N_{1n}^2(x)}$$
$$+ \frac{1}{\pi} \int_0^\infty \frac{J_{n+1/2}(x)\,J_{1n}(x) + N_{n+1/2}(x)\,N_{1n}(x)}{J_{1n}^2(x) + N_{1n}^2(x)}\,\cos(x\tau)\,\mathrm{d}x\,. \tag{4.36}$$

A comparison of the formulas (4.34) and (4.36) gives us an opportunity to obtain one more integral representation for the transition function:

$$\chi_n(\tau) = \frac{2}{\pi} \int_0^\infty \frac{J_{n+1/2}(x)\,J_{1n}(x) + N_{n+1/2}(x)\,N_{1n}(x)}{J_{1n}^2(x) + N_{1n}^2(x)}\,\cos(x\tau)\,\mathrm{d}x\,. \tag{4.37}$$

As $x \to \infty$, the numerator of the integrand in (4.37), contrary to the numerator of the integrand in (4.34), tends to zero; consequently, we should expect the rapid convergence of (4.37) as compared to (4.34).

Since the denominators of the integrands in (4.34) and (4.37) are proportional to polynomials, the integrals (4.34) and (4.37) can be calculated by means of the theory of residues. Besides, these integral representations can be useful for numerical calculation of $\chi_n(\tau)$ under a high n, when determination of the polynomial roots (the poles) turns out to be complicated.

In order to illustrate application of the formulas (4.34) and (4.37), let us derive $\chi_0(\tau)$ and $\chi_1(\tau)$. Taking into account the expressions for the Bessel functions of a semi-integer index via the elementary functions (Watson (1945), Kuznetsov (1965)) and making the rearrangements under the integral signs, we obtain (Gradshtein, Ryzhik (1971))

$$\chi_0(\tau) = \frac{2}{\pi} \int_0^\infty \frac{x \sin(x\tau)}{1 + x^2} \, \mathrm{d}x = \frac{2}{\pi} \int_0^\infty \frac{\cos(x\tau)}{1 + x^2} \, \mathrm{d}x = \mathrm{e}^{-\tau} \, ,$$

$$\chi_1(\tau) = \frac{2}{\pi} \int_0^\infty \frac{x^3 \sin(x\tau)}{4 + x^4} \, \mathrm{d}x = \frac{2}{\pi} \int_0^\infty \frac{2 + x^2}{4 + x^4} \cos(x\tau) \, \mathrm{d}x = \mathrm{e}^{-\tau} \cos\tau \, ;$$

these formulas correspond to the expressions presented in Kharkevich (1950) and Grigolyuk, Gorshkov (1976).

4.2 Propagation of Disturbances from a Cavity

Using the designations and dimensionless values introduced before, let us consider the following problem. A spherical cavity of the unit radius is located in an infinite linearly elastic space. The axially symmetric kinematic or force disturbances are applied to its surface. At the initial time $\tau = 0$, the medium is at rest.

As it follows from Sect. 1.1, the equation of motion of the medium with respect to the elastic potentials of the displacements φ_0 and ψ_0 and the corresponding initial conditions have the following form:

$$\gamma_0^2 \frac{\partial^2 \varphi_0}{\partial \tau^2} = \Delta \varphi_0 \, ,$$

$$\eta_0^2 \frac{\partial^2 \psi_0}{\partial \tau^2} = \Delta \psi_0 - \frac{\psi_0}{r^2 \sin \theta} \, ,$$

$$\varphi_0|_{\tau=0} = \left. \frac{\partial \varphi_0}{\partial \tau} \right|_{\tau=0} = \psi_0|_{\tau=0} = \left. \frac{\partial \psi_0}{\partial \tau} \right|_{\tau=0} = 0 \, . \tag{4.38}$$

We shall consider two types of boundary conditions.

Problem A (force conditions). The following stresses are predetermined at the cavity's surface:

$$\sigma_{rr}^{(0)}\Big|_{r=1} = p(\tau,\,\theta)\,, \qquad \sigma_{r\theta}^{(0)}\Big|_{r=1} = q(\tau,\,\theta)\,. \tag{4.39}$$

Problem B (kinematic conditions). The following displacements are predetermined at the cavity's surface:

$$u_0|_{r=1} = U(\tau,\,\theta)\,, \qquad v_0|_{r=1} = V(\tau,\,\theta)\,. \tag{4.40}$$

At infinity, as we did before, we state the absence of disturbances.

Representing the right-hand sides of the conditions (4.39) and (4.40) in the form of the series in terms of the Legendre polynomials and the Gegenbauer polynomials

$$p(\tau,\,\theta) = \sum_{n=0}^{\infty} p_n(\tau)\, P_n(\cos\theta)\,,$$

$$q(\tau,\,\theta) = -\sin\theta \sum_{n=1}^{\infty} q_n(\tau)\, C_{n-1}^{3/2}(\cos\theta)\,,$$

$$U(\tau,\,\theta) = \sum_{n=0}^{\infty} U_n(\tau)\, P_n(\cos\theta)\,,$$

$$V(\tau,\,\theta) = -\sin\theta \sum_{n=1}^{\infty} V_n(\tau)\, C_{n-1}^{3/2}(\cos\theta)\,, \tag{4.41}$$

and expanding the stresses $\sigma_{rr}^{(0)}$ and $\sigma_{r\theta}^{(0)}$, the displacements u_0 and v_0, and the potentials φ_0 and ψ_0 into the corresponding series, we obtain the following boundary value problem:

$$\gamma_0^2 \frac{\partial^2 \varphi_{0n}}{\partial \tau^2} = \Delta_n \varphi_{0n}\,,$$

$$\eta_0^2 \frac{\partial^2 \psi_{0n}}{\partial \tau^2} = \Delta_n \psi_{0n}\,,$$

$$\varphi_{0n}|_{\tau=0} = \frac{\partial \varphi_{0n}}{\partial \tau}\bigg|_{\tau=0} = \psi_{0n}|_{\tau=0} = \frac{\partial \psi_{0n}}{\partial \tau}\bigg|_{\tau=0} = 0\,; \tag{4.42}$$

the boundary conditions for Problem A are

$$\sigma_{rrn}^{(0)}\Big|_{r=1} = p_n(\tau)\,, \qquad \sigma_{r\theta n}^{(0)}\Big|_{r=1} = q_n(\tau) \tag{4.43}$$

and for Problem B are

$$u_{0n}|_{r=1} = U_n(\tau), \qquad v_{0n}|_{r=1} = V_n(\tau). \tag{4.44}$$

Let us note that Problem B can be solved using the transition functions and the convolution introduced in Sect. 4.1. However, for the sake of unification, we shall solve Problem B as we did in the case of Problem A, that is, using directly the concept of *generalized spherical waves*.

From the absence of disturbances at infinity it follows that the converging waves should not be represented in the solution. Using the representation of the general solution of the wave equation (1.114) and taking into account(3.38) and (3.45), let us represent the potentials and the components of the stress–strain state of the medium in the space of the Laplace transform in the following form ($\gamma_0 = 1$):

$$\varphi_{0n}^{L}(r, s) = \frac{1}{r^{n+1}} R_{n0}(rs) f_n^{L}(s) e^{-(r-1)s},$$

$$\psi_{0n}^{L}(r, s) = \frac{1}{r^{n+1}} R_{n0}(\eta_0 rs) g_n^{L}(s) e^{-\eta_0(r-1)s},$$

$$u_{0n}^{L}(r, s) = -\frac{1}{r^{n+2}} \Big[R_{n1}(rs) f_n^{L}(s) e^{-(r-1)s}$$
$$+ n(n+1) g_n^{L}(s) R_{n0}(\eta_0 rs) e^{-\eta_0(r-1)s} \Big],$$

$$v_{0n}^{L}(r, s) = \frac{1}{r^{n+2}} \Big[R_{n0}(rs) f_n^{L}(s) e^{-(r-1)s}$$
$$+ R_{n3}(\eta_0 rs) g_n^{L}(s) e^{-\eta_0(r-1)s} \Big],$$

$$\sigma_{rrn}^{(0)L}(r, s) = \frac{\beta_0}{r^{n+3}} \Big[Q_{n1}^{(0)}(rs) f_n^{L}(s) e^{-(r-1)s}$$
$$+ n(n+1) Q_{n2}^{(0)}(\eta_0 rs) g_n^{L}(s) e^{-\eta_0(r-1)s} \Big],$$

$$\sigma_{r\theta n}^{(0)L}(r, s) = \frac{\beta_0}{r^{n+3}} \Big[Q_{n2}^{(0)}(rs) f_n^{L}(s) e^{-(r-1)s}$$
$$+ Q_{n3}^{(0)}(\eta_0 rs) g_n^{L}(s) e^{-\eta_0(r-1)s} \Big]. \tag{4.45}$$

As it follows from Chap. 1, the relationships (4.45) determine a solution of (4.42) with the corresponding initial conditions. The arbitrary functions $f_n(\tau)$ and $g_n(\tau)$ should be derived from the boundary conditions (4.43) or (4.44). Substituting (4.45) into the boundary conditions (4.43), in the case of Problem A, we get

$$\beta_0 \Big[Q_{n1}^{(0)}(s) f_n^{L}(s) + n(n+1) Q_{n2}^{(0)}(\eta_0 s) g_n^{L}(s) \Big] = p_n^{L}(s),$$

$$\beta_0 \Big[Q_{n2}^{(0)}(s) f_n^{L}(s) + Q_{n3}^{(0)}(\eta_0 s) g_n^{L}(s) \Big] = g_n^{L}(s). \tag{4.46}$$

We can write the solution of this algebraic problem as follows:

$$f_n^{\rm L}(s) = \frac{Y_{n11}(s)}{X_{n0}(s)} \, p_n^{\rm L}(s) + \frac{Y_{n12}(s)}{X_{n0}(s)} \, q_n^{\rm L}(s) \,,$$

$$g_n^{\rm L}(s) = \frac{Y_{n21}(s)}{X_{n0}(s)} \, p_n^{\rm L}(s) + \frac{Y_{n22}(s)}{X_{n0}(s)} \, q_n^{\rm L}(s) \,. \tag{4.47}$$

At that, the polynomials $Y_{nij}(s)$ and $X_{n0}(s)$ have the following form:

$$X_{n0}(s) = \beta_0 D_{n2}^{(0)}(s, \eta_0 s) \,, \qquad Y_{n11}(s) = Q_{n3}^{(0)}(\eta_0 s) \,,$$

$$Y_{n12}(s) = -n(n+1) \, Q_{n2}^{(0)}(\eta_0 s) \,, \qquad Y_{n21}(s) = -Q_{n2}^{(0)}(s) \,,$$

$$Y_{n22}(s) = Q_{n1}^{(0)}(s) \,. \tag{4.48}$$

Here, the polynomials $Q_{ni}^{(0)}(s)$ and $D_{n2}^{(0)}(x, y)$ are already determined by the expressions (3.22) and (3.71).

Similarly, for Problem B, from the conditions (4.44) we obtain that the functions $f_n^{\rm L}(s)$ and $g_n^{\rm L}(s)$ can be determined as follows:

$$f_n^{\rm L}(s) = \frac{Y_{n11}(s)}{X_{n0}(s)} \, U_n(s) + \frac{Y_{n12}(s)}{X_{n0}(s)} \, V_n(s) \,,$$

$$g_n^{\rm L}(s) = \frac{Y_{n21}(s)}{X_{n0}(s)} \, U_n(s) + \frac{Y_{n22}(s)}{X_{n0}(s)} \, V_n(s) \,,$$

$$X_{n0}(s) = -D_{n3}(s, \eta_0 s) \,,$$

$$D_{n3}(y, z) = R_{n1}(y) \, R_{n3}(z) - n(n+1) \, R_{n0}(y) \, R_{n0}(z) \,,$$

$$Y_{n11}(s) = R_{n3}(\eta_0 s) \,, \qquad Y_{n12}(s) = n(n+1) \, R_{n0}(\eta_0 s) \,,$$

$$Y_{n21}(s) = -R_{n0}(s) \,, \qquad Y_{n22}(s) = -R_{n1}(s) \,. \tag{4.49}$$

Calculations of the originals of the functions $f_n^{\rm L}(s)$ and $g_n^{\rm L}(s)$ by the formulas (4.47) is not difficult since the coefficients at $p_n^{\rm L}(s)$ and $g_n^{\rm L}(s)$ are the rational functions. The originals of the latter, similarly to what we did in the previous chapters, can be derived directly by means of residues. The followed convolution of these functions and $p_n(\tau)$ and $q_n(\tau)$ allows us to determine $f_n(\tau)$ and $g_n(\tau)$. Using the formulas (4.45) and making summation of the corresponding series, we can determine the stress–strain state of the medium, when the disturbances propagate from a spherical cavity.

Let us, now, demonstrate that it is easy to derive a solution of the problem of propagation of the waves in an acoustic space having a spherical cavity using the solution obtained. In order to do this, it is necessary to set $q = 0$, $\kappa_0 = 1$ (or $\eta_0 \to \infty$), and $g_n(\tau) \equiv 0$.

Taking into account the definitions (1.126), (3.22), and (3.46) of the polynomials $R_{ni}(s)$ and $Q_{ni}^{(0)}(s)$, it is easy to prove that

$$R_{n0}(s) = s^{n+1} + O(s^n), \qquad R_{n3}(s) = s^{n+1} + O(s^n),$$

$$Q_{n1}^{(0)}(s) = s^{n+2} + A_{n1}s^{n+1} + A_{n2}s^n + O(s^{n-1}),$$

$$Q_{n2}^{(0)}(s) = \frac{2}{\eta_0^2} \left[s^{n+1} + (B_{n1} + 1) s^n + (B_{n2} + A_{n1}) s^{n-1} + O(s^{n-2}) \right],$$

$$Q_{n3}^{(0)}(s) = \frac{1}{\eta_0^3} \left\{ s^{n+2} + C_{n1}s^{n+1} \right.$$

$$\left. + \left[C_{n2} + (n + 2)(n - 1) \right] s^n + O(s^{n-1}) \right\}, \qquad s \to \infty. \tag{4.50}$$

Then, as it follows from (4.48) and (4.50), the following asymptotic estimates are valid:

$$Y_{n11}(s) = \eta_0^n s^{n+2} + O(\eta_0^{n-1}), \qquad Y_{n21}(s) = O(1),$$

$$Y_{n12}(s) = -2n(n+1)s^{n+1}\eta_0^{n-1} + O(\eta_0^{n-2}), \qquad Y_{n22}(s) = O(1),$$

$$D_{n2}^{(0)}(s, \eta_0 s) = \eta_0^n s^{n+4} R_{n0}(s) + O(\eta_0^{n+1}),$$

$$X_{n0}(s) = \beta_0 \eta_0^n s^{n+4} R_{n0}(s) + O(\eta_0^{n-1}), \qquad \eta_0 \to \infty. \tag{4.51}$$

Let us consider Problem A for an acoustic medium. Making a passage to the limit in (4.47), as $\eta_0 \to \infty$, we obtain

$$f_n^L(s) = \frac{p_n^L(s)}{\beta_0 s^2 R_{n0}(s)}, \qquad g_n^L(s) \equiv 0. \tag{4.52}$$

The last equality demonstrates the absence of shear waves; this fact corresponds to the model of an acoustic medium.

A similar passage to the limit for Problem B gives us the following relationships:

$$D_{n3}(s, \eta_0 s) = \eta_0^{n+1} s^{n+1} R_{n1}(s) + O(\eta_0^n),$$

$$X_{n12}(s) = -\eta_0^{n+1} s^{n+1} R_{n1}(s) + O(\eta_0^n),$$

$$Y_{n11}(s) = \eta_0^{n+1} s^{n+1} + O(\eta_0^n),$$

$$Y_{n12}(s) = n(n+1)\eta_0^n s^n + O(\eta_0^{n-1}),$$

$$Y_{n21}(s) = O(1), \qquad Y_{n22}(s) = O(1), \qquad \eta_0 \to \infty,$$

$$f_n^L(s) = -\frac{U_n(s)}{R_{n1}(s)}, \qquad g_n^L(s) \equiv 0. \tag{4.53}$$

A comparison of the formulas for $f_n^L(s)$ in (4.53) with the expressions (4.5) shows us that there is a relationship between Problem B and that considered in Sect. 4.1. It is evident that they coincide, when $U_n(\tau) = H(\tau)$.

In order to calculate the originals using (4.47) and (4.49), it is necessary to know the roots of the polynomials $X_{n0}(s)$. They can be calculated using any method known. Considering Problem B, let us demonstrate that the polynomial $X_{n0}(s)$ has the root $s = 0$ of the order 2. In fact, using the properties of the coefficients of the polynomials $R_{ni}(s)$, it is easy to obtain

$$X_{n0}(0) = -B_{n,\,n+1}(B_{n,\,n+1} - A_{nn}) + n(n+1)\,A_{nn}^2 = 0\,,$$
$$X'_{n0}(0) = -R'_{n1}(0)\,R_{n3}(0) - R_{n1}(0)\,R'_{n3}(0) + 2n(n+1)\,R_{n0}(0)\,R'_{n0}(0)$$
$$= -2R_{n1}(0)\,R_{n3}(0) + 2n(n+1)\,R_{n0}(0) = 0\,.$$

Let us note that the problems of the mixed kinematic and force conditions at a cavity's boundary can be studied similarly; however, the practical importance of such problems is quite minor. The possible variants of the mixed kinematic and force conditions are as follows:

$$\left.\sigma_{rr}^{(0)}\right|_{r=1} = p(\tau,\,\theta)\,, \qquad \left.v_0\right|_{r=1} = V(\tau,\,\theta)$$

or

$$\left.u_0\right|_{r=1} = U(\tau,\,\theta)\,, \qquad \left.\sigma_{r\theta}\right|_{r=1} = q(\tau,\,\theta)\,.$$

Let us consider, for example, propagation of the disturbances from a spherical cavity in the elastic medium (steel, $\gamma_0 = 1$ and $\eta_0 = 1.871$), when the radial stresses are predetermined at the boundary's surface $r = 1$ (Problem A):

$$\left.\sigma_{rr}^{(0)}\right|_{r=1} = \frac{1}{2}(1 + \cos\theta)\,f(\tau) = \frac{1}{3}\left[P_0(\cos\theta) + 2P_2(\cos\theta)\right]\,f(\tau)\,,$$
$$\left.\sigma_{r\theta}^{(0)}\right|_{r=1} = 0\,.$$

We shall choose the function $f(\tau)$ in the form of the triangle pulse

$$f(\tau) = -2\left[\tau_+ - 2\left(\tau - \frac{1}{2}\right)_+ + (\tau - 1)_+\right] = -\frac{2}{s}\left(1 - e^{-s/2} + e^{-s}\right)\,.$$

Solution of a similar problem is presented in Suzuki (1967).

The results of calculations of the stresses at the distance $r = 1.514$ from the cavity's center, when $\theta = 0$, $\theta = \pi/4$, and $\theta = \pi/2$, are presented in Fig. 4.1a, b, c, respectively. For $\theta = 0$ and $\theta = \pi/4$, the radial stresses σ_{rr} prevail. For $\theta = \pi/2$, the other components of the stress tensor get the higher absolute values.

4.3 Diffraction of Waves by a Cavity or by an Immovable Spherical Inclusion

Let us assume that in an infinite linearly elastic medium there exists a spherical cavity (Problem A) or an immovable spherical inclusion (Problem B). We shall consider a plane or spherical incoming compression wave interacting with the cavity or the inclusion. The potential of the wave $\varphi_{0s}(r,\,\theta,\,\tau)$ can be determined by the relationships (3.9) or (3.10), respectively (see Chap. 3).

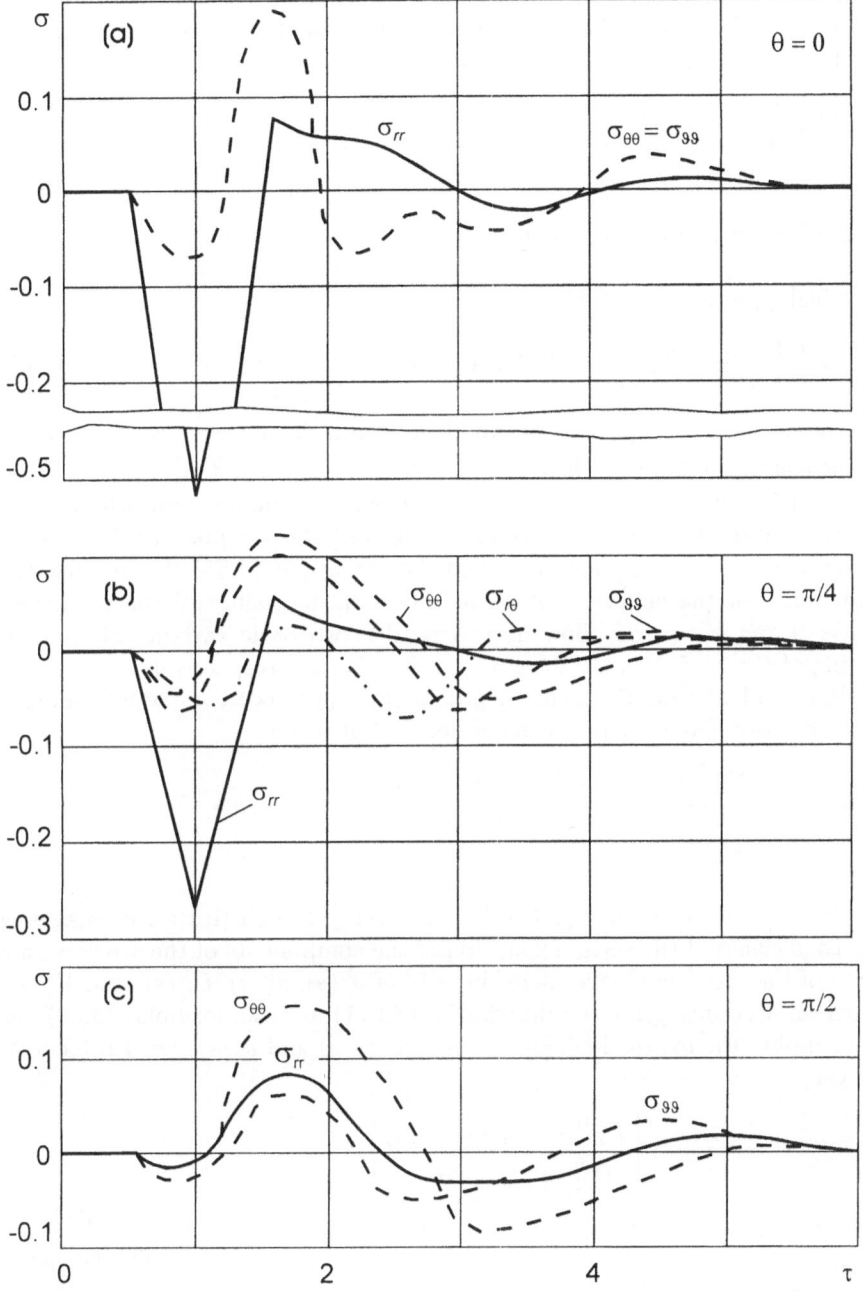

Fig. 4.1. The stresses at the point $r = 1.514$ for the axially symmetric Problem A of propagation of disturbances from a spherical cavity

The mathematical statement of the problem is determined by the equations and initial conditions (4.38), the condition of the absence of disturbances at infinity, and the boundary conditions at the sphere's surface $r = 1$; for Problem A, the boundary conditions are

$$\sigma_{rr}^{(0)}\Big|_{r=1} + \sigma_{rr\,s}^{(0)}\Big|_{r=1} = 0,$$

$$\sigma_{r\theta}^{(0)}\Big|_{r=1} + \sigma_{r\theta\,s}^{(0)}\Big|_{r=1} = 0; \tag{4.54}$$

in the case of Problem B, we get

$$u_0\big|_{r=1} + u_{0\,s}\big|_{r=1} = 0,$$

$$\sigma_{r\theta}^{(0)}\Big|_{r=1} + \sigma_{r\theta\,s}^{(0)}\Big|_{r=1} = k_{10}(v_0 + v_{0\,s})\big|_{r=1}. \tag{4.55}$$

Here, $\sigma_{rr\,s}^{(0)}$, $\sigma_{r\theta\,s}^{(0)}$, $u_{0\,s}$, and $v_{0\,s}$ are the components of the stress–strain state of the elastic medium produced by the incoming wave and $\sigma_{rr}^{(0)}$, $\sigma_{r\theta}^{(0)}$, u_0, and v_0 are the additional stresses and displacements of the medium produced by the wave diffraction. At that, the conditions (4.54) correspond to the absence of stresses at the cavity's surface and the conditions (4.55) characterize the immobility of the spherical obstacle. The limiting values of the coefficient $k_{10} = 0$ and $k_{10} = \infty$ allow us to consider two basic variants of contact: sliding of the surfaces and absolutely stiff adhesion, respectively.

It is evident that Problems A and B (for $k_{10} = \infty$) completely coincide with the corresponding problems of Sect. 4.2 if to set

$$p = -\sigma_{rr\,s}^{(0)}\Big|_{r=1}, \qquad q = -\sigma_{r\theta\,s}^{(0)}\Big|_{r=1},$$

$$U = -u_{0\,s}\big|_{r=1}, \qquad V = -v_{0\,s})\big|_{r=1}.$$

At that, in (4.47) and (4.49), it is necessary to substitute the images of the coefficients of the series expansion of the components of the stress–strain state of the incoming wave (3.25) instead of $p_n^L(s)$, $q_n^L(s)$, $U_n^L(s)$, and $V_n^L(s)$. Then, in accounting for the designations (3.71) and the formulas (3.68), we can simplify the expressions for the images $f_n^L(s)$ and $g_n^L(s)$. For Problem A, we get

$$f_n^L(s) = -E_0(s)\left[\frac{D_{n2}^{(0)}(-s, \eta_0 s)}{D_{n2}^{(0)}(s, \eta_0 s)} - e^{-2s}\right],$$

$$g_n^L(s) = 2\,(-1)^n(1 - \kappa_0)\,s^{2n+1}[s^2 + (1 - \kappa_0)\,(n - 1)\,(n + 2)]\,\frac{E_0(s)}{D_{n2}^{(0)}(s, \eta_0 s)}. \tag{4.56}$$

Under arbitrary values of the coefficients k_{10}, Problem B can be derived from the solutions obtained in Sect. 4.2; this problem requires a separate study. Deriving a solution by analogy with that in the previous section, for an arbitrary $n \geq 0$, we obtain

$$f_n^L(s) = -\frac{Y_{n11}(s)}{X_{n0}(s)} \left. u_{0ns}^L \right|_{r=1} - \frac{Y_{n12}(s)}{X_{n0}(s)} \left. (\sigma_{r\theta ns}^{(0)L} - k_{10}v_{0ns}^L) \right|_{r=1},$$

$$g_n^L(s) = -\frac{Y_{n21}(s)}{X_{n0}(s)} \left. u_{0ns}^L \right|_{r=1} - \frac{Y_{n22}(s)}{X_{n0}(s)} \left. (\sigma_{r\theta ns}^{(0)L} - k_{10}v_{0ns}^L) \right|_{r=1},$$

$$X_{n0}(s) = -\beta_0 D_{n1}^{(0)}(s, \eta_0 s) + k_{10}D_{n3}(s, \eta_0 s),$$

$$Y_{n11}(s) = \beta_0 Q_{n3}^{(0)}(\eta_0 s) - k_{10}R_{n3}(\eta_0 s), \qquad Y_{n12}(s) = n(n+1)R_{n0}(\eta_0 s),$$

$$Y_{n21}(s) = -\beta_0 Q_{n2}^{(0)}(s) + k_{10}R_{n0}(s), \qquad Y_{n22}(s) = -R_{n1}(s), \qquad (4.57)$$

where $D_{n1}^{(0)}(x, y)$ are determined by (3.71).

Substituting the functions u_{0ns}^L, v_{0ns}^L, and $\sigma_{r\theta ns}^{(0)}$ (3.25) into (4.57), in accounting for (3.86), we obtain

$$f_n^L(s) = E_0(s) \left[\frac{\beta_0 D_{n1}^{(0)}(-s, \eta_0 s) - k_{10}D_{n3}(-s, \eta_0 s)}{X_{n0}(s)} - e^{-2s} \right],$$

$$g_n^L(s) = 2(-1)^n s^{2n+1} E_0(s) \frac{\beta_0(1 - \kappa_0) - k_{10}}{X_{n0}(s)}. \qquad (4.58)$$

Let us consider two cases of contact conditions.

(1) A stiff adhesion ($k_{10} = \infty$):

$$f_n^L(s) = -E_0(s) \left[\frac{D_{n3}(-s, \eta_0 s)}{D_{n3}(s, \eta_0 s)} - e^{-2s} \right],$$

$$g_n^L(s) = 2(-1)^{n+1} s^{2n+1} \frac{E_0(s)}{D_{n3}(s, \eta_0 s)}. \qquad (4.59)$$

(2) A sliding ($k_{10} = 0$):

$$f_n^L(s) = -E_0(s) \left[\frac{D_{n1}^{(0)}(-s, \eta_0 s)}{D_{n1}^{(0)}(s, \eta_0 s)} - e^{-2s} \right],$$

$$g_n^L(s) = 2(-1)^{n+1} s^{2n+1} \frac{(1 - \kappa_0) E_0(s)}{D_{n1}^{(0)}(s, \eta_0 s)}. \qquad (4.60)$$

Making a passage to the limit, as $\eta \to \infty$, in (4.56) and taking into account the asymptotic relationship for $D_{n2}^{(0)}$ in (4.51), for a cavity in an acoustic medium (Problem A), we obtain

$$f_n^L(s) = -E_0(s) \left[\frac{R_{n0}(-s)}{R_{n0}(s)} - e^{-2s} \right], \qquad g_n^L(s) \equiv 0. \qquad (4.61)$$

For an immovable sphere surrounded by an ideal fluid, in accounting for the representation (4.53) for D_{n3} and the similarly derived formula for $D_{n1}^{(0)}$,

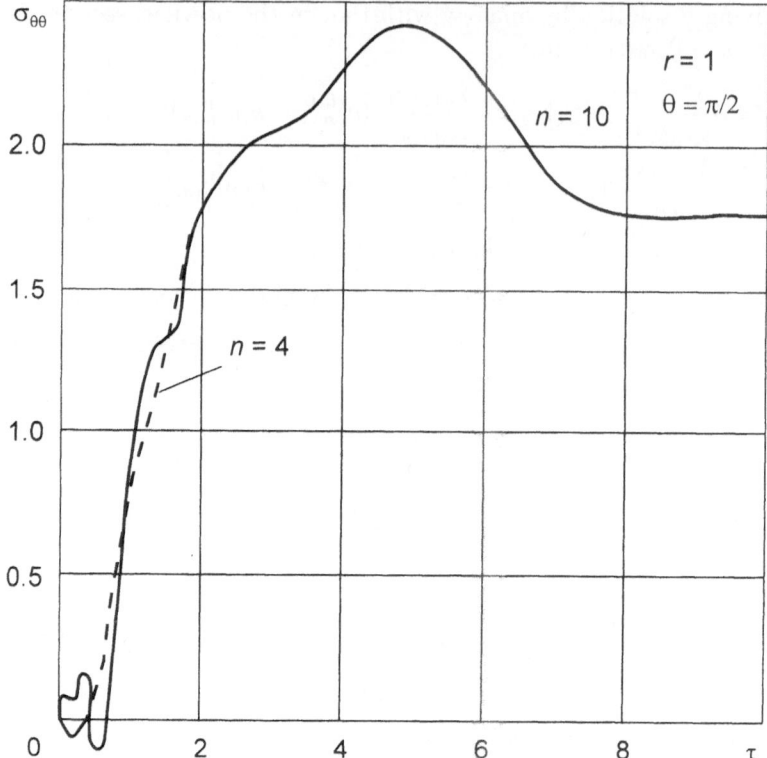

Fig. 4.2. Diffraction of a plane wave by a spherical cavity

$$D_{n1}^{(0)}(\pm s, \eta_0 s) = \eta_0^n s^{n+2} R_{n1}(\pm s) + O(\eta_0^{n-1}), \qquad \eta_0 \to \infty, \tag{4.62}$$

any one of the formulas (4.58)–(4.60) gives us the following solution:

$$f_n^{\mathrm{L}}(s) = -E_0(s)\left[\frac{R_{n1}(-s)}{R_{n1}(s)} - e^{-2s}\right], \qquad g_n^{\mathrm{L}}(s) \equiv 0. \tag{4.63}$$

Let us notice that inversion of the images of the components of the stress–strain state of the medium surrounding the obstacle should be made similarly to that as before.

Let us present the results of calculation of the total stress $\sigma_{\theta\theta} = \sigma_{\theta\theta}^{(0)} + \sigma_{\theta\theta\,s}^{(0)}$ at the cavity's wall $r = 1$, when the angle $\theta = \pi/2$ (Fig. 4.2). Here, we have assumed that the elastic medium is characterized by Poisson's ratio $\nu_0 = 1/3$ (at that, $\gamma_0 = 1$ and $\eta_0 = 2$). The law of variation in time of the stresses in the incoming wave has been assumed to have a form of the Heaviside function

$$f(\tau) = H(\tau), \qquad f^{\mathrm{L}}(s) = \frac{1}{s}.$$

A similar problem was studied in Huang, Wang (1972).

In Fig. 4.2, we plot the curves corresponded to the partial sums of the series (3.6) for $n = 4$ (the dashed curve) and for $n = 10$ (the solid curve). The maximum difference between the results takes place in the vicinity of the elastic wave front. When $\tau > 2$, the curves practically coincide. When farther away from the front, the rate of convergence of the series increases.

The stress $\sigma_{\theta\theta}$ takes its maximum value at $\tau \approx 5$, when the incoming wave front has already passed by. For $\tau > 8$, $\sigma_{\theta\theta}$ takes a practically steady state value. The calculations have shown that this stedy state value is ≈ 1.781.

4.4 The Resultant Force at an Immovable Sphere

Keeping within the framework of the problem of diffraction of the elastic waves by an immovable spherical obstacle considered in Sect. 4.3 (Problem B), let us determine a resultant force applied to the sphere by the surrounding medium. Because of the problem symmetry, only the projection of the resultant force onto the axis of symmetry R_x differs from zero. This dimensionless component

$$\tilde{R}_x = \frac{R_x}{(\lambda_* + 2\mu_*)\, b^2}$$

can be calculated as follows (a tilde is omitted):

$$R_x = -\iint_S (\mathbf{T}_n, \mathbf{e}_1)\, dS, \qquad \mathbf{n} = -\mathbf{e}_r. \tag{4.64}$$

Here, $\mathbf{e}_1$ is the unit vector of the x axis, $\mathbf{n}$ is the unit vector of a normal to the surface of spherical cavity, $\mathbf{e}_r$ is the unit vector of the spherical coordinate system, and $\mathbf{T}_n$ is the vector of contact pressure in the elastic medium; according to (1.14), we get

$$\mathbf{T}_n = -\left.\sigma_{rr}\right|_{r=1} \mathbf{e}_r - \left.\sigma_{r\theta}\right|_{r=1} \mathbf{e}_\theta - \left.\sigma_{r\vartheta}\right|_{r=1} \mathbf{e}_\vartheta,$$

$$\sigma_{\alpha\beta} = \sigma_{\alpha\beta}^{(0)} + \sigma_{\alpha\beta\, s}^{(0)}. \tag{4.65}$$

Taking into account the direction of the x axis and the direction of measurement of the angle θ, let us represent the relationship between the Cartesian and spherical coordinates of the position vector as follows (see Fig. 1.1):

$$\mathbf{r} = x\mathbf{e}_1 + y\mathbf{e}_2 + z\mathbf{e}_3,$$

$$x = -r\cos\theta, \qquad y = r\sin\theta\cos\vartheta, \qquad z = r\sin\theta\sin\vartheta. \tag{4.66}$$

At that, the basis in the Cartesian system $(\mathbf{e}_1, \mathbf{e}_2, \mathbf{e}_3)$ and the basis in the spherical coordinate system $(\mathbf{e}_r, \mathbf{e}_\theta, \mathbf{e}_\vartheta)$ are interrelated by the following formulas:

$$\mathbf{e}_1 = -\cos\theta\,\mathbf{e}_r + \sin\theta\,\mathbf{e}_\theta\,,$$

$$\mathbf{e}_2 = \sin\theta\cos\vartheta\,\mathbf{e}_r + \cos\theta\cos\vartheta\,\mathbf{e}_\theta - \sin\vartheta\,\mathbf{e}_\vartheta\,,$$

$$\mathbf{e}_3 = \sin\theta\sin\vartheta\,\mathbf{e}_r + \cos\theta\sin\vartheta\,\mathbf{e}_\theta + \cos\vartheta\,\mathbf{e}_\vartheta\,. \tag{4.67}$$

Taking into account (4.64), (4.65), and (4.67), we obtain

$$R_x = \iint_S \left[-\sigma_{rr}|_{r=1}\cos\theta + \sigma_{r\theta}|_{r=1}\sin\theta \right] dS\,. \tag{4.68}$$

Making a transition to a double integral in (4.68) in accounting for the fact that $dS = \sin\theta\,d\theta\,d\vartheta$, we finally get

$$R_x = \int_0^{2\pi} d\vartheta \int_0^\pi \left[-\sigma_{rr}|_{r=1}\cos\theta + \sigma_{r\theta}|_{r=1}\sin\theta \right] \sin\theta\,dS\,. \tag{4.69}$$

Let us substitute the expansions of the stresses into the integral (4.69). At that, let us also take into account the axial symmetry of the problem (that is, the independence of the integrands from the angle ϑ) and the orthogonality of the Legendre polynomials and the Gegenbauer polynomials (Gradshtein, Ryzhik (1971)), that is,

$$\int_0^\pi P_n(\cos\theta)\,P_m(\cos\theta)\,\sin\theta\,d\theta = \frac{2}{2n+1}\,\delta_{mn}\,,$$

$$\int_0^\pi C_n^{3/2}(\cos\theta)\,C_m^{3/2}(\cos\theta)\,\sin^3\theta\,d\theta = \frac{2\,(n+1)\,(n+2)}{2n+3}\,\delta_{mn}\,,$$

$$P_1(\cos\theta) = \cos\theta\,, \qquad C_0^{3/2}(\cos\theta) = 1\,, \tag{4.70}$$

where δ_{mn} is the Kronecker delta.

Then, we can find out that the force R_x is determined only by the first form of the series expansion of the functions sought; thus, we get

$$R_x = \frac{4\pi}{3}\left[-\sigma_{rr1}^{(0)} - \sigma_{rr1\,s}^{(0)} + 2\,(\sigma_{r\theta1}^{(0)} + \sigma_{r\theta1\,s}^{(0)}) \right]_{r=1}\,. \tag{4.71}$$

Substituting the values of the functions $f_n^L(s)$ and $g_n^L(s)$ from (4.58) into the expressions for the images of the stresses (4.45) for $n = 1$ and taking into account the formula for the stresses in the incoming wave (3.25), we obtain

$$\sigma_{rr1}^{(0)L}(1,\,s) + \sigma_{rr1\,s}^{(0)L}(1,\,s)$$

$$= \beta_0 E_0(s)\left[Q_{11}^{(0)}(-s) + Q_{11}^{(0)}(s)\,\frac{\beta_0 D_{11}^{(0)}(-s,\,\eta_0 s) - k_{10}D_{13}(-s,\,\eta_0 s)}{X_{10}(s)} \right.$$

$$\left. - 4Q_{12}^{(0)}(\eta_0 s)\,\frac{\beta_0(1 - \kappa_0) - k_{10}}{X_{10}(s)} \right],$$

$$\sigma_{r\theta 1}^{(0)\mathrm{L}}(1,\,s) + \sigma_{r\theta 1\,s}^{(0)\mathrm{L}}(1,\,s)$$

$$= \beta_0 E_0(s)\left[Q_{12}^{(0)}(-s) + Q_{12}^{(0)}(s)\,\frac{\beta_0 D_{11}^{(0)}(-s,\,\eta_0 s) - k_{10}D_{13}(-s,\,\eta_0 s)}{X_{10}(s)}\right.$$

$$\left. - 2s^3 Q_{13}^{(0)}(\eta_0 s)\,\frac{\beta_0(1-\kappa_0) - k_{10}}{X_{10}(s)}\right]. \quad (4.72)$$

Calculating now R_x by means of (4.71) and (4.72), taking into account the definition (4.53) of the polynomials $D_{n3}(y,\,z)$ and the formulas (3.68), and making the necessary Laplace transform, we finally get

$$R_x^{\mathrm{L}}(s) = \frac{8\pi}{3}\,\beta_0 E_0(s)\,\frac{s^3 P(s)}{\eta_0^2 R(s)}\,, \qquad P(s) = \eta_0^2 k_{10} P_1(s) - \beta_0 P_2(s)\,,$$

$$R(s) = \eta_0^2 k_{10} R_1(s) - \beta_0 R_2(s)\,, \qquad P_1(s) = \eta_0^4 s^2 + (2 + \eta_0^2)\,(\eta_0 s + 1)\,,$$

$$P_2(s) = \eta_0^5 s^3 + 3\eta_0^4 s^2 + 2\,(2 + \eta_0^2)\,(\eta_0 s + 1)\,,$$

$$R_1(s) = G_1(s;\,1,\,\eta_0)\,, \qquad R_2(s) = G_2(s;\,1,\,\eta_0)\,,$$

$$D_{13}(\alpha s,\,\beta s) = s^2 G_1(s;\,\alpha,\,\beta)\,, \qquad D_{11}^{(i)}(\alpha s,\,\beta s) = \eta_i^{-2} s^2 G_2(s;\,\alpha,\,\beta)\,,$$

$$G_1(s;\,\alpha,\,\beta) = \alpha^2\beta^2 s^2 + \alpha\beta(\alpha + 2\beta)\,s + \alpha^2 + 2\beta^2\,,$$

$$G_2(s;\,\alpha,\,\beta) = \alpha^2\beta^3 s^3 + \alpha\beta^2(3\alpha + 2\beta)\,s^2$$

$$+ 2\beta(\alpha^2 + 3\alpha\beta + \beta^2)\,s + 2\,(\alpha^2 + 3\beta^2). \quad (4.73)$$

The function $E_0(s)$ determines the form of the wave (plane or spherical) and the law of variation in time of the stress–strain state behind the front. This function is defined by (3.25) for $n = 1$. From (4.73) it is possible to obtain an expression for $R_x^{\mathrm{L}}(s)$ for the cases of stiff adhesion ($k_{10} = \infty$) and free sliding ($k_{10} = 0$) of the surfaces of contact. In order to do this, the polynomials $P(s)$ and $R(s)$ should be replaced by $P_1(s)$, $R_1(s)$ and $P_2(s)$, $R_2(s)$, respectively. Let us write the function $R_x^{\mathrm{L}}(s)$ in the explicit form for the following particular cases ($d = d_0$).

(1) An elastic medium and a spherical wave:

$$R_x^{\mathrm{L}}(s) = -4\pi f^{\mathrm{L}}(s)\,\beta_0\,\frac{d-1}{d^2}\,\frac{(sd+1)\,P(s)}{\eta_0^2 R(s)}\,. \qquad (4.74)$$

(2) An elastic medium and a plane wave:

$$R_x^{\mathrm{L}}(s) = -4\pi f^{\mathrm{L}}(s)\,\beta_0\,\frac{sP(s)}{R(s)}\,. \qquad (4.75)$$

(3) An acoustic medium and a spherical wave:

$$R_x^{\mathrm{L}}(s) = -4\pi f^{\mathrm{L}}(s)\,\beta_0\,\frac{d-1}{d^2}\,\frac{(sd+1)\,s^2}{s^2 + 2s + 2}\,. \qquad (4.76)$$

(4) An acoustic medium and a plane wave:

$$R_x^L(s) = -4\pi f^L(s)\, \beta_0 \frac{s^3}{s^2 + 2s + 2}\,. \tag{4.77}$$

It should be pointed out that transition to an acoustic medium should be provided by means of calculation of the limits, as $\eta_0 \to \infty$, in (4.74) and (4.75). The formulas for the plane waves (4.75) and (4.77) can be derived from (4.74) and (4.76) if $d \to \infty$ in the last ones.

Let us consider a problem of choice of the function f^L, which characterizes the source intensity. For a spherical wave, using the designations introduced in Sect. 3.1 for the spherical coordinate system whose origin coincides with the source, we can write the expression for the images of the radial stresses as follows:

$$\sigma_{ll}^{(0)L}(l, s) = \frac{d-1}{l^3}\, \beta_0 f^L(s)\, e^{-(l-d+1)\,s} Q_{01}^{(0)}(ls)\,. \tag{4.78}$$

Using the theorem on the limiting values of the Laplace transform, for $l = d - 1$, we obtain

$$\sigma_{ll}^{(0)}(d-1, +0) = \frac{1}{(d-1)^2}\, \beta_0 \lim_{s \to \infty} s f^L(s)\, Q_{01}^{(0)}[(d-1)\,s]$$

$$= \beta_0 \lim_{s \to \infty} s^3 f^L(s)\,. \tag{4.79}$$

Assuming that $\sigma_{ll}^{(0)} = -p_0$ at the instant when the incoming compression wave touches the sphere, we obtain

$$\beta_0 \lim_{s \to \infty} s^3 f^L(s) = -p_0\,, \qquad f^L(s) = -\frac{p_0}{\beta_0 P_3(s)}\,,$$

$$P_3(s) = s^3 + a_1 s^2 + a_2 s + a_3\,. \tag{4.80}$$

For an acoustic medium, it is usually assumed (Guz et al. (1978a)) that

$$f^L(s) = -\frac{p_0}{\beta_0 s^3}\,, \qquad f(\tau) = -\frac{p_0}{2\beta_0}\,\tau_+^2\,, \qquad \tau_+^2 = \tau^2 H(\tau)\,. \tag{4.81}$$

If we choose the law (4.81) for the elastic medium, then, as it follows from (4.74), (4.75), and (4.78), as $\tau \to \infty$, both the stresses and the resultant force will be unbounded. For boundedness of the solution, it is sufficient to assume $a_2 \neq 0$ in (4.80). Basically, a form of the function $f^L(s)$ should be chosen in accounting for the character of a particular problem, and should follow from the solution of the nonlinear problem on a point source of explosion in a continuous medium (which is beyond our context).

Under a particular choice of the function $f^L(s)$, inversion of the expressions (4.74)–(4.77) can be made without difficulty by means of the theory of residues. Choosing $f^L(s)$ in the form (4.81), we can represent

$$R_x(\tau) = \operatorname*{res}_{s=0} R_x^{\mathrm{L}}(s)\, e^{s\tau} + \sum_k \operatorname*{res}_{s=s_k} R_x^{\mathrm{L}}(s)\, e^{s\tau}\,. \tag{4.82}$$

At that, it is necessary to know the roots s_k of the equation $R(s) = 0$. In the case of stiff adhesion, we get

$$s_{1,2} = -\alpha \pm i\beta\,,$$
$$\alpha = 1 + \frac{1}{2\eta_0}\,, \qquad \beta^2 = 1 - \frac{1}{\eta_0} + \frac{3}{4\eta_0^2} > 0\,. \tag{4.83}$$

Since $0 \le \eta_0^{-2} \le 1/2$ for any linearly elastic medium, the possible range of α and β will be as follows:

$$1 \le \alpha \le 1 + \frac{1}{2\sqrt{2}}\,, \qquad \frac{11 - 4\sqrt{2}}{8} \le \beta^2 \le 1\,.$$

A passage to the limit, as $\eta \to \infty$, for an acoustic medium gives us $\alpha = \beta = 1$; this result corresponds completely to the roots of the denominators in (4.76) and (4.77).

Since a final expression for $R_x(\tau)$ in the case of elastic medium has a relatively sophisticated form, we confine ourselves to an analysis of the acoustic medium. When applying the formulas (4.76) for a spherical wave, we obtain

$$R_x(\tau) = 4\pi p_0 \left\{ e^{-\tau} \sin \tau \right.$$
$$\left. + \frac{1}{2d}\left[\left(1 - \frac{1}{d}\right)(1 - e^{-\tau}\cos\tau) - \left(3 - \frac{1}{d}\right)e^{-\tau}\sin\tau\right]\right\}\,. \tag{4.84}$$

A passage to the limit, as $d \to \infty$, in the solution (4.84) (the case of a plane disturbing wave) gives us the known expression presented, for example, in Grigolyuk, Gorshkov (1976)

$$R_x(\tau) = 4\pi p_0\, e^{-\tau} \sin \tau\,. \tag{4.85}$$

At the initial interactions of the wave and the obstacle, we obtain the similar result $R_x(+0) = 0$ in both cases.

However, in the steady-state regime (as $\tau \to \infty$), the situation turns out to be different. For a plane wave, from (4.85) we get

$$\lim_{\tau\to\infty} R_x = 0\,;$$

for a spherical wave, from (4.84) it follows that

$$\lim_{\tau\to\infty} R_x = 2\pi p_0\, \frac{d-1}{d^2}\,.$$

The fact that in the last case the limiting value of the resultant force differs from zero is related to the irregular distribution of pressure over the obstacle's surface.

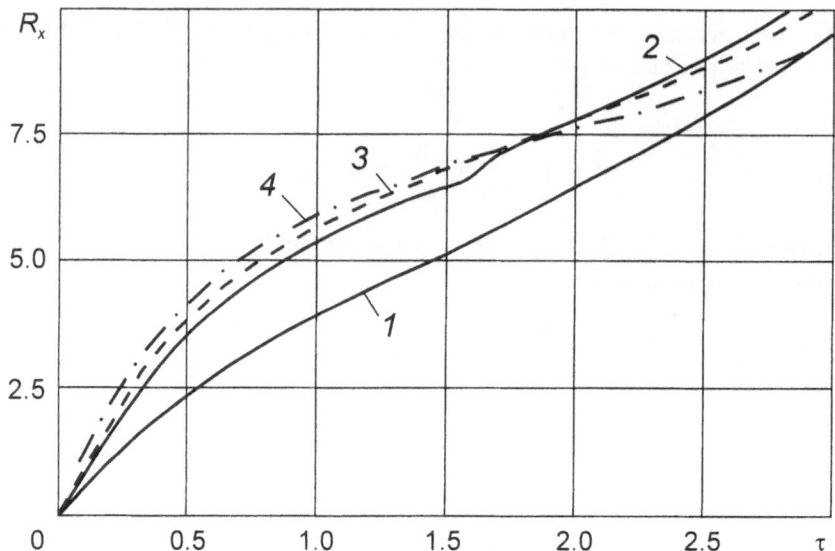

Fig. 4.3. Resultant force for an immovable sphere. Stiff cohesion ($k_{10} = \infty$)

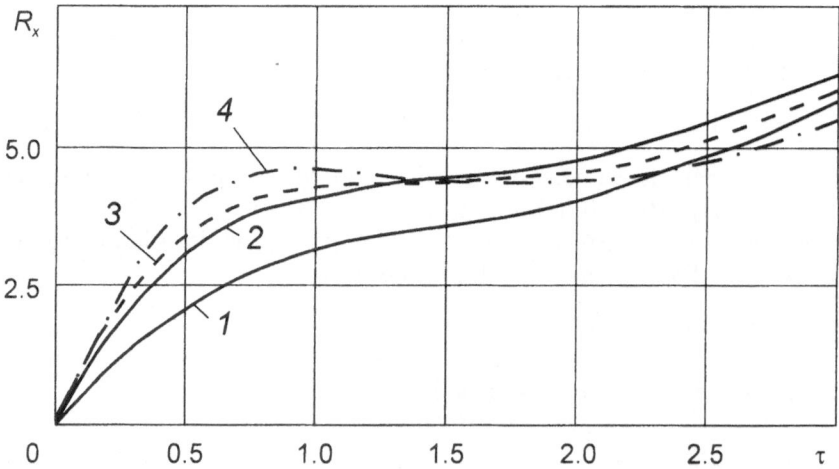

Fig. 4.4. Resultant force for an immovable sphere. Free sliding ($k_{10} = 0$)

To make an example, we have calculated a resultant force at an immovable sphere located in steel ($\gamma_0 = 1$; $\eta_0 = 1.871$) in the case of stiff adhesion of the sphere to the medium ($k_{10} = \infty$, Fig. 4.3) and in the case of free sliding ($k_{10} = 0$, Fig. 4.4). When performing calculations, we assumed that the law of variation of the potential in time had the following form:

$$f(\tau) = -\frac{1}{2}\tau_+^2, \qquad f^L(s) = -\frac{1}{s^3}.$$

The curves presented in Figs. 4.3 and 4.4 show a dependence of the resultant force on the incoming wave geometry and location of the point source. Curves *1–3* correspond to a spherical wave ($d = 2$, $d = 5$, and $d = 10$, respectively), and curve *4* corresponds to a plane wave ($d = \infty$). The reason why the force R_x increases without bound with an increase of τ was pointed out above. From the graphs it follows that a plane wave at the initial stage of interaction gives us a higher value of R_x as compared to a spherical wave. However, in succeeding instants of time, the resultant force value becomes smaller than that in the case of a point source. In the case of free sliding of surfaces of contact, the force R_x is smaller than that in the case of stiff adhesion. It is evident that for the real cases of contact, the value of the reaction R_x will be within the limiting values presented.

5. Diffraction of Plane (Spherical) Waves by a Spherical Barrier Supported by a Thin-Walled Shell

In Chap. 3, we have considered the problems of interaction of waves with a spherical barrier (a cavity, an absolutely stiff or elastic sphere) separated from a surrounding elastic medium by a thick-walled spherical shell. Though, we have derived an exact solution, the computations required are quite complicated. However, in many practically important cases, a shell to be analyzed is sufficiently thin so that its motion can be described by the equations of thin-walled shells (Sect. 1.3). When using such a statement, the solution can be simplified significantly since in the model of thin-walled shells we ignore the waves propagating through the barrier within its thickness. Thus, the difficulties of computations, when solving such problems, turn out to be similar to those when the shell is ignored completely (Chap. 4). Only the order of polynomials in the images of the functions sought in the resulting relationships will be different.

The problems of diffraction of waves by thin-walled barriers were studied sufficiently well. These problems were analyzed in Mnev, Pertsev (1970), Grigolyuk, Gorshkov (1974), Nigul et al. (1974), Nigul (1976), Volmir (1976), Kubenko (1979), Volmir (1979), Guz, Kubenko (1982), and Guz et al. (1984). The authors considered mainly the acoustic media surrounding and occupying a shell. Many works related to this problem were discussed in the reviews Gorshkov (1974), Gorshkov (1976), Guz et al. (1978b), Gorshkov (1979), Gorshkov (1980), Gorshkov (1981), Pao (1983), and Vestyak et al. (1983).

The axially symmetric problem of propagation of elastic waves in a space having a spherical cavity supported by a thin-walled shell was studied in Yanyutin (1984). The solution of this problem was reduced to the Volterra integral equations.

The main method of solution of the problem of diffraction of plane or spherical acoustic waves by a hollow spherical shell presumes expansion of the functions sought into the Legendre polynomials followed by application of the Laplace transform with respect to time. Inversion of each term should be performed exactly by means of the theory of residues. This method was used in Babaev et al. (1979), Grigolyuk et al. (1968), Grigolyuk et al. (1974), Huang (1969), Huang (1979), Kubenko (1972a), Kubenko (1972b), Veksler (1974), and Kubenko, Babaev (1983). A similar approach, when the Laplace transform was replaced by the Fourier one, was used in Mann–Nachbar (1957).

In Berger (1969), in order to improve the convergence of the series, the Cesaro mean method was used for summation. In Baidak, Torsky (1983), the authors applied the concept of *generalized* spherical waves and obtained a governing system of ordinary differential equations.

The simplified conditions for the problem be coupled (the hypothesis of plane reflection) were used in Lou, Klosner (1973). The excitation of the system 'shell – fluid' was produced by means of a concentrated force. In Lou, Klosner (1971), the authors considered a spherical shell supported by a ring.

The incidence of a plane wave on a hollow spherical shell in an elastic medium was studied in Huth, Cole (1955). The problem was solved in the uncoupled statement when the stresses produced by reflection and irradiation were not taken into account.

In Vasudevan, DiMaggio (1981), the authors estimated the accuracy of all three approximate schemes of interaction of fluid and shell, using, as an example, a solution of the diffraction of a plane pressure wave by a thin-walled elastic closed spherical shell. In Kenner, Goldsmith (1972), the authors presented the results of comparison of different approximations of shell models, numerical calculations, and experimental studies using, as an example, a thin-walled spherical shell filled with an acoustic medium and loaded by a given normal pressure.

The interaction of a spherical wave and a directed wave beam with a spherical shell surrounded by and filled with an ideal compressible fluid was studied in Grilitsky, Onischuk (1980a) and Porokhovsky (1982).

The problem of diffraction of a plane wave by a system of two concentrically located spherical shells surrounded by acoustic media was studied in Huang (1979). At that, a representation of the solution in the form of a superposition of elementary waves was used. The interaction of the acoustic waves with a shell in the case when the filler contains spherical inclusions was considered in Babaev, Kubenko (1976) and Babaev, Kubenko (1977a). For each term of the Legendre polynomial of the functions sought, a system of the Volterra equations was derived, which was solved numerically (see Kubenko (1975c)).

The motion of a structure composed of a three-layer thin-walled spherical shell elastically connected to the internal masses and subjected to the action of a plane acoustic wave was considered in Grigolyuk, Kuznetsov (1975). The integration of the equations of motion of the system was performed numerically.

The axially symmetric problems of deformation of a thin-walled spherical shell impinged by the wave generated by a point source located in an acoustic medium occupying the sphere were studied in Kubenko (1975a), Babaev, Kubenko (1977b), Babaev (1980), Kubenko, Stepanenko (1980), Babaev et al. (1980), and Aleksandrova (1982). The inversion of the Laplace transform was made either by reduction of the problem to the integral equations or by means of series expansion of the images in terms of exponential functions; the latter case corresponds to representation of the problem in the form of a superposition of generalized spherical waves. A similar internal problem for a system of concentrically located shells was solved in Saprykin (1984). The vibrations of a shell with a filler subjected to external pressure were studied in Kubenko et al. (1986).

An approximate solution of the problem of interaction of a plane wave of pressure with a shallow spherical shell fixed in a movable screen was derived in Kurochkin (1981); the author used the method of separation of variables and the integral transform with respect to time. The wave front was considered to be parallel to the supporting element. The same problem was solved for the case of immovable screen in Grigolyuk, Gorshkov (1968) and Grigolyuk, Gorshkov (1974).

The studies of the acoustic echo-signals (echo pulses) produced by a shell of spherical shape were presented in Nigul et al. (1974), Nigul (1976), and Veksler, Metsaveer (1981).

A method of numerical solution of the coupled nonstationary plane and axially symmetric problems of interaction of elastic-plastic thin-walled structural elements with the shock waves in ideal fluids, when the boundary conditions were predetermined at the deformable surfaces of contact, was developed in Bazhenov et al. (1977), Bazhenov, Mikhailov (1979), Bazhenov et al. (1979), and Bazhenov et al. (1981). The dynamic behavior of the physically and geometrically nonlinear shells was described within the framework of the theory of shells (accounting for shear and inertia of rotation) and the differential theory of plasticity in accounting for linear translational hardening. In order to facilitate integration of the equations of motion of a medium, it was proposed to apply the Godunov shock-capturing explicit finite difference scheme. A numerical solution of the equations, which describe dynamic deformation of elastic-plastic shells, was obtained by means of the variational-difference method using the *cross* explicit scheme. The geometrical nonlinearity was accounted for by a stepwise readjustment of the shell geometry. The plastic components of the deformation tensor were determined by iteration.

The problems of stability of four types of explicit difference schemes used for solution of the problems of interaction of shells with fluid were discussed in Neishlos et al. (1983). Application of different numerical methods was also considered in Akkas (1979) and Akkas, Engin (1980).

Recently, the finite element method (FEM) became popular in the mechanics of continuum. Application of this method to solution of a wide range of problems of interaction of fluids with structures (barriers) was discussed in Donea et al. (1977), Gross, Hofmann (1977), Belytschko (1980), Belytschko et al. (1980), Kulak (1981), and Neilson et al. (1981). The authors demonstrated mainly the technique of application of FEM (in the Euler and Lagrange coordinate systems) for solution of nonstationary problems of interaction of a fluid with a nuclear reactor.

The area of applications of the results obtained in the transient hydroelasticity of thin-walled structures increases continuously. Let us note only some of the applications: analysis of offshore platforms (fixed and free) and of the elements of nuclear reactors and oil and gas pipelines, biomechanics, studies of the behavior of structures in fluid (and having a fluid inside) during earthquakes (Taylor (1981)), etc.

In our days, much of attention in the area of transient aerohydroelasticity is paid to the analysis of various civil engineering structures subjected to wind and seismic loads, sea waves, high and low tides, and streams. Since these structures (buildings, nuclear electric plants, towers, bridges, pipelines, offshore platforms, reservoirs, dams, etc.) incorporate thin-walled elements, we can apply for their analysis the methods and techniques previously used in the aerohydroelasticity of plates and shells.

A numerical method for analysis of spherical reservoirs, which are supported by columns, are used for storage of liquefied gases, and can be impinged by weak air explosive waves, was developed in Kuranov et al. (1974), Gorshkov et al. (1979b), and Gorshkov et al. (1980). The algorithm proposed was based on a semi-analytical variant of FEM. The analysis of a stress–strain state of the structure's elements of a reservoir was made in a quasi-static statement.

A brief review of the theoretical and experimental works devoted to the analysis of shells of blasting chambers (which have plane, elliptic or spherical bottoms) was presented in Adischev, Kornev (1979). As a rule, the problem was stated as a problem of study of a stress–strain state of the shell and other structural elements of a chamber under given dynamic loads. The authors presented various idealized statements of the problems to be solved in the area of blasting chambers, and paid special attention to the presence of local effects in the vicinities of flanges.

The processes of nonstationary interaction of thin-walled structures and their elements with environment are some of the most complicated problems in mechanics. Rapid variation of the process parameters in time, presence of waves' fronts moving in time, presence of cavitation effects, formation of plastic zones in the barrier's material, and reflection and irradiation of the waves, – all these factors complicate the study significantly forcing the authors to apply simplifications which should be proven experimentally.

We should also note that, currently, some of the problems cannot be solved analytically at all. For this reason, the role of experimental studies, from which we can obtain reliable data on the behavior of structures and their elements, increases.

Some principles of organization of experimental–theoretical studies of dynamic behavior of thin-walled structures, as well as a description of the experimental arrangements used for this purpose were presented in Karmishin et al. (1979), Skurlatov (1979), and Karmishin et al. (1982).

The dynamic loading of plates, shells, and structures can be provided by different methods (Gorshkov (1979)).

Important problems related to damage accumulation in shells under repeated loading by shock waves were studied by an experimental approach in Skurlatov (1979) and Zaripov, Ilgamov (1979).

The results of experimental studies of the spherical blasting chambers were presented in Belov et al. (1984) and Efimova, Stepanenko (1984).

Let us note once again that the main purpose of this chapter is to present systematically the problems of dynamics of thin-walled spherical shells surrounded by linearly elastic or acoustic media, and to derive the solutions in the form of a superposition of generalized spherical waves; at that, we shall follow one general concept.

5.1 Propagation of Disturbances from a Supported Spherical Cavity

Let us consider a spherical cavity of unit dimensionless radius placed into an infinite elastic space. The cavity's surface is supported by an elastic spherical shell. The normal force p_s and the tangential force q_s are applied to the shell's internal surface. Assuming an axially symmetric character of the load and using the notation introduced in the previous chapters, we shall obtain a mathematical statement of the problem to be presented. At that, all the values indexed by 0 will belong to the surrounding sphere and those indexed by 1 will belong to the shell (Fig. 5.1).

The equations of motion of the elastic medium in terms of potentials have the following form (see (1.12)):

$$\frac{\partial^2 \varphi_0}{\partial \tau^2} = \Delta \varphi_0 \,,$$

$$\eta_0^2 \frac{\partial^2 \psi_0}{\partial \tau^2} = \Delta \psi_0 - \frac{\psi_0}{r^2 \sin^2 \theta} \,. \tag{5.1}$$

According to (1.56), we can write the equations of vibrations of the spherical shell as follows:

$$\gamma_1^2 \, \ddot{\mathbf{u}} = \mathbf{L}(\mathbf{u}) + \alpha_0 \, (\mathbf{p}_0 + \mathbf{p}_s) \,,$$

$$\mathbf{u} = (u_1, \, w_1, \, \chi_1)^{\mathrm{T}} \,, \qquad \mathbf{L} = (L_{ij})_{3 \times 3} \,,$$

$$\mathbf{p}_0 = (q_0, \, p_0, \, 0)^{\mathrm{T}} \,, \qquad \mathbf{p}_s = (q_s, \, p_s, \, 0)^{\mathrm{T}} \,,$$

$$c_* = c_1^{(0)} \,, \qquad \gamma_1 = \frac{c_1^{(0)}}{c_{01}} \,, \qquad c_{01} = \frac{E_0 (1 - \nu_0)}{\rho} \,. \tag{5.2}$$

Here, $\mathbf{u}$, $\mathbf{p}_0$, and $\mathbf{p}_s$ are the columns of the kinematic parameters of the shell, the contact forces, and the loads applied to the internal surface of the shell; L is the matrix of the differential operators L_{ij} defined by (1.56).

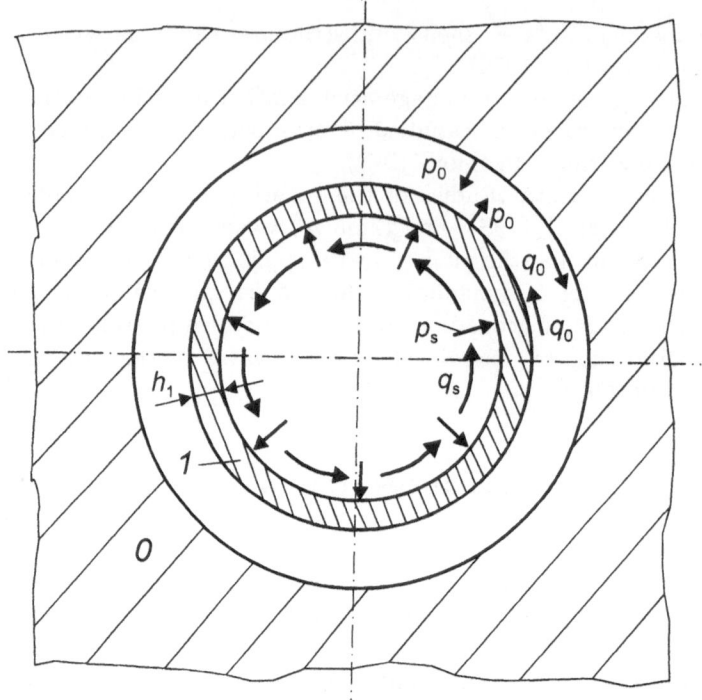

Fig. 5.1. Propagation of disturbances from a supported spherical cavity

Let us represent the conditions of contact of the elastic sphere with the shell in the following form:

$$\left.\sigma_{rr}^{(0)}\right|_{r=1} = \beta_1 p_0 , \qquad \left. u_0\right|_{r=1} = w_1 , \qquad \beta_1 = \frac{E_1}{\lambda_0 + 2\mu_0} ,$$

$$\left.\sigma_{r\theta}^{(0)}\right|_{r=1} = \beta_1 q_0 = k_{10}\left(\left. v_0\right|_{r=1} - u_{1+}\right), \qquad \beta_0 = 1 , \qquad u_{1+} = u_1 - \frac{\delta_1}{2}\chi_1 ;$$

$$(5.3)$$

here, u_{1+} is the tangential displacement of the shell at the external surface. Introduction of the coefficient k_{10} allows us to consider simultaneously two cases of contact of the surrounding medium with the shell. The case $k_{10} = 0$ corresponds to free sliding of the surfaces of contact and the case $k_{10} = \infty$ corresponds to stiff adhesion.

At infinity, as $r \to \infty$, we assume the absence of disturbances.

At the initial instant of time, the elastic medium and the shell are at rest, and this corresponds to the following initial conditions:

$$\left.\varphi_0\right|_{\tau=0} = \left.\dot\varphi_0\right|_{\tau=0} = \left.\psi_0\right|_{\tau=0} = \left.\dot\psi_0\right|_{\tau=0} = 0 ,$$

$$\left. u_1\right|_{\tau=0} = \left.\dot u_1\right|_{\tau=0} = \left. w_1\right|_{\tau=0} = \left.\dot w_1\right|_{\tau=0} = \left.\chi_1\right|_{\tau=0} = \left.\dot\chi_1\right|_{\tau=0} = 0 . \qquad (5.4)$$

We shall solve the mixed initial and boundary value problem (5.1)–(5.4) applying as before the method of incomplete separation of variables. Let us represent the potentials and the components of the stress–strain state of the surrounding medium and pressure p_0, p_s, q_0, and q_s in the form of the series in terms of the Legendre and Gegenbauer polynomials (see (3.6), (3.7), and (4.41)). Similarly, we can expand the displacements u_1 and w_1 and the angle of rotation of the shell's normal χ_1 into the series, that is,

$$u_1 = -\sin\theta \sum_{n=1}^{\infty} u_{1n} C_{n-1}^{3/2}(\cos\theta), \qquad w_1 = \sum_{n=0}^{\infty} w_{1n} P_n(\cos\theta),$$

$$\chi_1 = -\sin\theta \sum_{n=1}^{\infty} \chi_{1n} C_{n-1}^{3/2}(\cos\theta). \tag{5.5}$$

Taking into account the properties of the Legendre polynomials and their derivatives (1.79)–(1.82), as well as linearity of the operators L_{ij} (1.56), we get

$$L_{11}(u_1) = -\sin\theta \sum_{n=1}^{\infty} l_{11n} u_{1n} C_{n-1}^{3/2}(\cos\theta), \qquad l_{11n} = 1 - n(n+1) - \alpha_1,$$

$$L_{12}(w_1) = -\sin\theta \sum_{n=1}^{\infty} l_{12n} w_{1n} C_{n-1}^{3/2}(\cos\theta), \qquad l_{12n} = \alpha_2,$$

$$L_{13}(\chi_1) = -\sin\theta \sum_{n=1}^{\infty} l_{13n} \chi_{1n} C_{n-1}^{3/2}(\cos\theta), \qquad l_{13n} = -\alpha_7,$$

$$L_{21}(u_1) = \sum_{n=0}^{\infty} l_{21n} u_{1n} P_n(\cos\theta), \qquad l_{21n} = \alpha_2 n(n+1),$$

$$L_{22}(w_1) = \sum_{n=0}^{\infty} l_{22n} w_{1n} P_n(\cos\theta), \qquad l_{22n} = -\alpha_3 - \alpha_7 n(n+1),$$

$$L_{23}(\chi_1) = \sum_{n=0}^{\infty} l_{23n} \chi_{1n} P_n(\cos\theta), \qquad l_{23n} = \alpha_7 n(n+1),$$

$$L_{31}(u_1) = -\sin\theta \sum_{n=1}^{\infty} l_{31n} u_{1n} C_{n-1}^{3/2}(\cos\theta), \qquad l_{31n} = -\alpha_4,$$

$$L_{32}(w_1) = -\sin\theta \sum_{n=1}^{\infty} l_{32n} w_{1n} C_{n-1}^{3/2}(\cos\theta), \qquad l_{32n} = \alpha_4,$$

$$L_{33}(\chi_1) = -\sin\theta \sum_{n=1}^{\infty} l_{33n} \chi_{1n} C_{n-1}^{3/2}(\cos\theta), \qquad l_{33n} = 1 - n(n+1) - \alpha_5.$$

$$\tag{5.6}$$

Substituting (5.5) and (5.6) into (5.2), we obtain

$$\gamma_1^2 \, \ddot{\mathbf{u}} = l_n \mathbf{u}_n + \alpha_0 \left(\mathbf{p}_{0n} + \mathbf{p}_{s\,n} \right),$$

$$\mathbf{u}_n = (u_{1n}, \, w_{1n}, \, \chi_{1n})^{\cdot \mathrm{T}}, \qquad l_n = (l_{in})_{3 \times 3} \,,$$

$$\mathbf{p}_{0n} = (q_{0n}, \, p_{0n}, \, 0)^{\mathrm{T}}, \qquad \mathbf{p}_{s\,n} = (q_{s\,n}, \, p_{s\,n}, \, 0)^{\mathrm{T}}. \tag{5.7}$$

At that, in accounting for (1.57), the deformations, curvature changes, and forces in the shell can be represented by the following series:

$$\epsilon_i = \sum_{n=0}^{\infty} \epsilon_{in}^{(1)} P_n(\cos\theta) + \cos\theta \sum_{n=1}^{\infty} \epsilon_{in}^{(2)} C_{n-1}^{3/2}(\cos\theta) \,, \qquad i = 1, \, 2 \,,$$

$$\epsilon_{1n}^{(1)} = w_{1n} - n(n+1) \, u_{1n} \,, \qquad \epsilon_{1n}^{(2)} = u_{1n} \,,$$

$$\epsilon_{2n}^{(1)} = w_{1n} \,, \qquad \epsilon_{2n}^{(2)} = -u_{1n} \,,$$

$$\omega = -\sin\theta \sum_{n=1}^{\infty} \omega_n C_{n-1}^{3/2}(\cos\theta) \,, \qquad \omega_n = -u_{1n} + w_{1n} - \chi_{1n} \,,$$

$$\kappa_1 = \sum_{n=0}^{\infty} \kappa_{1n}^{(1)} P_n(\cos\theta) + \cos\theta \sum_{n=1}^{\infty} \kappa_{1n}^{(2)} C_{n-1}^{3/2}(\cos\theta) \,,$$

$$\kappa_{1n}^{(1)} = -n(n+1) \, \chi_{1n} \,, \qquad \kappa_{1n}^{(2)} = \chi_{1n} \,,$$

$$\kappa_2 = \cos\theta \sum_{n=1}^{\infty} \kappa_{2n} C_{n-1}^{3/2}(\cos\theta) \,, \qquad \kappa_{2n} = -\chi_{1n} \,,$$

$$N_i = \sum_{n=0}^{\infty} N_{in}^{(1)} P_n(\cos\theta) + \cos\theta \sum_{n=1}^{\infty} N_{in}^{(2)} C_{n-1}^{3/2}(\cos\theta) \,, \qquad i = 1, \, 2 \,,$$

$$N_{1n}^{(k)} = \epsilon_{1n}^{(k)} + \nu_1 \epsilon_{2n}^{(k)} \,, \qquad N_{2n}^{(k)} = \epsilon_{2n}^{(k)} + \nu_1 \epsilon_{1n}^{(k)} \,, \qquad k = 1, \, 2 \,,$$

$$H_i = \sum_{n=0}^{\infty} H_{in}^{(1)} P_n(\cos\theta) + \cos\theta \sum_{n=1}^{\infty} H_{in}^{(2)} C_{n-1}^{3/2}(\cos\theta) \,, \qquad i = 1, \, 2 \,,$$

$$H_{1n}^{(1)} = -\kappa_{1n}^{(1)} \,, \qquad H_{1n}^{(2)} = -\kappa_{1n}^{(2)} - \nu_1 \kappa_{2n} \,,$$

$$H_{2n}^{(1)} = -\nu_1 \kappa_{1n}^{(1)} \,, \qquad H_{2n}^{(2)} = -\kappa_{2n} - \nu_1 \kappa_{1n}^{(2)} \,,$$

$$S = -\sin\theta \sum_{n=1}^{\infty} S_n C_{n-1}^{3/2}(\cos\theta) \,, \qquad S_n = \alpha_7 \omega_n \,. \tag{5.8}$$

After expansion into the series, the equations of motion of the elastic medium, the conditions of contact, and the initial conditions take the following form:

$$\ddot{\varphi}_{0n} = \Delta_n \varphi_{0n} \,,$$

$$\eta_0^2 \ddot{\psi}_{0n} = \Delta_n \psi_{0n} \,, \tag{5.9}$$

$$\sigma_{rrn}^{(0)}\Big|_{r=1} = \beta_1 p_{0n}, \qquad u_{0n}\big|_{r=1} = w_{1n},$$

$$\sigma_{r\theta n}^{(0)}\Big|_{r=1} = \beta_1 q_{0n} = k_{10}\left(v_{0n}\big|_{r=1} - u_{1n} + \frac{\delta_1}{2}\chi_{1n}\right), \tag{5.10}$$

$$\varphi_{0n}\big|_{\tau=0} = \dot{\varphi}_{0n}\big|_{\tau=0} = \psi_{0n}\big|_{\tau=0} = \dot{\psi}_{0n}\Big|_{\tau=0} = 0,$$

$$u_{1n}(0) = w_{1n}(0) = \dot{u}_{1n}(0) = \dot{w}_{1n}(0) = \chi_{1n}(0) = \dot{\chi}_{1n}(0) = 0. \tag{5.11}$$

Let us analyze the mixed initial and boundary value problem (5.7), (5.9)–(5.11) by means of the Laplace transform with respect to time τ. If s is the transform parameter, then the relationships (5.7) in the space of images can be represented as follows:

$$(\gamma_1^2 s^2 E - 1_n)\,\mathbf{u}_n^L(s) = \alpha_0\left[\mathbf{p}_{0n}^L(s) + \mathbf{p}_{sn}^L(s)\right],$$

$$\mathbf{u}_n^L(s) = \left(u_{1n}^L(s),\, w_{1n}^L(s),\, \chi_{1n}^L(s)\right)^T,$$

$$\mathbf{p}_{0n}^L(s) = \left(q_{0n}^L(s),\, p_{0n}^L(s),\, 0\right)^T,$$

$$\mathbf{p}_{sn}^L(s) = \left(q_{sn}^L(s),\, p_{sn}^L(s),\, 0\right)^T; \tag{5.12}$$

here, E is the identity matrix.

Similarly to Chap. 4, we shall represent the potentials and components of the stress–strain state of the surrounding medium in the form of a sum of diverging waves (4.45). We can eliminate the contact pressure and radial displacements of the shell from the system of algebraic equations (5.12) using the boundary conditions (5.10),

$$p_{0n}^L = \frac{1}{\beta_1}\,\sigma_{rrn}^{(0)L}\Big|_{r=1}, \qquad q_{0n}^L = \frac{1}{\beta_1}\,\sigma_{r\theta n}^{(0)L}\Big|_{r=1}, \qquad w_{1n}^L = u_{0n}^L\big|_{r=1},$$

and add the last of the boundary conditions (5.10) to (5.12) in the form

$$\sigma_{r\theta n}^{(0)L}\Big|_{r=1} - k_{10}\left(v_{0n}^L\big|_{r=1} - u_{1n}^L\right) = 0.$$

Then, we obtain the following system of algebraic equations with respect to the images of the arbitrary functions $f_n^L(s)$ and $g_n^L(s)$, and the coefficients of series $u_{1n}^L(s)$ and $\chi_{1n}^L(s)$:

$$M_n F_n^L = \alpha_0 \mathbf{p}_{sn}, \qquad M_n = (M_{nij}(s))_{4\times 4},$$

$$F_n^L = \left(f_n^L(s),\, g_n^L(s),\, u_{1n}^L(s),\, \chi_{1n}^L(s)\right)^T; \ \cdot \tag{5.13}$$

here, $M_{nij}(s)$ are the polynomials of the argument s of the order not higher than $n + 2$, which can be determined as follows:

$$M_{n11}(s) = -\alpha_0 \frac{\beta_0}{\beta_1} Q_{n2}^{(0)}(s) + l_{12n} R_{n1}(s), \qquad M_{n13}(s) = -\gamma_1^2 s^2 - l_{11n},$$

$$M_{n12}(s) = -\alpha_0 \frac{\beta_0}{\beta_1} Q_{n3}^{(0)}(s) + l_{12n} n(n+1) R_{n0}(\eta_0 s), \qquad M_{n14}(s) = -l_{13n},$$

$$M_{n21}(s) = -(\gamma_1^2 s^2 - l_{22n}) R_{n1}(s) - \alpha_0 \frac{\beta_0}{\beta_1} Q_{n1}^{(0)}(s),$$

$$M_{n22}(s) = -n(n+1) \left[(\gamma_1^2 s^2 - l_{22n}) R_{n0}(\eta_0 s) + \alpha_0 \frac{\beta_0}{\beta_1} Q_{n2}^{(0)}(\eta_0 s) \right],$$

$$M_{n23}(s) = -l_{21n}, \qquad M_{n24}(s) = -l_{23n},$$

$$M_{n31}(s) = l_{32n} R_{n1}(s), \qquad M_{n32}(s) = n(n+1) l_{32n} R_{n0}(\eta_0 s),$$

$$M_{n33}(s) = -l_{31n}, \qquad M_{n34}(s) = \gamma_1^2 s^2 - l_{33n},$$

$$M_{n41}(s) = \beta_0 Q_{n2}^{(0)}(s) - k_{10} R_{n0}(s),$$

$$M_{n42}(s) = \beta_0 Q_{n3}^{(0)}(\eta_0 s) - k_{10} R_{n3}(\eta_0 s),$$

$$M_{n43}(s) = k_{10}, \qquad M_{n44}(s) = -k_{10} \frac{\delta_1}{2}. \qquad (5.14)$$

We should note once again that there is a logical link between (5.13) and a similar solution of the problem of propagation of disturbances from a non-supported cavity (Sect. 4.2). Though, the order of the system of equations (5.13) is two orders higher than that of the system of equations (4.46), their notional meanings are similar. In the problem under consideration, we have no reflected waves; hence, as it follows from the comparison with the solutions of Chap. 3, we should consider the recursive sequences of linear algebraic equations.

The solution (5.13) can be represented in the following form:

$$\mathbf{F}_n^{\mathrm{L}} = \frac{\alpha_0}{X_{n0}(s)} \mathbf{Y}_n \mathbf{p}_{sn}^{\mathrm{L}} = \frac{\alpha_0}{X_{n0}(s)} \left[q_{sn}^{\mathrm{L}}(s) \mathbf{Y}_{n1}(s) + p_{sn}^{\mathrm{L}}(s) \mathbf{Y}_{n2}(s) \right],$$

$$\mathbf{Y}_{nk}(s) = \left(Y_{n1k}(s), Y_{n2k}(s), Y_{n3k}(s), Y_{n4k}(s) \right)^{\mathrm{T}}, \qquad k = 1, 2,$$

$$\mathbf{Y}_n = (Y_{nij})_{4 \times 4}, \qquad X_{n0}(s) = \det \mathbf{M}_n(s), \qquad \mathbf{Y}_n = X_{n0} \mathbf{M}_n^{-1}. \qquad (5.15)$$

Here, $\mathbf{Y}_n$ is the polynomial matrix having the elements, which are the polynomials $Y_{nij}(s)$ of the order not higher than $2(n+3)$; at that, it is possible to prove that the order of its determinant (the polynomial $X_{no}(s)$) is also equal to $2(n+3)$.

As it follows from (5.15), the coefficients at $q_{sn}^{\mathrm{L}}(s)$ and $p_{sn}^{\mathrm{L}}(s)$ are the rational functions of the argument s. As we mentioned before, calculation of their originals does not make any difficulties. When the forces $p_{sn}(\tau)$ and $q_{sn}(\tau)$ are predetermined, the originals of the functions $f_n^{\mathrm{L}}(s)$ and $q_n^{\mathrm{L}}(s)$, and of the others, as well, can be calculated by means of the theorem on convolution of originals.

We can also note that a similar analysis of the determinant, which corresponds to the polynomial $X_{n0}(s)$, demonstrates that $X_{n0}(0) = X'_{n0}(0) = 0$, that is, this polynomial has two zero roots.

Let us consider two limiting cases of the solution obtained. We shall assume that the surrounding medium is of an acoustic type. Then, it is necessary to make the passage to the limit, as $\kappa_0 \to 1$ ($\eta_0 \to \infty$). Making the passage, we arrive at the simplified variant of the resulting relationships

$$\mathbf{F}_n^L(s) = \frac{\alpha_0}{X_{n0}(s)} p_{sn}^L(s) \, \mathbf{Y}_{n2}(s) \,, \qquad \mathbf{Y}_n = X_{n0} \mathbf{M}_n^{-1} \,,$$

$$M_{n11}(s) = l_{12n} R_{n1}(s) \,, \qquad M_{n13}(s) = \gamma_1^2 s^2 - l_{11n} \,,$$

$$M_{n14}(s) = -l_{13n} \,,$$

$$M_{n21}(s) = -(\gamma_1^2 s^2 - l_{22n}) R_{n1}(s) - \alpha_0 \frac{\beta_0}{\beta_1} s^2 R_{n0}(s) \,,$$

$$M_{n23}(s) = -l_{21n} \,, \qquad M_{n24}(s) = -l_{23n} \,,$$

$$M_{n31}(s) = l_{32n} R_{n1}(s) \qquad M_{n33}(s) = -l_{31n} \,,$$

$$M_{n34}(s) = \gamma_1^2 s^2 - l_{33n} \,, \tag{5.16}$$

where the matrix $\mathbf{M}_n$ of the dimensions 3×3 was obtained from the corresponding one in (5.13) by eliminating the fourth row and the second column of the matrix $\mathbf{Y}_n$.

The second limiting case presumes that the shell is absent. In order to pass to it, it is necessary to set $\beta_1 \to 0$ ($\gamma_1 \to \infty$), $u_{1n}^L \equiv 0$, and $\chi_{1n}^L \equiv 0$ in (5.13). Then, the matrix $\mathbf{M}_n$ becomes the matrix of the dimensions 2×2 (it is necessary to eliminate the third and fourth columns and the same rows). At that, the system of equations (5.13) turns to (4.46), that is, we obtain Problem A of Sect. 4.2.

5.2 Diffraction of Elastic Waves by a Thin-Walled Shell Occupied by an Elastic Medium

Let us now turn to the main and most complicated problem of this chapter. We shall assume that a thin-walled spherical shell occupied by an elastic medium is placed into an infinite linearly elastic medium; at that, the properties of these media are different. A plane or spherical compression wave, the front of which touches the shell's surface at the initial instant of time $\tau = 0$, propagates in the external medium. Then, using the dimensionless potentials and in accounting for the axial symmetry, we can write the equations of motion of the elastic media as follows:

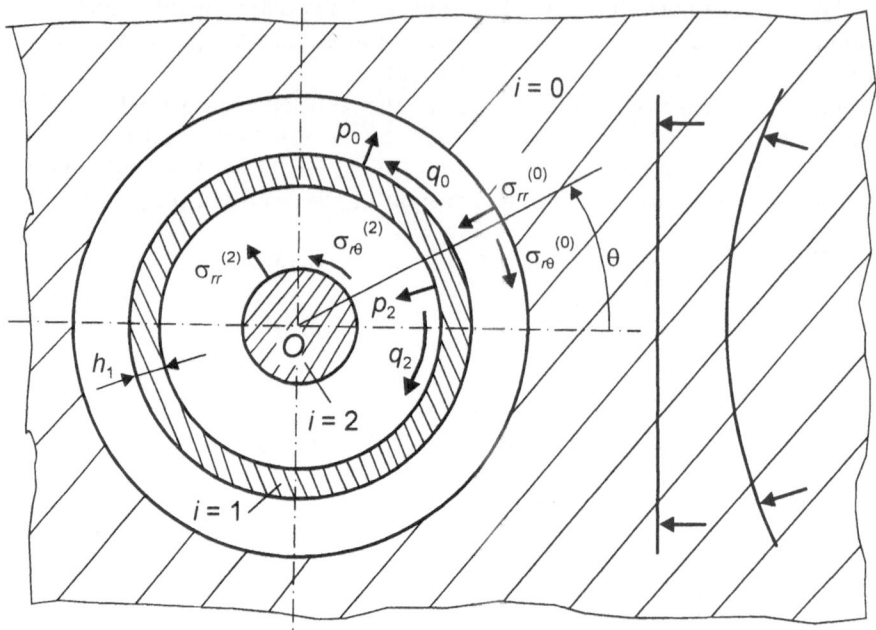

Fig. 5.2. Diffraction of waves by a spherical shell contacting the elastic media

$$\gamma_i^2 \frac{\partial^2 \varphi_i}{\partial \tau^2} = \Delta \varphi_i \,,$$

$$\eta_i^2 \frac{\partial^2 \psi_i}{\partial \tau^2} = \Delta \psi_i - \frac{\psi_i}{r^2 \sin^2 \theta} \,, \qquad i = 0, 2 \,. \tag{5.17}$$

The index $i = 0$ corresponds to the external medium and the index $i = 2$ corresponds to the internal one. We shall use the speed of compression waves in the external medium $c_1^{(0)}$ ($\gamma_0 = 1$) as the characteristic speed of sound c_*; the characteristic linear dimension is the radius of the shell's middle surface.

According to (1.56) and (5.2), the equations of motion of the shell take the form

$$\gamma_1^2 \ddot{\mathbf{u}} = \mathbf{L}(\mathbf{u}) + \alpha_0 (\mathbf{p}_0 - \mathbf{p}_2) \,,$$

$$\mathbf{p}_0 = (q_0, p_0, 0)^{\mathrm{T}} \,, \qquad \mathbf{p}_2 = (q_2, p_2, 0)^{\mathrm{T}} \,, \tag{5.18}$$

where q_i and p_i, $i = 1, 2$, are the tangential contact forces and the normal contact forces at the shell's interfaces with the external and internal media (Fig. 5.2).

We assume a condition of generalized contact: $k_{i0} = 0$ corresponds to free sliding and $k_{i0} = \infty$ corresponds to stiff adhesion. Under a bilateral adhesion, when lamination of the surfaces of contact is not accounted for, we can represent the boundary conditions at $r = 1$ as follows:

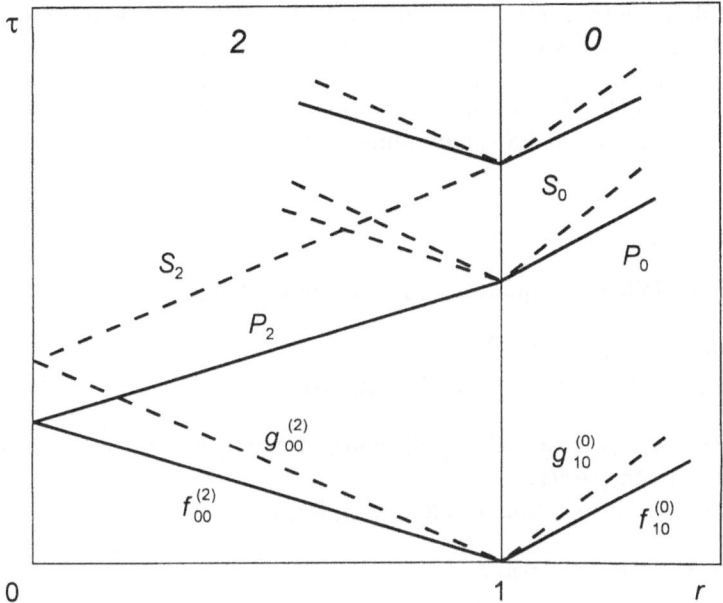

Fig. 5.3. The external and internal problems. The interface is a thin-walled shell

$$\sigma_{rr}^{(0)}\Big|_{r=1} + \sigma_{rrs}^{(0)}\Big|_{r=1} = \beta_1 p_0 \,,$$

$$u_0|_{r=1} + u_{0s}|_{r=1} = w_1 \,,$$

$$\sigma_{r\theta}^{(0)}\Big|_{r=1} + \sigma_{r\theta s}^{(0)}\Big|_{r=1} = \beta_1 q_0 = k_{10}(v_0|_{r=1} + v_{0s}|_{r=1} - u_{1+})\,,$$

$$\sigma_{rr}^{(2)}\Big|_{r=1} = \beta_1 p_2 \,, \qquad u_2|_{r=1} = w_1 \,,$$

$$\sigma_{r\theta}^{(2)}\Big|_{r=1} = \beta_1 q_2 = k_{12}(u_{1-} - v_2|_{r=1})\,,$$

$$u_{1-} = u_1 + \frac{\delta_1}{2}\chi_1 \,. \tag{5.19}$$

Here, u_{1-} are the tangential displacements of the shell at the internal surface, the other notation is similar to the introduced before. The index s is used to designate the components of the stress–strain state of the incoming wave. Because the shell thickness δ_1 is small, the boundary conditions corresponding to the external and internal surfaces of the shell are stated at the middle surface.

At infinity, as $r \to \infty$, the absence of disturbances is stated.

Before interaction, the system under consideration is at rest. As a result, we get the zero initial conditions

$$\varphi_i|_{\tau=0} = \dot{\varphi}_i|_{\tau=0} = \psi_i|_{\tau=0} = \dot{\psi}_i\Big|_{\tau=0} = u_1|_{\tau=0} = \dot{u}_1|_{\tau=0} = 0\,,$$

$$w_1|_{\tau=0} = \dot{w}_1|_{\tau=0} = \chi_1|_{\tau=0} = \dot{\chi}_1|_{\tau=0} = 0\,. \tag{5.20}$$

In order to derive a solution of the problem (5.17)–(5.20), let us represent the components of the stress–strain state of the elastic media and the shell in the form of the series in terms of the Legendre and Gegenbauer polynomials (see Chap. 3 and Sect. 5.1). Then, for an arbitrary n, we obtain the following equation of motion of the media surrounding and occupying the shell:

$$
\begin{aligned}
\gamma_i^2\, \ddot{\varphi}_{in} &= \Delta_n \varphi_{in} \,, \\
\eta_i^2\, \ddot{\psi}_{in} &= \Delta_n \psi_{in} \,, \qquad i = 0,\, 2 \,;
\end{aligned}
\tag{5.21}
$$

we also obtain the following equation of motion of the shell:

$$
\begin{aligned}
\gamma_i^2 \ddot{\mathbf{u}}_n &= \mathbf{L}\mathbf{u}_n + \alpha_0 \left(\mathbf{p}_{0n} - \mathbf{p}_{2n} \right) , \\
\mathbf{p}_{0n} &= \left(q_{0n},\, p_{0n},\, 0 \right)^{\mathrm{T}} , \qquad \mathbf{p}_{2n} = \left(q_{2n},\, p_{2n},\, 0 \right)^{\mathrm{T}} ;
\end{aligned}
\tag{5.22}
$$

here, q_{in} and p_{in}, $i = 0,\, 2$, are the coefficients of expansion of the contact forces q_i and p_i into the series.

The boundary conditions take the following form:

$$
\left. \sigma_{rrn}^{(0)} \right|_{r=1} + \left. \sigma_{rrsn}^{(0)} \right|_{r=1} = \beta_1 p_{0n} \,,
$$

$$
\left. u_{0n} \right|_{r=1} + \left. u_{0sn} \right|_{r=1} = w_{1n} \,,
$$

$$
\left. \sigma_{r\theta n}^{(0)} \right|_{r=1} + \left. \sigma_{r\theta sn}^{(0)} \right|_{r=1} = \beta_1 q_{0n}
$$

$$
= k_{10} \left(\left. v_{0n} \right|_{r=1} + \left. v_{0sn} \right|_{r=1} - u_{1n} + \frac{\delta_1}{2} \chi_{1n} \right) ,
$$

$$
\left. \sigma_{rrn}^{(2)} \right|_{r=1} = \beta_1 p_{2n} \,, \qquad \left. u_{2n} \right|_{r=1} = w_{1n} \,,
$$

$$
\left. \sigma_{r\theta n}^{(2)} \right|_{r=1} = \beta_1 q_{2n} = k_{12} \left(u_{1n} + \frac{\delta_1}{2} \chi_{1n} - \left. v_{2n} \right|_{r=1} \right) .
\tag{5.23}
$$

We get the zero initial conditions in the problem (5.21)–(5.23). In the case of interaction under consideration, contrary to the problems considered in the previous section, converging and diverging waves in the internal medium takes place. We can represent the elastic potentials in the form of the superposition of generalized spherical waves, like we did in Chap. 3 (3.36). However, because of a difference of the wave patterns (Fig. 5.3), we write them anew, as follows:

$$
\varphi_{in}(r,\tau) = \sum_{j,\,k} \left[\varphi_{0jk}^{(i)}(r,\tau)\, H(\tau - T_{0jk}^{(i)}) + \varphi_{1jk}^{(i)}(r,\tau)\, H(\tau - T_{1jk}^{(i)}) \right] ,
$$

$$
\psi_{in}(r,\tau) = \sum_{j,\,k} \left[\psi_{0jk}^{(i)}(r,\tau)\, H(\tau - S_{0jk}^{(i)}) + \psi_{1jk}^{(i)}(r,\tau)\, H(\tau - S_{1jk}^{(i)}) \right] ,
$$

$$
\varphi_{0jk}^{(i)}(r,\tau) = \frac{(-1)^n}{r^{n+1}} \sum_{p=0}^{n} (-\gamma_i r)^{n-p} A_{np} \frac{\mathrm{d}^{n-p}}{\mathrm{d}\tau^{n-p}} f_{0jk}^{(i)}(\tau - T_{0jk}^{(i)}) ,
$$

$$\varphi_{1jk}^{(i)}(r, \tau) = \frac{1}{r^{n+1}} \sum_{p=0}^{n} (\gamma_i r)^{n-p} A_{np} \frac{d^{n-p}}{d\tau^{n-p}} f_{1jk}^{(i)}(\tau - T_{1jk}^{(i)}),$$

$$\psi_{0jk}^{(i)}(r, \tau) = \frac{(-1)^n}{r^{n+1}} \sum_{p=0}^{n} (-\eta_i r)^{n-p} A_{np} \frac{d^{n-p}}{d\tau^{n-p}} g_{0jk}^{(i)}(\tau - S_{0jk}^{(i)}),$$

$$\psi_{1jk}^{(i)}(r, \tau) = \frac{1}{r^{n+1}} \sum_{p=0}^{n} (\eta_i r)^{n-p} A_{np} \frac{d^{n-p}}{d\tau^{n-p}} g_{1jk}^{(i)}(\tau - S_{1jk}^{(i)}),$$

$$T_{0jk}^{(i)} = \gamma_i(1 - r) + \tau_{jk}, \qquad T_{1jk}^{(i)} = \gamma_i(r - 1) + \tau_{jk},$$

$$S_{0jk}^{(i)} = \eta_i(1 - r) + \tau_{jk}, \qquad S_{1jk}^{(i)} = \eta_i(r - 1) + \tau_{jk},$$

$$\tau_{jk} = \gamma_2 k + \eta_2 j. \tag{5.24}$$

Here, the functions $f_{0jk}^{(i)}$ and $f_{1jk}^{(i)}$, $g_{0jk}^{(i)}$ and $g_{1jk}^{(i)}$ to be determined correspond to the converging and diverging compression waves and to the shear waves, respectively. The parameters $T_{mjk}^{(i)}$ and $S_{mjk}^{(i)}$, $i = 0, 2$, $m = 0, 1$, characterize the time delay of these waves.

Applying the Laplace transform with respect to time τ to the formulas (5.23), we obtain the formulas similar to (3.38) for the images of the potentials $\varphi_{in}^{L}(r, s)$ and $\psi_{in}^{L}(r, s)$. Because of the boundary condition at infinity, there are no converging waves in the external medium. It follows that $f_{0jk}^{(0)} = g_{0jk}^{(0)} \equiv 0$.

In the region occupied by the internal medium (the sphere $r \leq 1$), the solution should be bounded. Making an asymptotic analysis of the solution (5.24) of the equation of motion (5.21), as $r \to 0$, we obtain that the following equalities correspond to the limiting condition (see (3.39)–(3.43)):

$$(-1)^n f_{0j, k-1}^{(2)} + f_{1jk}^{(2)} = 0, \qquad (-1)^n g_{0, j-1, k}^{(2)} + g_{1jk}^{(2)} = 0. \tag{5.25}$$

An analysis of (5.25) shows that $f_{1jk}^{(2)} \equiv 0$ when j is odd or k is even, $g_{1jk}^{(2)} \equiv 0$ when j is even or k is odd, and $f_{1jk}^{(0)} = g_{1jk}^{(0)} = f_{0jk}^{(2)} = g_{0jk}^{(2)} \equiv 0$ when j is odd or k is odd.

Let us set

$$f_{jk}^{(2)} = f_{1, 2j, 2k+1}^{(2)} = (-1)^{n+1} f_{0, 2j, 2k}^{(0)}, \qquad f_{jk}^{(0)} = f_{1, 2j, 2k}^{(0)},$$

$$g_{jk}^{(2)} = g_{1, 2j+1, 2k}^{(2)} = (-1)^{n+1} g_{0, 2j, 2k}^{(0)}, \qquad g_{jk}^{(0)} = g_{1, 2j, 2k}^{(0)}. \tag{5.26}$$

Let us consider the notation (5.26) and the formulas for the relationships between the displacements, the stresses, and the potentials (3.7), as well as the formulas (5.24). Then, we can obtain the representations of the images of the components of the stress–strain state of the elastic media indexed by $i = 0, 2$; these representations coincide with (3.45) if to omit the indexes l and m and to substitute $r_1 = 1$ and $\tau_{jk} = 2(\gamma_2 k + \eta_2 j)$.

We shall represent the functions u_{1n}, w_{1n}, and χ_{1n} in the form of a sum of the terms with a retarded argument. At that, the delay corresponds to the arrival of elastic waves in the internal media at the shell's surface; in the space of the Laplace transforms, this corresponds to the following formulas:

$$u_{1n}^{L}(s) = \sum_{j,k} U_{jk}^{L}(s)\, e^{-\tau_{jk}s},$$

$$w_{1n}^{L}(s) = \sum_{j,k} W_{jk}^{L}(s)\, e^{-\tau_{jk}s},$$

$$\chi_{1n}^{L}(s) = \sum_{j,k} f_{jk}^{(1)L}(s)\, e^{-\tau_{jk}s}. \tag{5.27}$$

From the boundary conditions (5.25), we obtain

$$p_{0n} - p_{2n} = \frac{1}{\beta_1}\left(\sigma_{rrn}^{(0)} + \sigma_{rrsn}^{(0)} - \sigma_{rrn}^{(2)}\right)\Big|_{r=1},$$

$$q_{0n} - q_{2n} = \frac{1}{\beta_1}\left(\sigma_{r\theta n}^{(0)} + \sigma_{r\theta sn}^{(0)} - \sigma_{r\theta n}^{(2)}\right)\Big|_{r=1}, \qquad w_{1n} = u_{2n}\big|_{r=1}. \tag{5.28}$$

Substituting (5.28) into (5.22) and adding three equalities from the boundary conditions (5.23), we obtain

$$u_{0n}\big|_{r=1} + u_{0sn}\big|_{r=1} = u_{2n}\big|_{r=1},$$

$$\sigma_{r\theta n}^{(0)}\Big|_{r=1} + \sigma_{r\theta sn}^{(0)}\Big|_{r=1} = k_{10}\left(v_{0n}\big|_{r=1} + v_{0sn}\big|_{r=1} - u_{1n} + \frac{\delta_1}{2}\chi_{1n}\right),$$

$$\sigma_{r\theta n}^{(2)}\Big|_{r=1} = k_{12}\left(u_{1n} + \frac{\delta_1}{2}\chi_{1n} - v_{2n}\big|_{r=1}\right).$$

Then, equating the coefficients at the similar powers of the exponential functions and accounting for (5.27) and the representations of the displacements and stresses in the elastic media via the generalized spherical waves in the space of the Laplace transforms, we obtain the following recursive relationships for $j + k > 0$:

$$M_n \mathbf{f}_{jk}^{L}(s) = N_n \begin{pmatrix} f_{j,k-1}^{(2)L}(s) \\ g_{j-1,k}^{(2)L}(s) \end{pmatrix},$$

$$\mathbf{f}_{jk}^{L}(s) = \left(f_{jk}^{(0)L}(s),\, g_{jk}^{(0)L}(s),\, U_{jk}^{L}(s),\, f_{jk}^{(1)L}(s),\, f_{jk}^{(2)L}(s),\, g_{jn}^{(2)L}(s)\right)^{T},$$

$$M_n = (M_{nlm}(s))_{6\times 6}, \qquad N_n = (N_{nlm}(s))_{6\times 2}. \tag{5.29}$$

Here, M_n and N_n are the polynomial matrices such that

$$M_{n11}(s) = -\frac{\alpha_0}{\beta_0} \beta_0 Q_{n2}^{(0)}(s), \qquad M_{n12}(s) = -\frac{\alpha_0}{\beta_1} \beta_0 Q_{n3}^{(0)}(\eta_0 s),$$

$$M_{n13}(s) = \gamma_1^2 s^2 - l_{11n}, \qquad M_{n14}(s) = -l_{13n},$$

$$M_{n15}(s) = -\frac{\alpha_0}{\beta_1} \beta_2 Q_{n2}^{(2)}(-\gamma_2 s) - l_{12n} R_{n1}(-\gamma_2 s),$$

$$M_{n16}(s) = -\frac{\alpha_0}{\beta_1} \beta_2 Q_{n3}^{(2)}(-\eta_2 s) - n(n+1) l_{12n} R_{n0}(-\eta_2 s),$$

$$M_{n21}(s) = -\frac{\alpha_0}{\beta_1} \beta_0 Q_{n1}^{(0)}(s), \qquad M_{n22}(s) = -\frac{\alpha_0}{\beta_1} \beta_0 n(n+1) Q_{n2}^{(0)}(\eta_0 s),$$

$$M_{n23}(s) = -l_{21n}, \qquad M_{n24}(s) = -l_{23n},$$

$$M_{n25}(s) = (\gamma_1^2 s^2 - l_{22n}) R_{n1}(-\gamma_2 s) - \frac{\alpha_0}{\beta_1} \beta_2 Q_{n1}^{(2)}(-\gamma_2 s),$$

$$M_{n26}(s) = n(n+1) \left[(\gamma_1^2 s^2 - l_{22n}) R_{n0}(-\eta_2 s) - \frac{\alpha_0}{\beta_1} \beta_2 Q_{n2}^{(2)}(-\eta_2 s) \right],$$

$$M_{n31}(s) = M_{n32}(s) \equiv 0, \qquad M_{n33}(s) = -l_{31n},$$

$$M_{n34}(s) = \gamma_1^2 s^2 - l_{33n},$$

$$M_{n35}(s) = -l_{32n} R_{n1}(-\gamma_2 s), \qquad M_{n36}(s) = -l_{32n} R_{n0}(-\eta_2 s),$$

$$M_{n41}(s) = -R_{n1}(s), \qquad M_{n42}(s) = -n(n+1) R_{n0}(\eta_0 s),$$

$$M_{n43}(s) = M_{n44}(s) \equiv 0,$$

$$M_{n45}(s) = -R_{n1}(-\gamma_2 s), \qquad M_{n46}(s) = -n(n+1) R_{n0}(-\eta_2 s),$$

$$M_{n51}(s) = \beta_0 Q_{n2}^{(0)}(s) - k_{10} R_{n0}(s),$$

$$M_{n52}(s) = \beta_0 Q_{n3}^{(0)}(\eta_0 s) - k_{10} R_{n3}(\eta_0 s),$$

$$M_{n53}(s) = k_{10}, \qquad M_{n54}(s) = -k_{10} \frac{\delta_1}{2},$$

$$M_{n55}(s) = M_{n56}(s) = M_{n61}(s) = M_{n62}(s) \equiv 0,$$

$$M_{n63}(s) = -k_{12}, \qquad M_{n64}(s) = -k_{12} \frac{\delta_1}{2},$$

$$M_{n65}(s) = -\beta_2 Q_{n2}^{(2)}(-\gamma_2 s) - k_{12} R_{n0}(\gamma_2 s),$$

$$M_{n66}(s) = -\beta_2 Q_{n3}^{(2)}(-\eta_2 s) - k_{12} R_{n3}(-\eta_2 s),$$

$$N_{nlm}(s) = -M_{nl,\,m+4}(-s), \qquad l = 1, 2, \ldots, 6, \quad m = 1, 2. \qquad (5.30)$$

For $j = k = 0$, the boundary conditions at $r = 1$ bring us to the following initial conditions for the recursive relationships (5.29):

$$M_n \mathbf{f}_{00}^L(s) = \left(\frac{\alpha_0}{\beta_1} \sigma_{r\theta s n}^{(0)} \Big|_{r=1}, \; \frac{\alpha_0}{\beta_1} \sigma_{rr s n}^{(0)} \Big|_{r=1}, \; 0, \right.$$

$$\left. - u_{0 s n}|_{r=1}, \; 0, \; -\sigma_{r\theta s n}^{(0)} \Big|_{r=1} + k_{10} v_{0 s n}|_{r=1} \right)^T$$

$$= -E_0(s) \left[M_{n1}(-s) - e^{-2s} M_{n1}(s) \right]; \qquad (5.31)$$

here, $M_{n1}(s)$ is the first column of the matrix M_n.

The last equality in (5.31) is derived in accounting for the expressions (3.25) for the components of the stress–strain state of the elastic medium in the incoming wave.

We can write the recursive relationships (5.30) and (5.31) as follows:

$$\mathbf{f}_{jk}^{\mathrm{L}}(s) = \frac{Y_n(s)}{X_n(s)} \begin{pmatrix} f_{j,k-1}^{(2)\mathrm{L}}(s) \\ g_{j-1,k}^{(2)\mathrm{L}}(s) \end{pmatrix} , \qquad j+k > 0 ,$$

$$\mathbf{f}_{00}^{\mathrm{L}}(s) = E_0(s) \left[\frac{\mathbf{F}_0(s)}{X_n(s)} + \mathrm{e}^{-2s} \begin{pmatrix} 1 \\ 0 \\ \vdots \\ 0 \end{pmatrix} \right] ,$$

$$Y_n = X_n M_n^{-1} N_n , \qquad Y_n = (Y_{nij}(s))_{6 \times 2} , \qquad X_n = \det M_n ,$$

$$\mathbf{F}_0(s) = -X_n M_n^{-1} M_{n1}(-s) = \bigl(F_{01}(s), \ldots, F_{06}(s) \bigr)^{\mathrm{T}} ; \tag{5.32}$$

here, Y_n is the matrix whose elements are the polynomials $Y_{nij}(s)$; the order of these polynomials and the order of the matrix determinant $X_n(s)$, is not higher than $4n + 12$.

Since, as it follows from (5.32), the functions $f_{jk}^{(0)\mathrm{L}}(s)$, $g_{jk}^{(0)\mathrm{L}}(s)$, $U_{jk}^{\mathrm{L}}(s)$, and $f_{jk}^{(1)\mathrm{L}}(s)$ can be expressed via the functions $f_{jk}^{(2)\mathrm{L}}(s)$ and $g_{jk}^{(2)\mathrm{L}}(s)$ only, it is more convenient to use the recursive relationships in the following form:

$$f_{jk}^{(2)\mathrm{L}}(s) = \frac{Y_{n51}(s)}{X_n(s)} f_{j,k-1}^{(2)\mathrm{L}}(s) + \frac{Y_{n52}(s)}{X_n(s)} g_{j-1,k}^{(2)\mathrm{L}}(s) ,$$

$$q_{jk}^{(2)\mathrm{L}}(s) = \frac{Y_{n61}(s)}{X_n(s)} f_{j,k-1}^{(2)\mathrm{L}}(s) + \frac{Y_{n62}(s)}{X_n(s)} g_{j-1,k}^{(2)\mathrm{L}}(s) , \qquad i+j > 0 ,$$

$$f_{00}^{(2)\mathrm{L}}(s) = \frac{E_0(s)}{X_n(s)} F_{05}(s) , \qquad g_{00}^{(2)\mathrm{L}}(s) = \frac{E_0(s)}{X_n(s)} F_{06}(s) . \tag{5.33}$$

The coefficients in (5.32) and (5.33) are the rational functions of the argument s. For this reason, when calculating the originals $f_{jk}^{(0)}(\tau), \ldots, g_{jk}^{(2)}(\tau)$ and the components of the stress–strain state of the elastic media and the shell, we can use the algorithm described in Sect. 2.2.

5.3 Internal Problems on Interaction of Waves with a Thin-Walled Spherical Shell

Let us consider a mechanical system similar to that considered in the previous section that consisted of a thin-walled spherical shell surrounded and occupied by elastic media. However, in this case, we shall assume that the disturbance is produced by an elastic wave generated by a point source located in the internal medium $i = 2$ at the distance D from the sphere's center. The potential of a spherical P-wave produced by the source (in a dimensionless form) is determined by (3.13), where it is necessary to set $r_1 = 1$.

The mathematical statement of the problem is determined by (5.16), (5.17), and (5.19) under the modified boundary conditions at $r = 1$:

$$
\begin{aligned}
&\sigma_{rr}^{(0)}\Big|_{r=1} = \beta_1 p_0\,, \qquad u_0|_{r=1} = w_1\,, \\[2mm]
&\sigma_{r\theta}^{(0)}\Big|_{r=1} = \beta_1 q_0 = k_{10}\left(v_0|_{r=1} - u_1 + \frac{\delta_1}{2}\chi_1\right), \\[2mm]
&\sigma_{rr}^{(2)} + \sigma_{rr\,s}^{(2)}\Big|_{r=1} = \beta_1 p_2\,, \qquad u_2|_{r=1} + u_{2s}|_{r=1} = w_1\,, \\[2mm]
&\sigma_{r\theta}^{(2)}\Big|_{r=1} + \sigma_{r\theta\,s}^{(2)}\Big|_{r=1} = \beta_1 q_2 = k_{12}\left(u_1 + \frac{\delta_1}{2}\chi_1 - v_2|_{r=1} - v_{2s}|_{r=1}\right),
\end{aligned}
$$

$$(5.34)$$

where the index s belongs to the components of the stress–strain state of the incoming wave.

When deriving a solution of the problem under consideration, we use the same method as in Sect. 5.2. All the derivations already presented remain valid, including the resulting recursive equations (5.29) and the formulas for the elements of matrices (5.30). Only the initial conditions should be modified to take the following form:

$$
\mathbf{M}_n \mathbf{f}_{00}^{\mathrm{L}}(s) =
\begin{pmatrix}
-\dfrac{\alpha_0}{\beta_1}\,\sigma_{r\theta\,s\,n}^{(2)\mathrm{L}}\Big|_{r=1} + l_{12n} u_{2n\,s}^{\mathrm{L}}\big|_{r=1} \\[3mm]
-\dfrac{\alpha_0}{\beta_1}\,\sigma_{rr\,s\,n}^{(2)\mathrm{L}}\Big|_{r=1} - (\gamma_1^2 s^2 - l_{22n}) u_{2n\,s}^{\mathrm{L}}\big|_{r=1} \\[3mm]
l_{32n} u_{2n\,s}^{\mathrm{L}}\big|_{r=1} \\[3mm]
u_{2n\,s}^{\mathrm{L}}\big|_{r=1} \\[3mm]
0 \\[3mm]
-\sigma_{r\theta\,n}^{(2)\mathrm{L}} - k_{12} v_{2n\,s}^{\mathrm{L}}\big|_{r=1}
\end{pmatrix}
$$

$$
= -E_2(s)\,\mathbf{N}_{n1}\,; \qquad\qquad (5.35)
$$

here, the coefficients of the series expansion of the images of displacements and stresses in the incoming wave are determined by the expressions (3.25) at $r_1 = 1$, and $\mathbf{N}_{n1}$ is the first column of the matrix $\mathbf{N}_n$.

The recursive relationships for $j + k > 0$ are the same as in (5.32). For $j = k = 0$, in accounting for (3.35), the initial conditions can be represented as follows:

$$\mathbf{f}_{00}^{L}(s) = -\frac{E_2(s)}{X_n(s)}\, \mathbf{Y}_{n1}(s)\,;\tag{5.36}$$

here, $\mathbf{Y}_{n1}(s)$ is the first column of the matrix $\mathbf{Y}_n$ (5.32).

For $i + j > 0$, a scalar form of the recursive relationships coincides with (5.33), and the initial conditions have the following form:

$$f_{00}^{(2)L}(s) = -E_2(s)\,\frac{Y_{n51}(s)}{X_n(s)}\,,\qquad g_{00}^{(2)L}(s) = -E_2(s)\,\frac{Y_{n61}(s)}{X_n(s)}\,.\tag{5.37}$$

Let us note once more that the difficulties in solution of the external and internal problems are identical and the solutions of these problems differ only in the initial conditions of the recursive equations.

Let us now represent the solutions for two important cases of contact, sliding and stiff adhesion, in an explicit form. The same result will be valid for the external problem, as well. All the formulas presented in Sects. 5.2 and 5.3 remain valid in these particular cases. The only difference is in the simplification of some elements of the matrices $\mathbf{M}_n$ and $\mathbf{N}_n$.

(1) Sliding of the external medium relative to the shell ($k_{10} = 0$):

$$M_{51}(s) = \beta_0 Q_{n2}^{(0)}(s)\,,\qquad M_{n52}(s) = \beta_0 Q_{n3}^{(0)}(\eta_0 s)\,,$$
$$M_{n53}(s) = M_{n54}(s) \equiv 0\,.\tag{5.38}$$

(2) Stiff adhesion of the external medium to the shell ($k_{10} = \infty$):

$$M_{n51}(s) = -R_{n0}(s)\,,\qquad M_{n52}(s) = -R_{n3}(\eta_0 s)\,,$$
$$M_{n53}(s) = 1\,,\qquad M_{n54}(s) = -\frac{\delta_1}{2}\,.\tag{5.39}$$

(1′) Sliding of the internal medium relative to the shell ($k_{12} = 0$):

$$M_{n63}(s) = M_{n64}(s) \equiv 0\,,\qquad M_{n65}(s) = -\beta_2 Q_{n2}^{(2)}(-\gamma_2 s)\,,$$
$$M_{n66}(s) = -\beta_2 Q_{n3}^{(2)}(-\eta_2 s)\,.\tag{5.40}$$

(2′) Stiff adhesion of the shell's internal surface to the internal medium ($k_{12} = \infty$):

$$M_{n63}(s) = -1\,,\qquad M_{n64}(s) = -\frac{\delta_1}{2}\,,\qquad M_{n65}(s) = -R_{n0}(\gamma_2 s)\,,$$
$$M_{n66}(s) = -R_{n3}(-\eta_2 s)\,.\tag{5.41}$$

At that, the elements of the matrix $\mathbf{N}_n$, as before, are related to the elements of the matrix $\mathbf{M}_n$ (see (5.30)).

5.4 Interaction Between Shells and Acoustic Media

Let us consider an important particular case of the problems considered in Sect. 5.2 and 5.3. Namely, let us derive, by analogy to Sect. 3.4, a solution of the external and internal problems considered in this chapter in the case when the media surrounding the shell are of acoustic type. Similarly, we can derive the solutions for the cases when only one of the media (external or internal) is of acoustic type and the second one is of elastic type.

In order to consider the limiting case, it is necessary to set in the solutions presented in the previous sections

$$\kappa_0 = \kappa_2 = 1, \qquad (\eta_0 \to \infty, \qquad \eta_2 \to \infty),$$
$$k_{10} = k_{12} = 0, \qquad g_{jk}^{(0)} = g_{jk}^{(2)} \equiv 0,$$

to omit the index j in the designation of the other functions, and to retain the only summation over the index k in the formulas for summation with respect to generalized spherical waves.

Then, the recursive relationships (5.32) in both problems take the following form:

$$\mathbf{f}_k^{\mathrm{L}}(s) = \frac{\mathbf{Y}_{n1}(s)}{X_n(s)} f_{k-1}^{(2)\mathrm{L}}(s), \qquad k > 0,$$

$$\mathbf{Y}_{n1} = X_n \mathbf{M}_n^{-1} \mathbf{N}_{n1} = (Y_{n11}, Y_{n21}, Y_{n31}, Y_{n41})^{\mathrm{T}}, \qquad X_n = \det \mathbf{M}_n,$$

$$\mathbf{N}_{n1} = (N_{n11}, N_{n21}, N_{n31}, N_{n41})^{\mathrm{T}},$$

$$\mathbf{f}_k^{\mathrm{L}}(s) = \left(f_k^{(0)\mathrm{L}}(s), U_k^{\mathrm{L}}(s), f_k^{(1)\mathrm{L}}(s), f_k^{(2)\mathrm{L}}(s) \right)^{\mathrm{T}}. \tag{5.42}$$

Here, the matrix $\mathbf{M}_n$ is derived from a corresponding matrix presented in Sect. 5.2 by elimination of the second and third columns and the fifth sixth rows. The expressions for the other elements can be simplified as follows:

$$M_{n11}(s) \equiv 0, \qquad M_{n13}(s) = \gamma_1^2 s^2 - l_{11n},$$
$$M_{n14}(s) = -l_{13n}, \qquad M_{n15}(s) = -l_{12n} R_{n1}(-\gamma_2 s),$$
$$M_{n21}(s) = -\frac{\alpha_0}{\beta_1} \beta_0 s^2 R_{n0}(s),$$
$$M_{n23}(s) = -l_{21n}, \qquad M_{n24}(s) = -l_{23n},$$
$$M_{n25}(s) = (\gamma_1^2 s^2 - l_{22n}) R_{n1}(-\gamma_2 s) - \alpha_0 \frac{\beta_2}{\beta_1} \gamma_2^2 R_{n0}(-\gamma_2 s),$$
$$M_{n31}(s) \equiv 0, \qquad M_{n33}(s) = -l_{31n},$$
$$M_{n34}(s) = \gamma_1^2 s^2 - l_{33n}, \qquad M_{n35}(s) = -l_{32n} R_{n1}(-\gamma_2 s),$$
$$M_{n41}(s) = -R_{n1}(s), \qquad M_{n43}(s) = M_{n44}(s) \equiv 0,$$
$$M_{n45}(s) = -R_{n1}(-\gamma_2 s), \qquad N_{nl1}(s) = -M_{nl5}(-s), \qquad l = 1, 2, 3, 4.$$

$$\tag{5.43}$$

Taking into account (5.32) and (5.36), we can write the initial conditions for (5.42).

(1) The external problem:

$$
\mathbf{f}_0^{\mathrm{L}}(s) = E_0(s) \left[\frac{\mathbf{F}_0(s)}{X_n(s)} + e^{-2s} \begin{pmatrix} 1 \\ 0 \\ 0 \\ 0 \end{pmatrix} \right],
$$

$$
\mathbf{F}_0(s) = -X_n(s)\mathbf{M}_n^{-1}(s) \left(0,\ \mathrm{M}_{n21}(-s),\ 0,\ \mathrm{M}_{n41}(-s)\right)^{\mathrm{T}}
$$
$$
= \left(F_{01}(s),\ F_{02}(s),\ F_{03}(s),\ F_{04}(s)\right)^{\mathrm{T}}. \tag{5.44}
$$

(2) The internal problem:

$$
\mathbf{f}_0^{\mathrm{L}}(s) = -\frac{E_2(s)}{X_n(s)}\,\mathbf{Y}_{n1}(s).
$$

The recursive relationships with respect to the function $f_k^{(2)\mathrm{L}}(s)$ have the following scalar form:

$$
f_k^{(2)\mathrm{L}}(s) = \frac{Y_{n41}(s)}{X_n(s)}\, f_{k-1}^{(2)\mathrm{L}}(s), \qquad k > 0. \tag{5.45}
$$

At that, the initial conditions for (5.45) for the external and internal problems follow from (5.43) and (5.44), respectively. For the external problem, we get

$$
f_0^{(2)\mathrm{L}}(s) = E_0(s)\,\frac{F_{01}(s)}{X_n(s)} \tag{5.46}
$$

and for the internal one, we get

$$
f_0^{(2)\mathrm{L}}(s) = -E_2(s)\,\frac{Y_{n41}(s)}{X_n(s)}. \tag{5.47}
$$

Let us note that the rational functions of the argument s which enter into the resulting formulas include the polynomials of the order $2n + 8$.

Taking into consideration a relatively simple form of the recursive equations (5.45), it is possible to derive their general solution for any k by analogy to the problem of radial vibrations (Sects. 2.2 and 2.3). However, when making a practical analysis, it is more convenient to use the algorithm of calculation of originals presented in Sect. 3.5.

5.5 External Problem of Interaction of Waves with a Hollow Shell

This classical problem, as it follows from the review presented in the introduction to the current chapter, was considered in many publications. Let us demonstrate, how we can derive its solution from the general solution presented in Sect. 5.2.

The absence of internal medium corresponds to the following passage to the limit: $\beta_2 = 0$, $\gamma_2 = \eta_2 = \infty$, and $k_{12} = 0$. At that, because of the absence of converging and diverging waves, we have no need for the recursive equations (5.32) and the indexes j and k can be omitted. Only the initial conditions in (5.32) are saved.

However, it is impossible to represent immediately the resulting relationships by means of the elimination of the corresponding rows and columns of the matrices M_n and N_n, like we did it in Sect. 5.4, since, when deriving (5.32), we used the equality $w_{1n} = u_{2n}|_{r=1}$, which does not make sense in the case under consideration.

Let us consider the combination of arbitrary functions $f^{(2)L}(s)$ and $g^{(2)L}(s)$ from the fourth equation of (5.29)

$$M_{n45}(s)\, f^{(2)L}(s) + M_{n46}(s)\, g^{(2)L}(s) = -M_{n41}(s)\, f^{(0)L}(s)$$
$$- M_{n42}(s)\, g^{(0)L}(s) - E_0(s) \left[M_{n41}(-s) - e^{-2s} M_{n41}(s) \right]$$

and substitute it into the other equations of (5.29). We should note that this formal passage is equivalent to the application of the equality

$$w_{1n} = u_{0n}|_{r=1} + u_{0ns}|_{r=1} \ .$$

Discarding the sixth equation of (5.32), we obtain

$$\mathbf{K}_n(s)\, \mathbf{f}^L(s) = -E_0(s) \left[\mathbf{K}_{n1}(-s) - e^{-2s} \mathbf{K}_{n1}(s) \right] , \qquad \mathbf{K}_n = (K_{nij})_{4\times 4} ,$$
$$\mathbf{K}_{n1}(s) = \left(K_{n11}(s),\, K_{n21}(s),\, K_{n31}(s),\, K_{n41}(s) \right)^{\mathrm{T}} ,$$
$$\mathbf{f}^L(s) = \left(f^{(0)L}(s),\, g^{(0)L}(s),\, u_{1n}^L(s),\, \chi_{1n}^L(s) \right)^{\mathrm{T}} . \tag{5.48}$$

The elements of the polynomial matrix $\mathbf{K}_n$ have the following form ($M_{nij}(s)$ are taken from (5.30)):

$$K_{n11}(s) = -\frac{\alpha_0}{\beta_1} \beta_0 Q_{n2}^{(0)}(s) + l_{12n} R_{n1}(s) ,$$

$$K_{n12}(s) = -\frac{\alpha_0}{\beta_1} \beta_0 Q_{n3}^{(0)}(\eta_0 s) + (n+1)\, l_{12n} R_{n0}(\eta_0 s) ,$$

$$K_{n21}(s) = -\frac{\alpha_0}{\beta_1} \beta_0 Q_{n1}^{(0)}(s) - (\gamma_1^2 s^2 - l_{22n}) R_{n1}(s) ,$$

$$K_{n22}(s) = -n(n+1) \left[\frac{\alpha_0}{\beta_1} \beta_0 Q_{n2}^{(0)}(\eta_0 s) + (\gamma_1^2 s^2 - l_{22n}) R_{n0}(\eta_0 s) \right] ,$$

$$K_{n31}(s) = l_{32n} R_{n1}(s), \qquad K_{n32}(s) = n(n+1) l_{32n} R_{n0}(\eta_0 s),$$
$$K_{nlm}(s) = M_{nlm}(s), \qquad l = 1, 2, 3, \qquad m = 3, 4,$$
$$K_{n4m}(s) = M_{n5m}(s), \qquad m = 1, 2, 3, 4. \tag{5.49}$$

Resolving the relationships (5.48) with respect to the column $\mathbf{f}^{\mathrm{L}}(s)$, we obtain

$$\mathbf{f}^{\mathrm{L}}(s) = E_0(s) \left[\frac{\mathbf{F}_0(s)}{X_n(s)} + \mathrm{e}^{-2s} \begin{pmatrix} 1 \\ 0 \\ 0 \\ 0 \end{pmatrix} \right],$$

$$\mathbf{F}_0(s) = \left(F_{01}(s), F_{02}(s), F_{03}(s), F_{04}(s) \right)^{\mathrm{T}} = X_n(s) \mathrm{K}_n^{-1}(s) \, \mathbf{K}_{n1}(-s),$$

$$X_n(s) = \det \mathrm{K}_n(s). \tag{5.50}$$

It is easy to see that the matrix $\mathrm{K}_n(s)$ coincides with the matrix M_n (5.14) obtained for the problem of propagation of disturbances from a cavity supported by a shell. Thus, the difference between the solutions of these problems is only in the difference between the right-hand sides of (5.15) and (5.50).

The resulting relationships (5.50) take a more simple form in the case of diffraction of a pressure wave by a thin-walled shell located in an infinite acoustic medium. The passage to the limit, as $\kappa_0 \to 1$ ($\eta_0 \to \infty$), yields

$$\mathbf{f}^{\mathrm{L}}(s) = E_0(s) \left[\frac{\mathbf{F}_0(s)}{X_n(s)} + \mathrm{e}^{-2s} \begin{pmatrix} 1 \\ 0 \\ 0 \end{pmatrix} \right],$$

$$\mathbf{f}^{\mathrm{L}}(s) = \left(f^{(0)\mathrm{L}}(s), u_{1n}^{\mathrm{L}}(s), \chi_{1n}^{\mathrm{L}}(s) \right)^{\mathrm{T}},$$

$$\mathbf{F}_0(s) = \left(F_{01}(s), F_{02}(s), F_{03}(s) \right)^{\mathrm{T}} = X_n(s) \mathrm{K}_n^{-1}(s) \, \mathbf{K}_{n1}(-s),$$

$$\mathbf{K}_{n1}(s) = \left(K_{n11}(s), K_{n31}(s), K_{n41}(s) \right)^{\mathrm{T}}. \tag{5.51}$$

Here, K_n is derived from a corresponding matrix (5.49) by eliminating the fourth row and the second column. At that, some elements $K_{nij}(s)$ are simplified, that is,

$$K_{n11}(s) = l_{12n} R_{n1}(s),$$
$$K_{n12}(s) = -\frac{\alpha_0}{\beta_1} \beta_0 s^2 R_{n0}(s) - (\gamma_1^2 s^2 - l_{22n}) R_{n1}(s). \tag{5.52}$$

Let us also note that we can obtain the solutions of the problems considered before from the general solution (5.50). In the case of diffraction of waves by a spherical cavity, there is no shell; this case corresponds to the passage to the limit as $\beta_1 \to 0$ and $\gamma_1 \to \infty$. At that, $u_{1n} = \chi_{1n} \equiv 0$ and $k_{10} = 0$. Eliminating additionally the third and fourth rows and columns in (5.50), we obtain the relationships presented in Sect. 4.3.

In order to analyze the problem of diffraction of waves by an immovable spherical barrier, it is necessary to set $\gamma_1 = 0$, $\beta_1 = \infty$, $k_{10} = \infty$, and $u_{1n} = \chi_{1n} \equiv 0$. Eliminating the third and fourth columns and the second and third rows of the matrix K_n, dividing the first equations (5.50) by l_{12n} and the fourth one by (-1), we obtain a solution of the corresponding problem of Sect. 4.3.

5.6 Internal Problems in the Absence of Surrounding Medium

The problems of nonstationary behavior of the reservoirs that contain a filler, and can be subjected to different emergency conditions, are of a great practical importance. It is reasonable to use the model of a thin-walled spherical shell occupied by an elastic material in the absence of external medium. In this case, we can simulate the disturbances by an eccentrically located point source of waves.

The solution of this problem can be derived from the more general problem considered in Sect. 5.3. In order to do this, it is necessary to set $\beta_0 = 0$, $\gamma_0 = \eta_0 = \infty$, and $k_{10} = 0$. At that, $f_{jk}^{(0)L}(s) = g_{jk}^{(0)L}(s) \equiv 0$. The recursive relationships for calculation of the four arbitrary functions are as follows:

$$\mathbf{f}_{jk}^L(s) = \frac{Y_n(s)}{X_n(s)} \begin{pmatrix} f_{j,\,k-1}^{(2)L}(s) \\ g_{j-1,\,k}^{(2)L}(s) \end{pmatrix}, \qquad j + k > 0, \tag{5.53}$$

$$Y_n(s) = X_n(s)M_n^{-1}(s)N_n(s), \qquad X_n(s) = \det M_n(s),$$

$$\mathbf{f}_{jk}^L(s) = \left(U_{jk}^L(s),\, f_{jk}^{(1)L}(s),\, f_{jk}^{(2)L}(s),\, g_{jk}^{(2)L}(s) \right)^T,$$

$$Y_n = (Y_{nij})_{4 \times 2},$$

$$\mathbf{f}_{00}^L(s) = -\frac{E_2(s)}{X_n(s)} Y_{n1}(s). \tag{5.54}$$

Here, the matrices M_n and N_n are derived from the corresponding matrices presented in Sect. 5.2 by eliminating the fourth and fifth rows and the first two columns of the matrix M_n.

Let us also mark out four particular cases of solution of this problem defined by the relationships (5.53) and (5.54).

5.6.1 An Elastic Shell Occupied by an Acoustic Medium

This case presumes a passage to the limit, as $\kappa_2 \to 1$ ($\gamma_2 \to \infty$) and $k_{12} = 0$. Eliminating the last rows and columns in the matrices M_n and N_n, omitting the index j, and assuming $g_{jk}^{(2)L}(s) \equiv 0$, we arrive at

$$\mathbf{f}_k^{\mathrm{L}}(s) = \frac{\mathbf{Y}_{n1}(s)}{X_n(s)} f_{k-1}^{(2)\mathrm{L}}(s), \qquad k > 0.$$

$$M_n = (M_{nij})_{3\times 3},$$

$$N_n = \mathbf{N}_{n1} = (N_{nij})_{3\times 1}, \qquad \mathbf{Y}_n = \mathbf{Y}_{n1} = (Y_{nij})_{3\times 1}; \tag{5.55}$$

$$\mathbf{f}_k^{\mathrm{L}}(s) = \left(U_k^{\mathrm{L}}(s), f_k^{(1)\mathrm{L}}(s), f_k^{(2)\mathrm{L}}(s) \right)^{\mathrm{T}},$$

$$\mathbf{f}_0^{\mathrm{L}}(s) = -\frac{E_2(s)}{X_n(s)} \mathbf{Y}_{n1}(s). \tag{5.56}$$

5.6.2 An Absolutely Stiff Spherical Shell Occupied by an Elastic Medium

It is necessary to set $\gamma_1 = 0$, $\beta_1 \to \infty$, and $U_{1n}^{\mathrm{L}}(s) = \chi_{1n}^{\mathrm{L}}(s) \equiv 0$. Then, eliminating the second and third rows in the matrices M_n and N_n of (5.53) and (5.54) and eliminating additionally the third and fourth columns in the matrix M_n, we obtain

$$\begin{pmatrix} f_{jk}^{(2)\mathrm{L}}(s) \\ g_{jk}^{(2)\mathrm{L}}(s) \end{pmatrix} = \frac{\mathbf{Y}_n(s)}{X_n(s)} \begin{pmatrix} f_{j,k-1}^{(2)\mathrm{L}}(s) \\ g_{j-1,k}^{(2)\mathrm{L}}(s) \end{pmatrix}, \qquad j + k > 0, \tag{5.57}$$

$$M_n = (M_{nij})_{2\times 2}, \qquad N_n = (N_{nij})_{2\times 2}, \qquad Y_n = (Y_{nij})_{2\times 2},$$

$$\begin{pmatrix} f_{00}^{(2)\mathrm{L}}(s) \\ g_{00}^{(2)\mathrm{L}}(s) \end{pmatrix} = -\frac{E_2(s)}{X_n(s)} \begin{pmatrix} Y_{n11}(s) \\ Y_{n21}(s) \end{pmatrix}. \tag{5.58}$$

At that, as it can be proven by direct calculation of the matrix Y_n using the relationships (3.68), we can obtain the expression for the polynomials $Y_{n12}(s)$ and $Y_{n21}(s)$ in the following simple form:

$$Y_{n12}(s) = (-1)^{n+1} 2n(n+1) (\eta_2 s)^{2n+1} [\beta_2 (1 - \kappa_2) + k_{12}],$$

$$Y_{n21}(s) = (-1)^{n+1} 2 (\gamma_2 s)^{2n+1} [\beta_2 (1 - \kappa_2) + k_{12}]. \tag{5.59}$$

This mathematical model corresponds to the case when the stiffness of reservoir's walls is so high that the elastic properties cannot be accounted for.

5.6.3 A Stiff Reservoir Occupied by an Acoustic Medium

In this case, it is necessary to set $\kappa_2 = 1$, $\eta_2 = \infty$, and $k_{10} = 0$.

Considering (5.57)–(5.59), similarly to the passages to the acoustic medium, let us set $g_{jk}^{(2)\mathrm{L}}(s) \equiv 0$ and omit the index j. Then, we obtain the following simple recursive equations:

$$f_k^{(2)L}(s) = -\frac{R_{n1}(\gamma_2 s)}{R_{n1}(-\gamma_2 s)} f_{k-1}^{(2)L}(s), \qquad k > 0.$$

$$f_0^{(2)L}(s) = E_2(s) \frac{R_{n1}(\gamma_2 s)}{R_{n1}(-\gamma_2 s)}. \tag{5.60}$$

5.6.4 An Elastic Sphere Having a Free External Surface

This case makes a model to be used for the study of propagation of seismic waves in the Earth. Assuming that the shell is absolutely soft, we set $\beta_1 = 0$, $\gamma_1 = \infty$, and $k_{12} = 0$ in (5.53) and (5.54). We also set $U_{1n}^L(s) = \chi_{1n}^L(s) \equiv 0$, eliminate the first and third rows and the third and fourth columns in the matrices M_n and N_n, multiply the second equation by $\beta_1/(\alpha_0\beta_2)$ before making a passage to the limit, and divide the last equation by β_2. Then, we get the recursive relationships, which coincide by form with (5.57) and (5.58). At that, the polynomials $Y_{n12}(s)$ and $Y_{n21}(s)$ can be represented as follows:

$$Y_{n12}(s) = (-1)^{n+1} n(n+1)(1-\kappa_2)^2(\eta_2 s)^{2n+1}[\eta_2^2 s^2 + 2(n-1)(n+2)],$$
$$Y_{n21}(s) = (-1)^n \beta_2(\gamma_2 s)^{2n+1}(1-\kappa_2)[\gamma_2^2 s + (1-\kappa_2)(n-1)(n+2)].$$

Let us note that the problems with special boundary conditions at the shell's external surface considered in this section bring us to the more simple resulting formulas as compared to Sect. 5.3. A comparison of the results of calculations and the corresponding solutions, which take into account interaction with the surrounding medium, gives us an opportunity to estimate the effect of the media in each particular case.

6. Translational Motion of a Sphere in Elastic and Acoustic Media

In many cases of interaction of waves with obstacles, the kinematic parameters that characterize a displacement of the body's center of mass are of major interest. According to such a statement, we may often neglect the body's elasticity and consider it to be perfectly rigid. At that, as we shall demonstrate below, we obtain a coupled problem too, but in this case, the problem turns out to be significantly simpler, since the equations of motion of the obstacle are the ordinary differential equations of motion of a system with a finite number of degrees of freedom (in particular, we can obtain one equation).

The problem of nonstationary straight-line motion of a perfectly rigid sphere in an acoustic medium at a predetermined velocity was first analyzed in Kirchhoff (1876). Love (1905) studied the translational motion of a sphere in an acoustic medium under the action of elastic force (spherical pendulum) and at a predetermined velocity; at that, a general solution of the wave equation in the spherical coordinate system was used. The same solution was presented in Lamb (1931).

The determination of loads under a predetermined law of motion of a solid body is a first stage of solution of the problem of interaction of a movable obstacle with a surrounding medium. On the bases of the Laplace transform with respect to time, this problem was considered in Filippov, Egorychev (1977), Pao, Mow (1973), Grigolyuk, Gorshkov (1974), Zamyshlyaev, Yakovlev (1967), and Guz et al. (1978a). In Babichev (1966a), the author studied the straight-line motion of a sphere (which follows the law $z = z(t)$) in an infinite homogeneous isotropic linearly elastic medium in two cases of contact of the body and the medium: stiff adhesion and sliding. The elastic potentials of the waves of expansion and deformation were represented in the form of diverging waves. The ordinary differential equations were derived for arbitrary functions. This and other transient problems were considered in Babichev, Sarimsakov (1986).

The formulas for solution of the case when the law of motion was $z = t^2/2$ were derived in Babichev (1968), Babichev et al. (1967), and Saidov, Khidoyatov (1969); a detailed parametric study was performed. In Babichev et al. (1966) and Saidov, Khidoyatov (1969), the integration of the equations of motion of the elastic medium was made by means of the numerical method of characteristics. A comparison with an exact solution was presented, the resultant force R_z applied to the moving sphere was determined.

The response of an elastic medium to the motion of a sphere of radius r_0 was determined in Babichev (1973). The author considered a limiting case, as $r_0 \to 0$, and presented a comparison with the solution by Stokes. The *method of particles* was applied to reduce the problem of straight-line motion of a sphere at the predetermined velocity $v = v(t)$ to a system of ordinary differential equations (Sadykov (1972)). The boundary conditions at infinity were replaced by the condition of the absence of displacements in the medium on the spherical surface of a sufficiently large radius R. The translational motion of a perfectly rigid spherical reservoir filled with an ideal fluid was studied in Akkas (1979) and Pečinka et al. (1983).

The formula for a limiting value (as $t \to \infty$) of displacement of a solid body having two planes of symmetry impinged by a plane acoustic wave, whose front was perpendicular to the planes of symmetry, was derived in Novozhilov (1959).

The problem of motion of a perfectly rigid dynamically asymmetric sphere stiffly bonded to a surrounding infinite elastic medium, when the sphere is impinged by a plane compression or shear wave, was studied in Kovshov, Simonov (1967). The solution was derived in series form in terms of the spherical functions. In order to determine the coefficients of the series, the author used the Fourier transform with respect to time, the inversion of which was made by means of the theory of residues. The particular cases when the inertia tensor was of a diagonal type and when the sphere motion was translational and rotational were studied.

When solving the problem of translational motion of a perfectly rigid cylinder or a sphere impinged by a plane wave, the method of characteristics was used in Babichev (1966b). The ordinary differential equations describing approximately the process of straight-line motion of a perfectly rigid sphere impinged by a plane compression wave in an elastic medium were derived in Babichev (1969a) and Sarimsakov (1976).

A similar problem was studied in Mow (1965) and Mow (1966). In the latter one, the following inverse problem was also solved: the determination of the parameters of an incoming wave in the case when the motion of inclusion was known. The inverse problem was studied in Maaz (1964) applied to the problem of operation of a seismic detector. The author considered a perfectly rigid spherical shell, which contained an elastically constrained point particle. The solution was reduced to the integration of an ordinary differential equation, the order of which depended on the number of the coefficients of the series in terms of the Legendre polynomials $P_n(\cos \theta)$.

In Wolf (1945), the method of complete separation of variables was applied for solution of the problem of interaction of a plane wave in an acoustic or elastic medium with a perfectly rigid sphere.

A similar problem for an acoustic medium was considered in Tang (1968), Grigolyuk, Gorshkov (1967), Kreiser (1968), and Temkin (1972) using the Laplace integral transform and the Fourier integral transform.

The problem of diffraction of a plane acoustic wave by a sphere of finite mass having a soft coating was studied in Poruchikov et al. (1981). In addition to the traditional statement, one more degree of freedom (the reduction of the thickness of coating h) was introduced. The authors assumed that the pressure at the obstacle's surface was proportional to the reduction h.

Some problems of motion of perfectly rigid bodies impinged by acoustic pressure waves were studied in Israilov (1977).

6.1 Response of a Surrounding Medium to a Sphere's Motion at a Predetermined Law

Let us consider a perfectly rigid sphere of dimensionless radius $r = 1$ located in an infinite linearly elastic medium. The center of the sphere coincides with the origin O of the Cartesian and spherical coordinate systems. We assume that the sphere motion along the x axis follows the law $x(\tau)$. In order to determine the stress–strain state of the medium, we shall study the initial value problem (4.38) supplemented by the boundary conditions

$$u_0|_{r=1} = -x \cos \theta , \qquad \sigma_{r\theta}^{(0)}\Big|_{r=1} = k_{10} \left(x \sin \theta - v_0|_{r=1} \right) , \qquad (6.1)$$

where the meaning of the coefficient k_{10} is as before. At infinity, as $r \to \infty$, we state the absence of disturbances. After separation of the angle θ, for any $n \geq 0$, we obtain the initial value problem (4.42) and the following boundary conditions:

$$u_{01}|_{r=1} = -x(\tau) , \qquad \sigma_{r\theta 1}^{(0)}\Big|_{r=1} = k_{10} \left(v_{01}|_{r=1} + x(\tau) \right) ,$$

$$u_{0n}|_{r=1} = 0 , \qquad \sigma_{r\theta n}^{(0)}\Big|_{r=1} = k_{10} \, v_{0n}|_{r=1} , \qquad n \neq 1 . \qquad (6.2)$$

Let us note that for $n \neq 1$ the problem is homogeneous, consequently, it has only the trivial solution

$$u_{0n} = v_{0n} = \sigma_{\alpha\beta n}^{(0)} \equiv 0 , \qquad n \neq 1 . \qquad (6.3)$$

For $n = 1$, the solution to the problem should be derived by the methods considered earlier. Comparing the boundary conditions (4.55) and (6.2), it is easy to prove that the arbitrary functions $f_1^L(s)$ and $g_1^L(s)$ can be derived using the results obtained in Sect. 4.3; to do this, it is sufficient to set

$$u_{01\,s}^L\Big|_{r=1} = x^L(s) , \qquad v_{01\,s}^L\Big|_{r=1} = x^L(s) , \qquad \sigma_{r\theta 1\,s}^{(0)}\Big|_{r=0} \equiv 0$$

in (4.57).

It follows that in the problem under consideration, we have

$$f_1^{\mathrm{L}}(s) = \frac{Y_1(s)}{X_{10}(s)} \, x^{\mathrm{L}}(s) \,,$$

$$g_1^{\mathrm{L}}(s) = \frac{Y_2(s)}{X_{10}(s)} \, x^{\mathrm{L}}(s) \,,$$

$$Y_1(s) = -Y_{111}(s) + k_{10} Y_{112}(s) = -\beta_0 Q_{13}^{(0)}(\eta_0 s) + k_{10} R_{20}(\eta_0 s) \,,$$

$$Y_2(s) = -Y_{121}(s) + k_{10} Y_{122}(s) = \beta_0 Q_{12}^{(0)}(s) - k_{10} R_{20}(s) \,. \tag{6.4}$$

Taking into account (4.45) and (6.3), we obtain the following formulas for the images of displacements and stresses of an elastic medium at any point of space:

$$u_\theta^{\mathrm{L}}(r,\,s) = u_{01}^{\mathrm{L}}(r,\,s)\cos\theta \,, \qquad v_\theta^{\mathrm{L}}(r,\,s) = -v_{01}^{\mathrm{L}}(r,\,s)\sin\theta \,,$$

$$\sigma_{rr}^{(0)\mathrm{L}}(r,\,s) = \sigma_{rr1}^{(0)\mathrm{L}}(r,\,s)\cos\theta \,, \qquad \sigma_{r\theta}^{(0)\mathrm{L}}(r,\,s) = \sigma_{r\theta1}^{(0)\mathrm{L}}(r,\,s)\sin\theta \,,$$

$$u_{01}^{\mathrm{L}}(r,\,s) = -\frac{Z(s)}{r^3}\left[R_{11}(rs)\,Y_1(s)\,\mathrm{e}^{-(r-1)s}\right.$$

$$\left. + 2R_{10}(\eta_0 rs)\,Y_2(s)\,\mathrm{e}^{-\eta_0(r-1)s}\right] \,,$$

$$v_{01}^{\mathrm{L}}(r,\,s) = \frac{Z(s)}{r^3}\left[R_{10}(rs)\,Y_1(s)\,\mathrm{e}^{-(r-1)s}\right.$$

$$\left. + R_{13}(\eta_0 rs)\,Y_2(s)\,\mathrm{e}^{-\eta_0(r-1)s}\right] \,,$$

$$\sigma_{rr1}^{(0)\mathrm{L}}(r,\,s) = \frac{\beta_0 Z(s)}{r^4}\left[Q_{11}^{(0)}(rs)\,Y_1(s)\,\mathrm{e}^{-(r-1)s}\right.$$

$$\left. + 2Q_{12}^{(0)}(\eta_0 rs)\,Y_2(s)\,\mathrm{e}^{-\eta_0(r-1)s}\right] \,,$$

$$\sigma_{r\theta1}^{(0)\mathrm{L}}(r,\,s) = \frac{\beta_0 Z(s)}{r^4}\left[Q_{12}^{(0)}(rs)\,Y_1(s)\,\mathrm{e}^{-(r-1)s}\right.$$

$$\left. + Q_{13}^{(0)}(\eta_0 rs)\,Y_2(s)\,\mathrm{e}^{-\eta_0(r-1)s}\right] \,,$$

$$Z(s) = \frac{\eta_0^2 x^{\mathrm{L}}(s)}{s^2 R(s)} \,. \tag{6.5}$$

Inversion of (6.5) is as before when applying the theory of residues and the theorem on retarded argument for the Laplace transform.

An important characteristic of the sphere motion in a medium is the force of resistance R_x. Using the expression for resultant force (4.71), where we should set $\sigma_{rr\,s1}^{(0)} = \sigma_{r\theta\,s1}^{(0)} \equiv 0$, and the formulas (6.5), after some rearrangements, we obtain

$$R_x^{\mathrm{L}}(s) = -\frac{4\pi}{3}\,\beta_0 x^{\mathrm{L}}(s)\,\frac{Q(s)}{R(s)} \,, \qquad Q(s) = k_{10}\eta_0^2 Q_1(s) - \beta_0 Q_2(s) \,,$$

$$Q_1(s) = H_1(s; 1, \eta_0), \qquad Q_2(s) = H_2(s; 1, \eta_0),$$
$$H_1(s; \alpha, \beta) = \alpha\beta(\beta + 2\alpha)\, s^3 + (\beta^2 + 9\alpha\beta + 2\alpha^2)\, s^2 + 9\,(\alpha + \beta)\, s + 9,$$
$$H_2(s; \alpha, \beta) = \alpha\beta^3 s^4 + \beta(\beta^2 + 3\alpha\beta + 4\alpha^2)\, s^3$$
$$+ (3\beta^2 + 18\alpha\beta + 4\alpha^2)\, s^2 + 18\,(\alpha + \beta)\, s + 18. \quad (6.6)$$

The polynomial $R(s)$ can be determined by (4.73), as before.

Let us consider two main cases of contact of the sphere's surface with the surrounding medium: stiff adhesion ($k_{10} = \infty$) and free sliding ($k_{10} = 0$). As it follows from the analysis of (6.4) and (6.6), we should replace $Q(s)$ and $R(s)$ by $Q_1(s)$ and $R_1(s)$, and $Q_2(s)$ and $R_2(s)$, respectively.

The expression for resultant force has its simplest form in the case of an acoustic medium. Making a passage to the limit, as $\eta_0 \to \infty$, in (6.6) and taking into account (4.50), (4.51), (4.53), and (4.62), we arrive at

$$R_x^{\mathrm{L}}(s) = -\frac{4\pi}{3}\, \beta_0 x^{\mathrm{L}}(s)\, \frac{s^2(s+1)}{s^2 + 2s + 2}\,; \qquad (6.7)$$

this formula coincides with the known solutions presented, for instance, in Grigolyuk, Gorshkov (1976).

Let us consider a linear law of the sphere's motion

$$x(\tau) = \tau_+ = \tau H(\tau), \qquad x^{\mathrm{L}}(s) = \frac{1}{s^2}\,.$$

Then, in accounting for the roots of the denominator of (4.83), as $\eta_0 = \infty$, the original of the function $R_x^{\mathrm{L}}(s)$ is

$$R_x(\tau) = -\frac{4\pi}{3}\, \beta_0\, e^{-\tau} \cos\tau\,; \qquad (6.8)$$

that coincides, to an accuracy of a multiplier, with the transition function $\chi_1(\tau)$ presented in Sect. 4.1.

In Fig. 6.1, we present the results of the response of the surrounding medium (steel, $\gamma_0 = 1$, $\eta_0 = 1.871$), when the sphere's motion is translational and follows the law $x(\tau) = \tau_+^2/2$ (solid lines) or $x(\tau) = \tau_+$ (dashed lines). Curves *1* correspond to stiff contact of the sphere and the surrounding medium ($k_{10} = \infty$) and curves *2* correspond to the case of free sliding.

When the sphere's motion follows the law $x(\tau) = \tau_+$, the reaction R_x at the initial instant $\tau = 0$ differs from zero; this result can be obtained from the analysis of the image (6.7). Thus, in the case of stiff contact, we get

$$R_x(+0) = \lim_{s \to \infty} sR_x^{\mathrm{L}}(s) = -\frac{4\pi}{3}\left(1 + \frac{2}{\eta_0}\right), \qquad x^{\mathrm{L}}(s) = \frac{1}{s^2}\,,$$

and in the case of free sliding, we get

$$R_x = -\frac{4\pi}{3}\,.$$

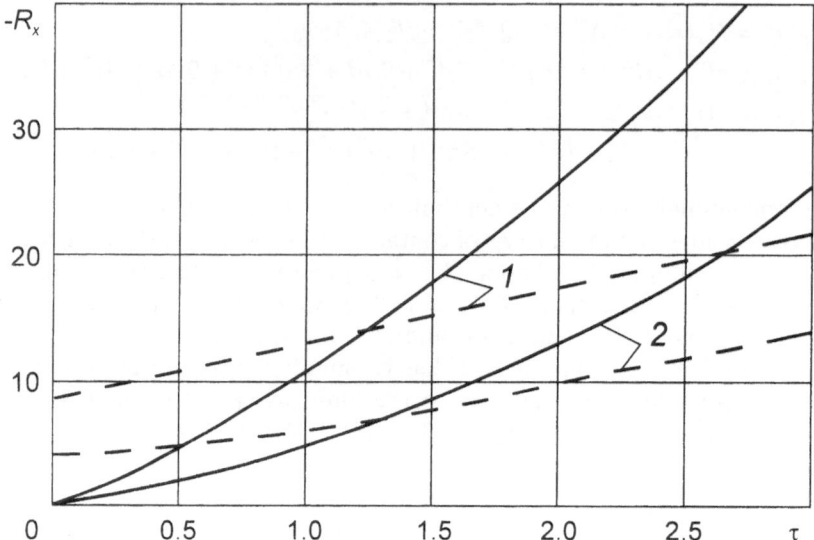

Fig. 6.1. Response of the medium to the translational motion of the sphere

As it follows from Fig. 6.1, the curves presented preserve qualitatively the form of the predetermined law of variation $x(\tau)$. This can be explained by the fact that the image of the reaction R_x can be represented in the form

$$R_x^{\mathrm{L}}(s) = -\frac{4\pi}{3}\left[r(s) + \frac{a_1}{s} + \frac{a_2}{s^2} + \frac{a_3}{s^3}\right],$$

where $r(s)$ is a proper rational fraction. For example, in the case of stiff adhesion, the denominator's roots are

$$s_{1,\,2} \approx -1.267 \pm \mathrm{i}\,0.824\,.$$

Hence, the original of the first term is damped rapidly and affects the result significantly only at the initial instant of time.

6.2 Motion of a Perfectly Rigid Spherical Inclusion Under Action of Predetermined Forces

Let us consider a perfectly rigid sphere of radius b and mass M and a linearly elastic medium surrounding the sphere. We shall consider symmetrical distribution of mass in the sphere. The force $R_e(\tau)$ acts on the sphere in such a manner that the line of action coincides with the x axis. At the initial instant of time, the sphere and the infinite surrounding medium are at rest.

Because of the problem symmetry, the sphere will move along the x axis. In a dimensionless form, the equation of the motion is as follows:

$$M_0 \ddot{x} = \tilde{R}_e + \tilde{R}_x , \qquad x(0) = \dot{x}(0) = 0 ,$$

$$M_0 = \frac{M}{\rho_* b^3} = \frac{4\pi}{3} m_0 , \qquad m_0 = \frac{M}{\frac{4\pi}{3} \rho_* b^3} ,$$

$$\tilde{R}_e = \frac{R_e}{(\lambda_* + 2\mu_*) b^2} , \qquad \lambda_* = \lambda_0 , \qquad \mu_* = \mu_0 , \qquad c_* = c_1^{(0)} . \tag{6.9}$$

Here, the dots are used to designate differentiation with respect to time τ, the dimensionless parameters are introduced as before, $\tilde{R}_x$ is the dimensionless reaction of the surrounding medium (from here on, we omit tildes in dimensionless values). By analogy to the problems of fluid mechanics (Grigolyuk, Gorshkov (1976)), the parameter m_0 is known as a *buoyancy*.

Applying the Laplace transform with respect to time τ to the problem (6.9) and taking into account the formula (6.6) for the reaction of the surrounding medium, we obtain

$$x^{\mathrm{L}}(s) = \frac{3}{4\pi} \frac{R_e^{\mathrm{L}}(s) R(s)}{m_0 s^2 R(s) + \beta_0 Q(s)} . \tag{6.10}$$

When the function $R_e(\tau)$ is predetermined, inversion of (6.10) is not difficult and can be done as usual.

From a practical point of view, it is more interesting to study the problem when the external force $R_e(\tau)$ is determined by the wave coming up against an obstacle. Let us assume that at the initial instant of time the immovable sphere's surface is impinged by the front of a plane or spherical compression wave, whose potential $\varphi_{0s}(r, \tau, \theta)$ is determined by (3.9) or (3.10). The corresponding mixed initial and boundary value problem is determined by the equations and initial conditions (4.38) and (6.9), the conditions at infinity, and the generalized boundary conditions at the sphere's surface:

$$u_0|_{r=1} + u_{0s}|_{r=1} = -x(\tau) \cos\theta ,$$

$$\sigma_{r\theta}^{(0)}\Big|_{r=1} + \sigma_{r\theta s}^{(0)}\Big|_{r=1} = k_{10} (x \sin\theta - v_0|_{r=1} - v_{0s}|_{r=1}) . \tag{6.11}$$

Because of the problem linearity, it can be represented in the form of the superposition of two problems with the boundary conditions of the following form:

$$u_0|_{r=1} = - u_{0s}|_{r=1} ,$$

$$\sigma_{r\theta}^{(0)}\Big|_{r=1} + k_{10} v_0|_{r=1} = - \left(\sigma_{r\theta s}^{(0)}\Big|_{r=1} + k_{10} v_{0s}|_{r=1} \right) ; \tag{6.12}$$

$$u_0|_{r=1} = -x \cos\theta ,$$

$$\sigma_{r\theta}^{(0)}\Big|_{r=1} + k_{10} v_0|_{r=1} = k_{10} x \sin\theta . \tag{6.13}$$

It is evident that the conditions (6.12) correspond to the problem of diffraction of the wave by an immovable obstacle (Sects. 4.2 and 4.3), and the conditions (6.13) correspond to the problem considered in Sect. 6.1.

Thus, in the case of interaction of an elastic wave with an immovable sphere, we should determine the force R_x in (6.9) by (6.6), as we did before, and R_e coincides with the reaction of a surrounding medium in the case of immovable sphere (4.73). Substituting the expressions mentioned into (6.9), we obtain the following image of the sphere displacement $x(\tau)$ in the space of the Laplace transform with respect to τ:

$$x^{\mathrm{L}}(s) = \frac{2\beta_0 E_0(s)}{\eta_0^2} \frac{s^3 P(s)}{m_0 s^2 R(s) + Q(s)}. \tag{6.14}$$

From (6.14) we can get the expressions for $x^{\mathrm{L}}(s)$ for two basic cases of adhesion of surfaces of contact. When the condition of stiff contact takes place ($k_{10} = \infty$), it is necessary to replace $P(s)$, $Q(s)$, and $R(s)$ by $P_1(s)$, $Q_1(s)$, and $R_1(s)$. In the case of free sliding ($k_{10} = 0$), the polynomials $P(s)$, $Q(s)$, and $R(s)$ should be replaced by $P_2(s)$, $Q_2(s)$, and $R_2(s)$, respectively.

Similarly to (4.74)–(4.77), let us write the formulas for four basic particular cases of the problem under consideration ($\beta_0 = 1$).

(1) An elastic medium and a spherical wave:

$$x^{\mathrm{L}}(s) = -3f^{\mathrm{L}}(s)\frac{d-1}{\eta_0^2 d^2} \frac{(sd+1)\,P(s)}{m_0 s^2 R(s) + Q(s)}. \tag{6.15}$$

(2) An elastic medium and a plane wave:

$$x^{\mathrm{L}}(s) = -\frac{3}{\eta_0^2}\,f^{\mathrm{L}}(s)\frac{sP(s)}{m_0 s^2 R(s) + Q(s)}. \tag{6.16}$$

(3) An acoustic medium and spherical wave:

$$x^{\mathrm{L}}(s) = -3f^{\mathrm{L}}(s)\frac{d-1}{d^2}\frac{sd+1}{m_0(s^2 + 2s + 2) + s + 1}. \tag{6.17}$$

(4) An acoustic medium and a plane wave:

$$x^{\mathrm{L}}(s) = -3f^{\mathrm{L}}(s)\frac{s}{m_0(s^2 + 2s + 2) + s + 1}. \tag{6.18}$$

Calculation of the originals of the functions (6.15)–(6.18) can be reduced to calculation of the residues of corresponding functions in the points coinciding with the denominator's roots and the singularities of the function $f^{\mathrm{L}}(s)$. At that, we can prove that for any type of contact and under any value of the parameter η_0 (including the case of acoustic medium $\eta_0 = \infty$), the denominator's roots are located in the left-hand half-plane.

In the case of acoustic medium, the solution can be represented in an explicit form. Let us analyze a case of spherical wave at $f(\tau) = -\tau_+^2/2$ $(f^L(s) = -s^{-3})$. Then, the sphere displacement can be represented as follows:

$$x(\tau) = \operatorname*{res}_{s=0} x^L(s)\, e^{s\tau} + \sum_{k=1}^{2} \operatorname*{res}_{s=s_k} x^L(s)\, e^{s\tau}\,,$$

$$m_0(s_k^2 + 2s_k + 2) + s_k + 1 = 0\,. \tag{6.19}$$

The form of the roots s_k depends on the buoyancy of obstacles m_0; in particular, as it follows from (6.19), we get

$$s_{1,2} = -\alpha \pm \beta < 0\,, \qquad m_0 < \frac{1}{2}\,,$$

$$s_1 = s_2 = -\alpha\,, \qquad m_0 = \frac{1}{2}\,,$$

$$s_{1,2} = -\alpha \pm \mathrm{i}\beta\,, \qquad m_0 > \frac{1}{2}\,, \tag{6.20}$$

where

$$\alpha = 1 + \frac{1}{2m_0}\,, \qquad \beta = \sqrt{\left|1 - \frac{1}{4m_0^2}\right|}\,.$$

Performing the necessary calculations in (6.19) and taking into account (6.17) and (6.20), we arrive at

$$x(\tau) = 3\left[A_0\tau^2 + A_1\tau - A_2 + e^{-\alpha\tau}\left(A_3\,\frac{\sinh\beta\tau}{\beta} + A_2\cosh\beta\tau\right)\right]\,,$$

$$m_0 < \frac{1}{2}\,,$$

$$x(\tau) = 3\left[A_0\tau^2 + A_1\tau - A_2 + e^{-\alpha\tau}\left(A_3\tau + A_2\right)\right]\,, \qquad m_0 = \frac{1}{2}\,,$$

$$x(\tau) = 3\left[A_0\tau^2 + A_1\tau - A_2 + e^{-\alpha\tau}\left(A_3\,\frac{\sin\beta\tau}{\beta} + A_2\cos\beta\tau\right)\right]\,,$$

$$m_0 > \frac{1}{2}\,, \tag{6.21}$$

where

$$A_0 = \frac{d-1}{2d^2(2m_0+1)}\,, \qquad A_1 = \frac{(d-1)^2}{d^2(2m_0+1)}\,,$$

$$A_2 = \frac{d(2m_0+1) - m_0 - 1}{d^2(2m_0+1)^2}\,(d-1)\,,$$

$$A_3 = \frac{d + m_0 - 1}{2d^2 m_0(2m_0+1)}\,(d-1)\,.$$

In the case of plane wave, the expressions for the displacement $x(\tau)$ can be derived either applying (6.19) to (6.18) or making a passage to the limit, as $d \to \infty$, in (6.21). Taking into account that

$$\lim_{d\to\infty} A_0 = 0, \qquad \lim_{d\to\infty} A_1 = \lim_{d\to\infty} A_2 = \frac{1}{2m_0 + 1},$$

$$\lim_{d\to\infty} A_3 = \frac{1}{2m_0(2m_0 + 1)}, \qquad\qquad\qquad (6.22)$$

we obtain

$$x(\tau) = \frac{3}{2m_0 + 1}\left[\tau - 1 + e^{-\alpha\tau}\left(\frac{1}{2m_0\beta}\sinh\beta\tau + \cosh\beta\tau\right)\right],$$

$$m_0 < \frac{1}{2},$$

$$x(\tau) = \frac{3}{2}\left(\tau - 1 + \frac{\tau + 1}{2}e^{-2\tau}\right), \qquad m_0 = \frac{1}{2},$$

$$x(\tau) = \frac{3}{2m_0 + 1}\left[\tau - 1 + e^{-\alpha\tau}\left(\frac{1}{2m_0\beta}\sin\beta\tau + \cos\beta\tau\right)\right],$$

$$m_0 > \frac{1}{2}. \qquad (6.23)$$

The last formulas coincide with the known results presented, for instance, in Grigolyuk, Gorshkov (1976).

An approximate solution of the problem of diffraction of waves by an immovable perfectly rigid sphere can be derived for the cases when the obstacle's buoyancy m_0 is small. For $m_0 = 0$, from (6.14) we get

$$x^L(s) = \frac{2\beta_0 E_0(s)}{\eta_0^2}\frac{s^3 P(s)}{Q(s)}. \qquad (6.24)$$

In the case of acoustic medium (spherical wave), the expression for the image of $x(\tau)$ takes the following simple form ($m_0 = 0$):

$$x^L(s) = -3f^L(s)\frac{d - 1}{d^2}\frac{sd + 1}{s + 1}. \qquad (6.25)$$

At that, for $f(\tau) = -\tau_+^2/2$, the original takes the form

$$x(\tau) = 3\frac{d - 1}{d^2}\left[\frac{1}{2}\tau^2 + (d - 1)\left(\tau - 1 + e^{-\tau}\right)\right]. \qquad (6.26)$$

The same result can be obtained from (6.21) for $m_0 < 1/2$ if account is taken of:

$$\lim_{m_0\to 0}(\alpha + \beta) = \infty, \qquad \lim_{m_0\to 0}(\alpha - \beta) = 1.$$

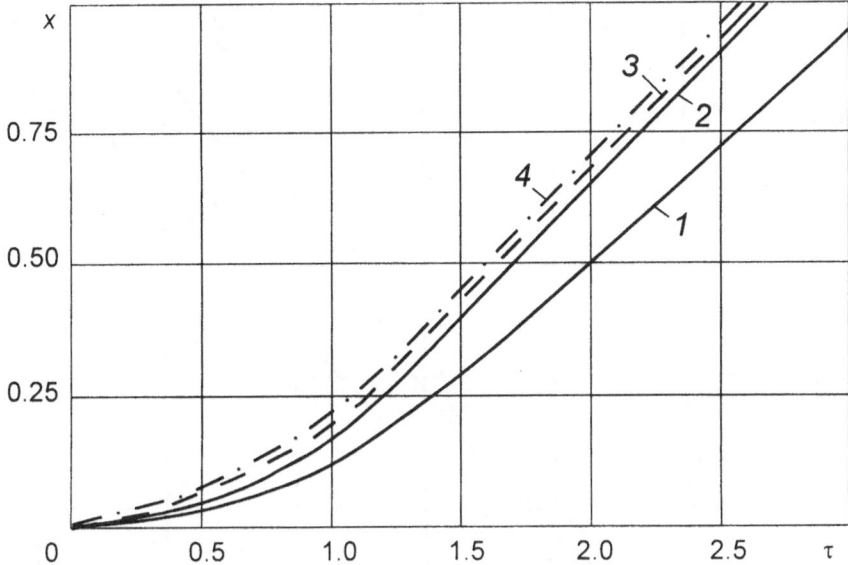

Fig. 6.2. The displacements of a perfectly rigid sphere ($m_0 = 1$) produced by an elastic wave. Stiff adhesion ($k_{10} = \infty$)

Let us now consider the determination of the medium's stress–strain state. When the force $R_e(\tau)$ is predetermined arbitrary, it is necessary to substitute (6.10) into (6.5).

Let us consider the problem of a wave impinging an immovable sphere. For $n \neq 1$, the images $f_n^L(s)$ and $g_n^L(s)$ are determined by (4.58). Then, the displacements and stresses should be determined by (4.45). For $n = 1$, it is necessary to sum the right-hand sides of (4.58) and (6.4) in accounting for (6.14).

The results for the displacement $x(\tau)$ of the center of mass of a spherical obstacle of *neutral buoyancy* ($m_0 = 1$) placed into an elastic medium ($\gamma_0 = 1$, $\eta_0 = 1.871$) and impinged by an incoming wave are presented in Figs. 6.2 and 6.3. In the first case, it was assumed that a stiff contact of the sphere with the surrounding medium ($k_{10} = \infty$) takes place, and in the second case, it was assumed that a free sliding of the surfaces of contact ($k_{10} = 0$) takes place. The following law of variation of pressure in the incoming wave was assumed:

$$f(\tau) = -\frac{\tau_+^2}{2}.$$

Curves *1*, *2*, *3*, and *4* correspond to $d = 2$, $d = 5$, $d = 10$, and $d = \infty$ (a plane wave), respectively.

The curves presented demonstrate that $x(\tau)$ tends to the results derived for a plane wave with an increase in the distance d between the point source

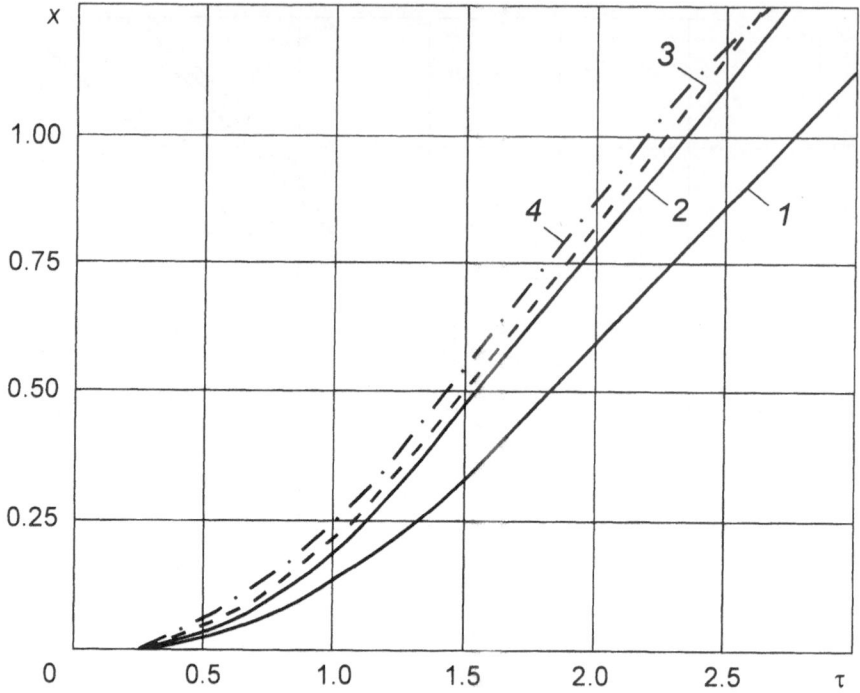

Fig. 6.3. The displacements of a perfectly rigid sphere ($m_0 = 1$) produced by an elastic wave. Free sliding ($k_{10} = 0$)

and the sphere's center. Comparison of the curves in Figs. 6.2 and 6.3 brings us to the conclusion that the conditions of contact do not affect the qualitative character of the variation of the displacement $x(\tau)$. However, in the case of free sliding, the displacement $x(\tau)$ is slightly larger than in the case of stiff adhesion.

6.3 Diffraction of Elastic Waves by a Perfectly Rigid Sphere Having Internal Elements

Let us consider the problem discussed in the previous section in a more complicated situation: we shall place N oscillators of mass M_l inside the perfectly rigid sphere. Each of the oscillators has one degree of freedom and can make only translational motion along the x axis. We shall designate the stiffness of suspension members of oscillators by k_l and the coefficient of linear damping by E_l (the Voigt model). Then, we can represent the system of equations of motion of the mechanical system under consideration along the x axis under the zero initial conditions in the following dimensionless form:

$$m_0 \ddot{x} = \frac{3}{4\pi} R_x + \sum_{l=1}^{N} m_l \omega_l^2 (x_l - x) + \sum_{l=1}^{N} \epsilon_l m_l (\dot{x}_l - \dot{x}) ,$$

$$\ddot{x}_l + \epsilon_l (\dot{x}_l - \dot{x}) + \omega_l^2 (x_l - x) = 0 ,$$

$$x(0) = \dot{x}(0) = x_l(0) = \dot{x}_l(0) = 0 ,$$

$$\omega_l^2 = \frac{\Omega_l^2 b^2}{c_*^2} , \qquad \epsilon_l = \frac{E_l b}{M_l c_*} , \qquad \Omega_l^2 = \frac{k_l}{M_l} , \qquad \frac{4}{3}\pi m_l = \frac{M_l}{\rho_* b^3} . \qquad (6.27)$$

Here, x_l is the absolute displacement of the lth oscillator along the x axis and R_x is the resultant force applied to the sphere when moving in an elastic medium. In the problem under consideration, when a wave propagating along the x axis impinges the mechanical system, we can determine the resultant force like we did deriving (6.14).

Applying the Laplace transform with respect to time τ to the system of equations (6.27) and taking into account the last note, we obtain

$$x^L(s) = 2\beta_0 E_0(s) \frac{s^3 P(s) T(s)}{\eta_0^2 X(s)} ,$$

$$x_l^L(s) = \frac{\epsilon_l s + \omega_l^2}{s^2 + \epsilon_l s + \omega_l^2} x^L(s) ,$$

$$X(s) = s^2 R(s) [m_0 T(s) + S(s)] + \beta_0 Q(s) T(s) ,$$

$$T(s) = \prod_{l=1}^{N} (s^2 + \epsilon_l s + \omega_l^2) , \qquad S(s) = \sum_{l=1}^{N} S_l(s) ,$$

$$S_l(s) = m_l (\epsilon_l s + \omega_l^2) \sum_{\substack{j=1 \\ j \neq l}}^{N} (s^2 + \epsilon_j s + \omega_j^2) . \qquad (6.28)$$

We can see that introduction of additional degrees of freedom does not complicate qualitatively the structure of expressions for the images of displacements of the sphere. Only the order of polynomials in the numerator and denominator of (6.28) increases.

Let us represent the solutions of the problem for the particular cases considered before ($\beta_0 \equiv 1$).

(1) An elastic medium and a spherical wave:

$$x^L(s) = -3 f^L(s) \frac{d-1}{d^2} \frac{(sd+1) P(s) T(s)}{\eta_0^2 X(s)} . \qquad (6.29)$$

(2) An elastic medium and a plane wave:

$$x^L(s) = -3 f^L(s) \frac{s P(s) T(s)}{\eta_0^2 X(s)} . \qquad (6.30)$$

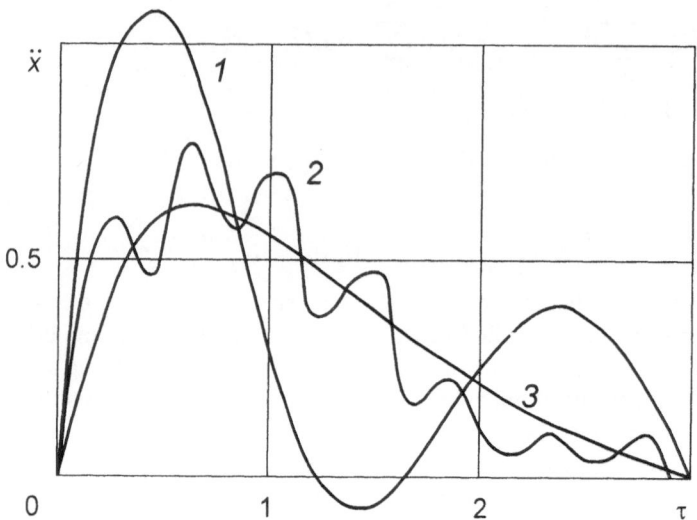

Fig. 6.4. Acceleration of the spheres with an oscillator

(3) An acoustic medium and a spherical wave:

$$x^L(s) = -3f^L(s)\,\frac{d-1}{d^2}\,\frac{(sd+1)\,T(s)}{[m_0T(s) + S(s)]\,(s^2+2s+2) + (s+1)\,T(s)}\,.$$

(6.31)

(4) An acoustic medium and a plane wave:

$$x^L(s) = -3f^L(s)\,\frac{sT(s)}{[m_0T(s) + S(s)]\,(s^2+2s+1) + (s+1)\,T(s)}\,.$$

(6.32)

Inversion of (6.29)–(6.32) under a predetermined law $f^L(s)$ can be made by means of residues. When making an analysis in practice, as a rule, it is important to determine the velocities and accelerations of the sphere and oscillators rather than their displacements. To calculate the velocities and accelerations, it is sufficient to derive the originals of corresponding expressions multiplied by s and s^2.

As an example, let us consider interaction of a plane wave with a sphere containing one oscillator ($N = 1$). The sphere is placed into an acoustic medium. We set

$$f(\tau) = -\frac{\tau_+^2}{2}\,, \qquad m_0 = m_1 = \frac{1}{2}\,, \qquad \epsilon_1 = 0\,.$$

Figure 6.4 illustrates the variation of the sphere acceleration $\ddot{x}(\tau)$ under various values of the frequency ω_1^2: curve *1* corresponds to $\omega_1^2 = 5$, curve *2* corresponds to $\omega_1^2 = 100$, and curve *3* corresponds to $\omega_1^2 = 1000$. Similar

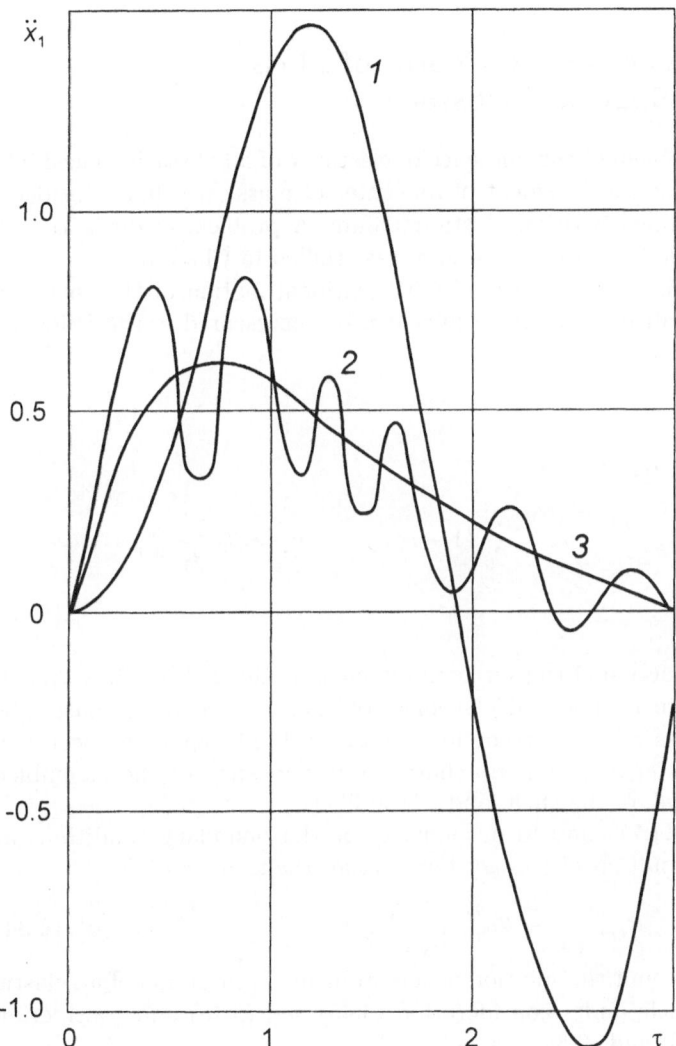

Fig. 6.5. Acceleration of an oscillator fastened inside a sphere

curves for the oscillator's acceleration $\ddot{x}_1(\tau)$ are presented in Fig. 6.5. The curves show that the variation of the eigenfrequency of oscillator affects significantly the parameters of motion of the system. As $\omega_1^2 \to \infty$, the curve $\ddot{x}_1(\tau)$ tends to the curve $\ddot{x}(\tau)$; this fact can be explained analytically by the relationship between $x_l^{\mathrm{L}}(s)$ and $x^{\mathrm{L}}(s)$ determined by (6.28).

6.4 Displacement of the Center of Mass of an Elastic Spherical Obstacle

Let us study a problem of the influence of elasticity of an obstacle placed into a continuum on the displacement of its center of mass. We shall consider a thin-walled shell filled with an elastic medium. A problem of diffraction of waves by an elastic body of such a kind was studied in Chap. 3.

Similarly to (6.9), the equation of translational motion of the center of mass of a thick-walled shell with a filler can be represented in the following dimensionless form:

$$m_0 \ddot{x} = \frac{3}{4\pi} R_x \,,$$

$$x(0) = \dot{x}(0) = 0 \,,$$

$$m_0 = m_1 + m_2 \,, \qquad m_1 = \frac{\gamma_1^2}{\beta_0} (1 - r_1^3) \,, \qquad m_2 = \beta_2 \frac{\gamma_1^2}{\beta_0} r_1^3 \,,$$

$$\beta_i = \frac{\lambda_i + 2\mu_i}{\lambda_1 + 2\mu_1} \,. \tag{6.33}$$

Here, R_x is the reaction of the surrounding medium divided by $(\lambda_0 + 2\mu_0) b^2$ and $x(\tau)$ is the dimensionless displacement of the shell's center of mass. The other dimensionless values correspond to those used in Chap. 3 and Sect. 6.2. The parameters m_0, m_1, and m_2 characterize buoyancy of the assembled structure in general, of the shell, and of the filler.

According to (4.71) and in accounting for the boundary conditions at $r = 1$ (3.3), the resultant of contact forces takes the form

$$\beta_0 R_x = \frac{4\pi}{3} \left(-\sigma_{rr1}^{(1)} \Big|_{r=1} + 2\sigma_{r\theta1}^{(1)} \Big|_{r=1} \right) \,. \tag{6.34}$$

Taking into account the solution of the problem of diffraction of an elastic wave by the obstacle under consideration (3.45) and (3.48) and considering the formulas (6.33) and (6.34), we get

$$x^{\mathrm{L}}(s) = \frac{\gamma_1^2}{\beta_0 m_0} \sum_{j,k,l,m} \left[R_{10}(-\gamma_1 s) f_{0jklm}^{(1)\mathrm{L}}(s) \right.$$

$$- R_{10}(\gamma_1 s) f_{1jklm}^{(1)\mathrm{L}}(s) \, e^{-\gamma_1 \delta s} - 2 R_{10}(-\eta_1 s) g_{0jklm}^{(1)\mathrm{L}}(s)$$

$$\left. + 2 R_{10}(\eta_1 s) g_{1jklm}^{(1)\mathrm{L}}(s) \, e^{-\eta_1 \delta s} \right] e^{-\tau_{jklm} s} \,. \tag{6.35}$$

At that, the functions $f_{0jklm}^{(1)L}(s)$, $f_{1jklm}^{(1)L}(s)$, $g_{0jklm}^{(1)L}(s)$, and $g_{1jklm}^{(1)L}(s)$ can be determined by a solution of the recursive system of equations (3.48) for $n = 1$.

Comparison of (3.45) for $r = 1$ and (6.35) shows that the displacement of the center of mass of an elastic obstacle does not coincide with the displacement determined by the first form of the series expansion in terms of the Legendre polynomials. The radial and tangential purely elastic displacements of the spherical shell u_1^e and v_1^e relative to the center of mass can be represented in the following form:

$$u_1^e = u_{10} + (u_{11} + x) \cos\theta + \sum_{n=2}^{\infty} u_{1n} P_n(\cos\theta),$$

$$v_1^e = -(v_{11} + x) \sin\theta + \sum_{n=2}^{\infty} v_{1n} C_{n-1}^{3/2}(\cos\theta). \tag{6.36}$$

As a particular case, let us consider the motion of the center of mass of a solid elastic sphere impinged by an elastic wave in the external medium.

We should note that the solution of the general problem obtained in Sect. 3.3 remains valid in this case when to set $\gamma_1 = \gamma_2$, $\eta_1 = \eta_2$, $\kappa_1 = \kappa_2$, and $\beta_2 = 1$. However, it can be simplified to reduce the number of unknown functions in the recursive relationships (3.48). At that, despite the fact that this simplification can be provided directly from (3.48), it is easier to perform anew the procedure of solution done in Sect. 3.3 for a solid sphere.

Taking into consideration the obstacle's homogeneity (i.e., arbitrary functions do not depend on the indexes j and k) and (3.45), we can write the images of displacements and stresses of the elastic sphere as follows:

$$u_{2n}^L(r, s) = -\frac{1}{r^{n+2}} \sum_{l,m} \left\{ \left[-R_{n1}(-\gamma_2 rs) e^{-\gamma_2(1-r)s} \right. \right.$$

$$\left. + R_{n1}(\gamma_2 rs) e^{-\gamma_2(1+r)s} \right] f_{lm}^L(s)$$

$$+ n(n+1) \left[-R_{n0}(-\eta_2 rs) e^{-\eta_2(1-r)s} \right.$$

$$\left. \left. + R_{n0}(\eta_2 rs) e^{-\eta_2(1+r)s} \right] g_{lm}^L(s) \right\} e^{-2(l\eta_2 + m\gamma_2)s},$$

$$v_{2n}^L(r, s) = \frac{1}{r^{n+2}} \sum_{l,m} \left\{ \left[-R_{n0}(-\gamma_2 rs) e^{-\gamma_2(1-r)s} \right. \right.$$

$$\left. + R_{n0}(\gamma_2 rs) e^{-\gamma_2(1+r)s} \right] f_{lm}^L(s)$$

$$+ \left[-R_{n3}(-\eta_2 rs) e^{-\eta_2(1-r)s} \right.$$

$$\left. \left. + R_{n3}(\eta_2 rs) e^{-\eta_2(1+r)s} \right] g_{lm}^L(s) \right\} e^{-2(l\eta_2 + m\gamma_2)s},$$

$$\sigma_{rrn}^{(2)L}(r, s) = \frac{\beta_2}{r^{n+3}} \sum_{l,m} \left\{ \left[-Q_{n1}^{(2)}(-\gamma_2 rs) e^{-\gamma_2(1-r)s} \right. \right.$$

$$\left. + Q_{n1}^{(2)}(\gamma_2 rs)\, e^{-\gamma_2(1+r)\,s} \right] f_{lm}^{L}(s)$$

$$+ n(n+1)\left[-Q_{n2}^{(2)}(-\eta_2 rs)\, e^{-\eta_2(1-r)\,s}\right.$$

$$\left.\left. + Q_{n2}^{(2)}(\eta_2 rs)\, e^{-\eta_2(1+r)\,s} \right] g_{lm}^{L}(s)\right\}\, e^{-2\,(l\eta_2+m\gamma_2)\,s}\,,$$

$$\sigma_{r\theta n}^{(2)L}(r,\,s) = \frac{\beta_2}{r^{n+3}} \sum_{l,\,m}\left\{\left[-Q_{n2}^{(2)}(-\gamma_2 rs)\, e^{-\gamma_2(1-r)\,s}\right.\right.$$

$$\left. + Q_{n2}^{(2)}(\gamma_2 rs)\, e^{-\gamma_2(1+r)\,s} \right] f_{lm}^{L}(s)$$

$$+ \left[-Q_{n3}^{(2)}(-\eta_2 rs)\, e^{-\eta_2(1-r)\,s}\right.$$

$$\left.\left. + Q_{n3}^{(2)}(\eta_2 rs)\, e^{-\eta_2(1+r)\,s} \right] g_{lm}^{L}(s)\right\}\, e^{-2\,(l\eta_2+m\gamma_2)\,s}\,. \quad (6.37)$$

As it was demonstrated in Sect. 3.3, the conditions of boundedness of the solution are satisfied in this case. Following the boundary conditions on the surface of solid sphere $r = 1$ and taking into account the representations of the components of the stress–strain state in the external medium (3.45), we arrive at the recursive relationships

$$f_{lm}^{L}(s) = \frac{Y_{n11}(s)}{X_{n0}(s)}\, f_{l,\,m-1}^{L}(s) + \frac{Y_{n12}(s)}{X_{n0}(s)}\, g_{l-1,\,m}^{L}(s)\,,$$

$$g_{lm}^{L}(s) = \frac{Y_{n21}(s)}{X_{n0}(s)}\, f_{l,\,m-1}^{L}(s) + \frac{Y_{n22}(s)}{X_{n0}(s)}\, g_{l-1,\,m}^{L}(s)\,, \quad (6.38)$$

which are similar to (3.48); we also obtain the initial conditions

$$f_{00}^{L}(s) = \frac{E_0(s)}{X_{n0}(s)}\, F_{01}(s)\,, \qquad g_{00}^{L}(s) = \frac{E_0(s)}{X_{n0}(s)}\, F_{02}(s)\,. \quad (6.39)$$

Here, the polynomials $Y_{nij}(s)$, $i, j = 1, 2$, $X_{n0}(s)$, $F_{01}(s)$, $F_{02}(s)$, and the function $E_0(s)$ have the same meaning as in (3.48) and (3.51). The elements of the matrices M_n and N_n are determined by (3.50) when to make the following changes:

$$M_{n11}(s) = -\beta_2 Q_{n1}^{(2)}(-\gamma_2 s)\,, \qquad M_{n12}(s) = -n(n+1)\,\beta_2 Q_{n2}^{(2)}(-\eta_2 s)\,,$$

$$M_{n21}(s) = -R_{n1}(-\gamma_2 s)\,, \qquad M_{n22}(s) = -n(n+1)\,R_{n0}(-\eta_2 s)\,,$$

$$M_{n31}(s) = -\beta_2 Q_{n2}^{(2)}(-\gamma_2 s)\,, \qquad M_{n32}(s) = -\beta_2 Q_{n3}^{(2)}(-\eta_2 s)\,,$$

$$M_{n41}(s) = -\beta_2 Q_{n2}^{(2)}(-\gamma_2 s) - k_{10} R_{n0}(-\gamma_2 s)\,,$$

$$M_{n42}(s) = -\beta_2 Q_{n3}^{(2)}(-\eta_2 s) - k_{10} R_{n3}(-\eta_2 s)\,. \quad (6.40)$$

Considering (6.33), (6,34), and (6.37), we finally obtain the Laplace images of the displacement of the center of mass of the solid elastic sphere along the x axis

$$x^L(s) = \sum_{l,m} \left\{ \left[R_{10}(-\gamma_2 s) - R_{10}(\gamma_2 s) e^{-2\gamma_2 s} \right] f_{lm}^L(s) \right.$$

$$+ 2 \left[-R_{10}(-\eta_2 s) + R_{10}(\eta_2 s) e^{-2\eta_2 s} \right] g_{lm}^L(s) \right\} e^{-2(l\eta_2 + m\gamma_2) s}.$$

$$(6.41)$$

In this case, the purely elastic components of displacements can also be represented by (6.36).

Inversion of (6.35) and (6.41) should be provided jointly with the inversion of corresponding recursive relationships using the methods proposed in Sect. 3.5.

6.5 A Perfectly Rigid Spherical Shell Filled with an Elastic Medium

Let us consider a perfectly rigid spherical shell of radius b, which makes a translational motion along the x axis following the predetermined law $x(\tau)$. The shell is filled with a linearly elastic medium, which is at rest at the initial instant of time. We shall assume that no external forces are applied to the shell. The displacement $x(\tau)$ is considered to be small within the hypothesis assumed in the linear theory of elasticity.

Using the dimensionless values introduced before and considering (1.12), we can write the equations of motion of the elastic medium (the index 2) with respect to the nonzero components of elastic potentials as follows:

$$\gamma_2^2 \varphi_2 = \Delta \varphi_2, \qquad \eta_2^2 \psi_2 = \Delta \psi_2 - \frac{\psi_2}{r^2 \sin^2 \theta}, \qquad r \le 1. \qquad (6.42)$$

The initial conditions are zeros, that is,

$$\varphi_2|_{\tau=0} = \dot{\varphi}_2|_{\tau=0} = \psi_2|_{\tau=0} = \dot{\psi}_2\Big|_{\tau=0} = 0. \qquad (6.43)$$

Let us represent the boundary conditions on the surface $r = 1$ in the generalized form. Then, we get

$$u_2|_{r=1} = -x \cos \theta, \qquad \sigma_{r\theta}^{(2)}\Big|_{r=1} = k_{12}(v_2|_{r=1} - x \sin \theta). \qquad (6.44)$$

We shall search for a solution of the problem (6.42)–(6.43) on the set of bounded functions, and is analogous to the condition at infinity for the external problems.

Expanding the functions sought into the series in terms of the Legendre and Gegenbauer polynomials (3.6), we obtain the following boundary conditions for the coefficients of these series:

$$u_{21}|_{r=1} = -x(\tau), \qquad \sigma_{r\theta 1}^{(2)}\Big|_{r=1} = -k_{12}(v_{21}|_{r=1} + x(\tau)),$$

$$u_{2n}|_{r=1} = 0, \qquad \sigma_{r\theta n}^{(2)}\Big|_{r=1} = -k_{12}\, v_{2n}|_{r=1}, \qquad n \ne 1. \qquad (6.45)$$

Because of homogeneity of the boundary conditions for $n \neq 1$, the mixed initial and boundary value problem has a nonzero solution only if $n = 1$.

Let us assume that $l + m > 0$. Then, applying the Laplace transform with respect to time τ, using the representations of images of displacements and stresses (6.37), and taking into account the boundary conditions (6.45), we obtain the system of recursive equations with respect to the functions $f_{lm}^L(s)$ and $g_{lm}^L(s)$

$$M_1 \begin{pmatrix} f_{lm}^L(s) \\ g_{lm}^L(s) \end{pmatrix} = N_1 \begin{pmatrix} f_{l,m-1}^L(s) \\ g_{l-1,m}^L(s) \end{pmatrix},$$

$$M_1 = (M_{1ij})_{2\times 2}, \qquad N_1 = (N_{1ij})_{2\times 2} \tag{6.46}$$

and the system of initial conditions

$$M_1 \begin{pmatrix} f_{00}^L(s) \\ g_{00}^L(s) \end{pmatrix} = - \begin{pmatrix} 1 \\ k_{12} \end{pmatrix} x^L(s). \tag{6.47}$$

At that, the elements of the polynomial matrices M_1 and N_1 can be calculated as follows:

$$
\begin{aligned}
&M_{111}(s) = R_{11}(-\gamma_2 s), &&M_{121}(s) = -\beta_2 Q_{12}^{(2)}(-\gamma_2 s) - k_{10} R_{10}(-\gamma_2 s), \\
&M_{112}(s) = 2R_{10}(-\eta_2 s), &&M_{122}(s) = -\beta_2 Q_{13}^{(2)}(-\eta_2 s) - k_{12} R_{13}(-\eta_2 s), \\
&N_{1ij}(s) = M_{1ij}(-s).
\end{aligned} \tag{6.48}
$$

Solving the relationships (6.46) and (6.47) for $f_{lm}^L(s)$ and $g_{lm}^L(s)$, we get

$$
\begin{pmatrix} f_{lm}^L(s) \\ g_{lm}^L(s) \end{pmatrix} = \frac{Y_1}{X_{10}(s)} \begin{pmatrix} f_{l,m-1}^L(s) \\ g_{l-1,m}^L(s) \end{pmatrix}, \qquad l + m > 0,
$$

$$
\begin{pmatrix} f_{00}^L(s) \\ g_{00}^L(s) \end{pmatrix} = \frac{x^L(s)}{X_{10}(s)} \begin{pmatrix} F_{01}(s) \\ F_{02}(s) \end{pmatrix},
$$

$$
\begin{pmatrix} F_{01}(s) \\ F_{02}(s) \end{pmatrix} = -X_{10} M_1^{-1} \begin{pmatrix} 1 \\ k_{12} \end{pmatrix},
$$

$$X_{10}(s) = \det M_1, \qquad Y_1 = (Y_{1ij})_{2\times 2} = X_{10} M_1^{-1} N_1. \tag{6.49}$$

Using (3.68) and applying the definitions of the functions $D_{n1}^{(i)}(y, z)$ (3.71) and $D_{n3}(y, z)$ (4.49), we obtain the following expressions for the elements of the matrix Y_1, the determinant X_{10}, and the function $F_{0k}(s)$ in explicit form:

$$X_{10}(s) = -\beta_2 D_{11}^{(2)}(-\gamma_2 s, -\eta_2 s) - k_{12} D_{13}(-\gamma_2 s, -\eta_2 s),$$

$$Y_{111}(s) = -\beta_2 D_{11}^{(2)}(\gamma_2 s, -\eta_2 s) - k_{12} D_{13}(\gamma_2 s, -\eta_2 s),$$

$$Y_{112}(s) = -4\eta_2 s^3(\eta_2^2 k_{12} + 2\beta_2),$$

$$Y_{121}(s) = -2\gamma_2^3 \eta_2^{-2} s^3(\eta_2^2 k_{12} + 2\beta_2),$$

$$Y_{122}(s) = -\beta_2 D_{11}^{(2)}(-\gamma_2 s, \eta_2 s) - k_{12} D_{13}(-\gamma_2 s, \eta_2 s),$$

$$F_{01}(s) = \beta_2 Q_{13}^{(2)}(-\eta_2 s) + k_{12} R_{20}(-\eta_2 s),$$

$$F_{02}(s) = -\beta_2 Q_{12}^{(2)}(-\gamma_2 s) - k_{12} R_{20}(-\gamma_2 s). \tag{6.50}$$

Transforming the polynomials (6.50) in accounting for the designations introduced in (4.73), we can write the recursive relationships (6.49) in scalar form

$$f_{lm}^{\mathrm{L}}(s) = \frac{Z_{11}(s)}{R(s)} f_{l,m-1}^{\mathrm{L}}(s) + \frac{Z_{12}(s)}{R(s)} g_{l-1,m}^{\mathrm{L}}(s),$$

$$g_{lm}^{\mathrm{L}}(s) = \frac{Z_{21}(s)}{R(s)} f_{l,m-1}^{\mathrm{L}}(s) + \frac{Z_{22}(s)}{R(s)} g_{l-1,m}^{\mathrm{L}}(s), \qquad l+m>0,$$

$$f_{00}^{\mathrm{L}}(s) = \frac{Z_{01}(s)}{s^2 R(s)} x^{\mathrm{L}}(s), \qquad g_{00}^{\mathrm{L}}(s) = \frac{Z_{02}(s)}{s^2 R(s)} x^{\mathrm{L}}(s). \tag{6.51}$$

Here, the polynomials $Z_{ij}(s)$ and $R(s)$ take the form

$$Z_{11}(s) = k_{12}\eta_2^2 G_1(s; \gamma_2, -\eta_2) + \beta_2 G_2(s; \gamma_2, -\eta_2),$$

$$Z_{12}(s) = 4\eta_2^3 s(\eta_2^2 k_{12} + 2\beta_2), \qquad Z_{21}(s) = 2\gamma_2^3 s(\eta_2^2 k_{12} + 2\beta_2),$$

$$Z_{22}(s) = k_{12}\eta_2^2 G_1(s; -\gamma_2, \eta_2) + \beta_2 G_2(s; -\gamma_2, \eta_2),$$

$$Z_{01}(s) = -k_{12}\eta_2^2 R_{20}(-\eta_2 s) - \beta_2 R_{12}(-\eta_2 s),$$

$$Z_{02}(s) = (k_{12}\eta_2^2 + 2\beta_2) R_{20}(-\gamma_2 s),$$

$$R(s) = k_{12}\eta_2^2 R_1(s) + \beta_2 R_2(s), \qquad R_1(s) = G_1(s; -\gamma_2, -\eta_2),$$

$$R_2(s) = G_2(s; -\gamma_2, -\eta_2). \tag{6.52}$$

Thus, using the algorithm of inversion of the Laplace transform presented in Sect. 3.5, the formulas (6.37) and the relationships (6.51), we can determine the components of the stress–strain state of the medium, which fills the sphere. In the problem under consideration, these components are as follows:

$$u_2(\tau, r, \theta) = u_{21}(r, \tau)\cos\theta, \qquad v_2(\tau, r, \theta) = -v_{21}(r, \tau)\sin\theta,$$

$$\sigma_{rr}^{(2)}(\tau, r, \theta) = \sigma_{rr1}^{(2)}(r, \tau)\cos\theta, \qquad \sigma_{r\theta}^{(2)}(\tau, r, \theta) = \sigma_{r\theta1}^{(2)}(r, \tau)\sin\theta. \tag{6.53}$$

When deriving the equations of motion of a perfectly rigid spherical shell, it is necessary to know the resultant force. Using (4.71) and taking into account the direction of the external normal to the surface bounding the elastic medium, we can represent the reaction R_x as follows:

$$R_x = \frac{4\pi}{3} \left(\sigma_{rr1}^{(2)} - 2\sigma_{r\theta1}^{(2)} \right) \Big|_{r=1} . \tag{6.54}$$

Substituting (6.37) into (6.54) and taking into account (3.22) in the space of the Laplace transforms, we obtain

$$R_x^L(s) = \frac{4\pi}{3} \beta_2 \gamma_2^2 s^2 \sum_{\cdot l, m} \left\{ \left[-R_{10}(-\gamma_2 s) + R_{10}(\gamma_2 s) \, e^{-2\gamma_2 s} \right] f_{lm}^L(s) \right.$$

$$\left. + 2 \left[R_{10}(-\eta_2 s) - R_{10}(\eta_2 s) \, e^{-2\eta_2 s} \right] g_{lm}^L(s) \right\} e^{-2 \, (l\eta_2 + m\gamma_2) \, s} . \tag{6.55}$$

The original $R_x(\tau)$ can be determined by (6.55) and (6.51) similarly to the originals of displacements and stresses.

In many cases, including the problems of contact dynamics, it is possible to confine our studies to the initial stages of motion. When $\tau < 2\gamma_2$, the exact value of the reaction R_x is determined by the first term of the series for $l = m = 0$ (see (6.55),). Then, in accounting for the designations introduced in (6.6), the image of the resultant force in this interval of time can be represented as follows:

$$R_x^L(s) = \frac{4\pi}{3} \beta_2 \frac{Q(s)}{R(s)} x^L(s) , \qquad Q(s) = k_{12} \eta_2^2 Q_1(s) + \beta_2 Q_2(s) ,$$

$$Q_1(s) = H_1(s; \, -\gamma_2, \, -\eta_2) , \qquad Q_2(s) = H_2(s; \, -\gamma_2, \, -\eta_2) . \tag{6.56}$$

We can see the similarity of (6.56) and the expression (6.6) in the case when the sphere moves in an infinite elastic medium. The difference is only in the signs of the formulas and in the signs of the polynomials $R_k(s)$ and $Q_k(s)$.

Two particular cases of contact follow from (6.56). When adhesion of the medium and the shell is perfectly stiff ($k_{12} = \infty$) or when sliding takes place ($k_{12} = 0$), it is necessary to replace the polynomials $Q(s)$ and $R(s)$ by $Q_1(s)$ and $R_1(s)$ or by $Q_2(s)$ and $R_2(s)$, respectively.

The results presented are in their simplest form for an acoustic medium ($\eta_2 \to \infty$). Then, it is necessary to set $g_{lm} \equiv 0$ in the recursive relationships and to omit the index l. Making a passage to the limit, as $\eta_2 \to \infty$, we obtain

$$f_m^L(s) = \frac{\gamma_2^2 s^2 + 2\gamma_2 s + 2}{\gamma_2^2 s^2 - 2\gamma_2 s + 2} f_{m-1}^L(s) , \qquad m > 0 ,$$

$$f_0^L(s) = \frac{x^L(s)}{\gamma_2^2 s^2 - 2\gamma_2 s + 2} . \tag{6.57}$$

At that, the images of the resultant force take the form

$$R_x^L(s) = \frac{4\pi}{3} \beta_2 \gamma_2^2 s^2 \left[-R_{10}(-\gamma_2 s) + R_{10}(\gamma_2 s) \, e^{-2\gamma_2 s} \right] \sum_{m=0}^{\infty} f_m^L(s) \, e^{-2m\gamma_2 s} . \tag{6.58}$$

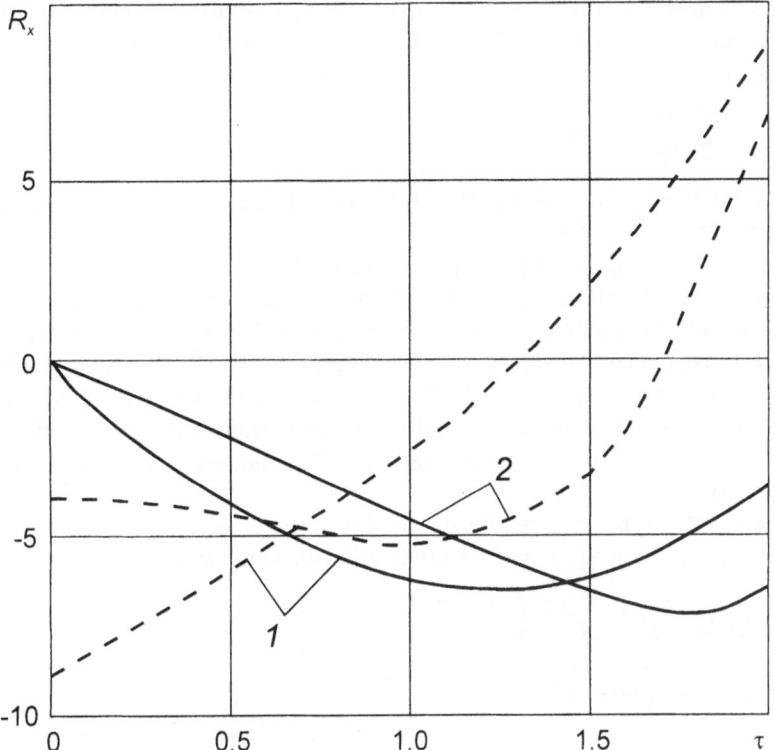

Fig. 6.6. The reaction of the filler under a translational motion of a perfectly rigid spherical shell

For $\tau < 2\gamma_2$, from (6.56), as $\eta_2 \to \infty$, we obtain

$$R_x^L(s) = -\frac{4\pi}{3}\,\beta_2\,\frac{s^2(\gamma_2 s - 1)}{\gamma_2^2 s^2 - 2\gamma_2 s + 2}\,x^L(s)\,. \tag{6.59}$$

Let us derive the roots of the denominator of (6.56) for the case of perfectly stiff contact (note that when sliding takes place, the degree of denominator is equal to three). Similarly to (4.83), we get

$$R(s_k) = 0\,, \qquad s_{1,2} = \alpha \pm \mathrm{i}\,\beta\,,$$

$$\alpha = \frac{1}{\gamma_2} + \frac{1}{2\eta_2}\,, \qquad \beta^2 = \frac{1}{\gamma_2^2} + \frac{3}{4\eta_2^2} - \frac{1}{\gamma_2\eta_2}\,,$$

$$1 \le \alpha\gamma_2 \le 1 + \frac{1}{2\sqrt{2}}\,, \qquad \frac{11 - 4\sqrt{2}}{8} \le \gamma_2^2\beta^2 \le 1\,. \tag{6.60}$$

As compared to the problem considered in Sect. 6.1, the roots s_1 and s_2 are located in the right-hand half-plane and the function $R_x(\tau)$ is of undamped type; this fact is typical for the internal problems (Babaev et al. (1974)).

Let us consider a linear law of variation of the displacement of shell $x(\tau) = \tau_+$. In accounting for (6.59), the original $R_x(\tau)$ can be represented as follows:

$$R_x = -\frac{4\pi}{3} \frac{\beta_2}{\gamma_2} e^{\tau/\gamma_2} \cos \frac{\tau}{\gamma_2} . \tag{6.61}$$

Let us point out once again the difference in the signs of exponential functions in (6.61) and (6.8).

For example, let us calculate the reaction of an elastic medium (steel, $\gamma_2 = 1$, $\eta_2 = 1.871$) filling the sphere under a translational motion of the shell. The results of calculations in the half-interval $0 \le \tau < 2$ are presented in Fig. 6.6. Here, curves 1 correspond to stiff adhesion of the shell to the filler ($k_{12} = \infty$) and curves 2 correspond to free sliding of the surfaces of contact ($k_{12} = 0$). The dashed lines correspond to the case when the motion follows the law $x(\tau) = \tau_+$ and the solid lines correspond to the case when the motion follows the law $x(\tau) = \tau_+^2/2$.

Similarly to Sect. 6.1, we can derive the limiting values of the reaction R_x, as $\tau \to +0$, in the case $x(\tau) = \tau_+$. Under stiff adhesion, we obtain

$$R_x(+0) = -\frac{4\pi}{3} \left(1 + \frac{2}{\eta_0}\right) ;$$

when sliding takes place, we get

$$R_x(+0) = -\frac{4\pi}{3} .$$

As it follows from Fig. 6.6, the type of contact of the filler with the shell has no marked effect on the resultant force R_x.

7. Penetration of Spherical Bodies into a Fluid Half-Space

In this chapter, we study the problem of impact of a deformable spherical body against a fluid half-space and the problem of penetration of such a body into the half-space. The fluid is assumed to be ideal. The problems of such a kind are of practical importance. The results of their studies can be applied in different areas of modern engineering. The state of the art is sufficiently well analyzed in the reviews Gorshkov (1976), Gorshkov (1981), and Vestyak et al. (1983) and in the books Kubenko (1981) and Sagomonyan (1986).

Below, we pay our main attention to the problem of the vertical entry of thin-walled elastic spherical shells into compressible fluids. Because of the complexity of derivation of the analytic solutions (this is mainly caused by considerable deformations of contacting and free surfaces, appearance and development of cavitation zones in the fluid, and plastic deformations in the structure's material), we consider the numerical methods to be the most promising ones.

In this chapter, we also study a problem of entry of a rigid sphere (a ball) into a fluid. The results of this study can be used for approximate analysis of submersion of deformable bodies and structures.

7.1 Penetration of Spherical Shells into a Compressible Fluid

Let us consider the initial stage of vertical entry of a thin-walled spherical shell connected with a rigid body of mass M_0 into a half-space occupied by an ideal compressible fluid (Fig. 7.1).

Before the instant of time when the shell will touch the fluid, the system 'shell – fluid' makes a translational motion at the speed V_0; the direction of the motion coincides with the axis of symmetry of the shell and with the normal to the fluid's undisturbed (plane) free surface, as well. The fluid motion caused by the shell submersion is axially symmetric. We shall neglect the viscosity, weight, and variation of internal energy of the fluid particles.

In order to describe the fluid behavior, let us apply the Lagrangian coordinates. Using the Lagrangian representation, we can determine the displacement of fluid's free surface directly in process of solution and state the

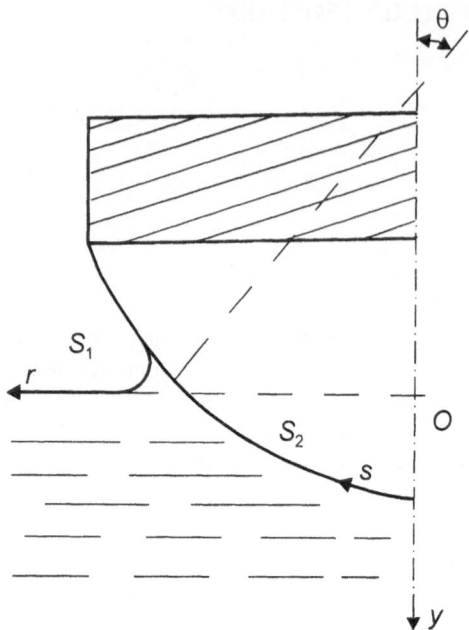

Fig. 7.1. Penetration of a spherical shell into a compressible fluid

boundary conditions at the shell's wetted surface exactly. Recall that we consider an axially symmetric motion of the fluid; then, using the immovable cylindrical coordinate system $r0y$ (Fig. 7.1), we can write the equations in the Lagrange form (Drobyshevsky (1980))

$$\frac{\partial u_1}{\partial \tau} = -\frac{1}{\rho}\frac{\partial p}{\partial r}, \qquad \frac{\partial u_2}{\partial \tau} = -\frac{1}{\rho}\frac{\partial p}{\partial y},$$

$$\rho r J = \text{const}, \qquad p = \frac{(\rho)^l - 1}{l},$$

$$\frac{\partial r}{\partial \tau} = u_1, \qquad \frac{\partial y}{\partial \tau} = u_2, \qquad J = \frac{\partial(r, y)}{\partial(r_0, y_0)}, \qquad (7.1)$$

where

$$\tilde{u}_1 = \frac{u_1}{c_1}, \qquad \tilde{u}_2 = \frac{u_2}{c_1}, \qquad \tilde{V} = \frac{V}{c_1}, \qquad \tilde{V}_0 = \frac{V_0}{c_1},$$

$$\tilde{r} = \frac{r}{R}, \qquad \tilde{y} = \frac{y}{R}, \qquad \tau = \frac{tc_1}{R}, \qquad \tilde{p} = \frac{p}{\rho_1 c_1^2}, \qquad \tilde{\rho} = \frac{\rho}{\rho_1}.$$

A tilde is used to designate the dimensionless values. From here on, we omit tildes in dimensionless values.

Here, u_1 and u_2 are the velocity components of the fluid particles in the directions r and y, respectively, p and ρ are the pressure and density of the fluid, c_1 and ρ_1 are the speed of sound and density of the undisturbed fluid,

J is the Jacobian, r_0 and y_0 are the initial coordinates of the fluid particles, R is the shell radius, V is the speed of submersion of the body at the instant of time t, and $l = 7.15$ is the constant of the state equation of water.

We should note that the equation of continuity (7.1) is represented in the form which differs from that presented in Chap. 1. This form of the equation of continuity is convenient for numerical calculations (it can be obtained if representing the equation of continuity in the integral form).

In order to describe the transient vibrations of the shell, let us use the geometrically nonlinear equations (1.53); recall that the model (1.53) takes into account the shear deformations and inertia of rotation of cross-sectional elements. Unlike (1.53), we set $b = R$, $c_* = c_1$, $\beta = \theta$, and $k_1 = k_2 = 1$, where θ is the angle in the spherical coordinate system (Fig. 7.1). Finally, the equations of axially symmetric motion of the shell take the form

$$\frac{\delta}{\xi^2}\frac{\partial^2 U}{\partial \tau^2} = \frac{\partial}{\partial \theta}(N_1 - \chi S) + f(N_1 - N_2 - \chi S)$$

$$+ (\eta N_1 + S) + \chi\,\frac{1-\nu^2}{\delta}\,\zeta p + \frac{\partial V}{\partial \tau}\,\frac{\sin\theta}{\xi^2},$$

$$\frac{\delta}{\xi^2}\frac{\partial^2 W}{\partial \tau^2} = \frac{\partial}{\partial \theta}(S + \eta N_1) + f(S + \eta N_1)$$

$$- N_1 + \chi S - N_2 - \frac{1-\nu^2}{\delta}\,\zeta p - \frac{\partial V}{\partial \tau}\,\frac{\cos\theta}{\xi^2},$$

$$\frac{1}{\xi^2}\frac{\partial^2 \chi}{\partial \tau^2} = -\frac{\partial H_1}{\partial \theta} - f(H_1 - H_2) - \eta H_1 + 12\,\frac{S}{\delta}, \qquad (7.2)$$

where

$$U = \frac{u}{h}, \qquad W = \frac{w}{h}, \qquad \eta = \delta\left(\frac{\partial W}{\partial \theta} - U\right),$$

$$\xi = \frac{c_0}{c_1} = \frac{1}{\gamma_1}, \qquad \zeta = \frac{\rho_1 c_1^2}{E}.$$

Replacing the shell action to the body by an equivalent system of forces distributed over the supporting contour, we obtain the following equation for determination of the speed of the body of mass M_0:

$$m_0\frac{dV}{d\tau} = \left[N_1|_{\theta=\theta_0}\sin\theta - S|_{\theta=\theta_0}\cos\theta\right]\frac{2\pi\sin\theta_0\delta}{(1-\nu^2)\,\xi}, \qquad m_0 = \frac{M_0}{\rho_1 R^3};$$

$$(7.3)$$

here, θ_0 is the half-angle of the shell.

The initial conditions are

$$V|_{\tau=0} = V_0, \qquad u_1|_{\tau=0} = u_2|_{\tau=0} = 0,$$

$$p|_{\tau=0} = 0, \qquad r|_{\tau=0} = r_0,$$

$$U|_{\tau=0} = W|_{\tau=0} = \chi|_{\tau=0} = \left.\frac{\partial U}{\partial \tau}\right|_{\tau=0} = \left.\frac{\partial W}{\partial \tau}\right|_{\tau=0} = \left.\frac{\partial \chi}{\partial \tau}\right|_{\tau=0} = 0. \quad (7.4)$$

The boundary conditions are as follows:

$$p|_{S_1} = 0, \qquad u_1|_{S_\infty} = u_2|_{S_\infty} = 0,$$

$$[u_1 \sin\theta + u_2 \cos\theta]|_{S_2} = V\cos\theta + \delta\frac{\partial W}{\partial \tau}; \qquad (7.5)$$

$$U|_{\theta=\theta_0} = W|_{\theta=\theta_0} = \chi|_{\theta=\theta_0} = 0; \qquad (7.6)$$

here, S_1 is the fluid's free surface, S_2 is the surface of contact of the shell with the fluid, and S_∞ is the fluid's undisturbed surface.

From the conditions of symmetry at the shell's pole, we get

$$U|_{\theta=0} = \chi|_{\theta=0} = \left.\frac{\partial W}{\partial \theta}\right|_{\theta=0} = 0. \qquad (7.7)$$

Thus, the problem of penetration of a spherical shell with an attached weight into a compressible fluid reduces to simultaneous integration of the system of equations (7.1)–(7.3) under corresponding initial and boundary conditions (7.4)–(7.7). The nonlinear problems of hydroelasticity formulated above make a complicated mathematical problem. For this reason, a rich variety of idealized and simplified schemes of penetration have been used (Vestyak et al. (1983), Kubenko (1981)).

The problem of submersion of deformable bodies (in particular, the bodies of spherical shape) can be divided conventionally into two interrelated parts:

(1) the determination of the fluid motion caused by submerging of deformable body (the *hydrodynamic problem*);

(2) the analysis of the structure dynamics under the action of the known hydrodynamic forces (the *elastic problem*).

At that, the hydrodynamic problem is a very complicated one from a mathematical point of view. In order to solve it, both the analytical and numerical methods are widely applied.

The analytical methods of determination of the hydrodynamic forces, which act on a blunt-nosed body while submerging either into a compressible fluid or into an incompressible fluid, are based mainly on a replacement of the problem of a body submergence by a problem of flow around a flat expanding plate or an expanding rigid disk (Grigolyuk, Gorshkov (1976),

Gorshkov (1981), Vestyak et al. (1983)). The solutions obtained based on such approaches are valid for small depths of submergence only; therefore, it does not always happen that the reaction forces in the submerging body reach their maximum values during the periods of time under consideration.

The usage of numerical methods (mainly the finite difference ones) gives us an opportunity to solve the problem under a more exact statement (Bazhenov, Mikhailov (1979), Sagomonyan (1986)). At that, the researchers employ the finite difference methods using both movable and immovable meshes. The solutions derived using immovable meshes (Sagomonyan (1986)) have the same limitations as the analytical methods due to the replacement of the problem of the body submergence by the determination of the flow around an expanding plate (disk). In the case when the movable meshes are used, the solution should be derived under a more exact statement when the simplification mentioned is not required (Bazhenov, Mikhailov (1979)); as a result, we have an opportunity to study the process of interaction of the shell with the fluid for the higher depths of submersion. When using the movable meshes, we can make the problem statement in the Lagrangian coordinates (Noh (1964), Butler (1971)), in the mixed Lagrange–Euler coordinates (Hirt (1971)), or in the Euler coordinates (Marchuk (1977)). At that, in all of these cases, the typical features of the Lagrangian representation are present.

To derive solutions using the mixed Lagrange–Euler coordinates or the Euler coordinates, it is necessary to make an additional reorganization of the mesh at each time step. Hence, the methods based on a pure Lagrangian representation are more efficient (they require less computing) but impose the limitations on the depth of submersion.

When studying the dynamics of shells during submersion (the elastic problem), the combined method is often used. This method consists of two stages: first, we make the transformation of the system of partial differential equations into the system of ordinary differential equations applying the Bubnov method or the method of lines; second, we seek a solution of the system of ordinary differential equations by means of the Runge–Kutta numerical procedure (Gorshkov (1981)).

In the case when we analyze the fluid behavior using the finite difference methods, it is more convenient to apply the same methods for the analysis of the shell dynamics, as well. Such an approach simplifies the procedure of synchronization of the solutions of equations of motion for the fluid and for the shell at each time step. The finite difference methods are also more efficient as compared to the Runge–Kutta method. Though the finite difference methods are of a lower order of approximation in time, they do not involve a considerable loss of accuracy. This can be explained by the fact that the shell's discretization gives us a higher error in the solution and, hence, we have no opportunity to obtain a more accurate solution by increasing the order of approximation in time.

The development of the finite element method (FEM) and its application to the problems of penetration of deformable structures into fluids was presented in Ershov, Shakhverdi (1984). At that, the finite elements were used both for the description of the structure motion and for the description of the fluid motion.

Below, we solve the system of equations (7.1)–(7.7) by means of the finite difference method. The fluid region to be analyzed is covered by a mesh whose cells are the quadrangular Lagrangian elements moving jointly with the fluid. The radius vectors and velocities are determined at the quadrangle's vertices while the pressure and density are assumed to be constant within a cell. The finite difference equations for the velocity are derived by the integration of both sides of the relationships (7.1) over some small volume followed by the transformation of the corresponding volume integrals into the curvilinear ones. The contour of integration for a given cell's vertex is formed by the segments of straight lines drawn through the centers of neighboring cells. For a given cell, the density is determined from the condition of conservation of the fluid mass and the pressure is determined from the equation of state (Drobyshevsky (1980), Drobyshevsky (1984)).

In order to determine the surface of contact of the shell with the fluid, the following method can be used. We introduce two additional nodes into the scheme of numerical calculation and calculate their coordinates and velocities. The first node corresponds to the boundary of the shell's wetted surface; at that, for this node, we derive the velocity component normal to the body's surface from the boundary condition (7.5) on S_2 and we determine the corresponding tangential component by means of the linear interpolation based on two neighboring (regular) nodes of the shell.

The second node is located on the fluid's free surface at the distance $\delta_0 \leq V_* \Delta\tau$ from the first one (here, V_* is the speed of expansion of the shell's wetted surface and $\Delta\tau$ is the time step). The velocity of the second node is equal to the velocity of the nearest node located on the body's wetted surface at the last time step, when this regular node was still on the fluid's free surface.

The radius of the shell's wetted surface is calculated at each time step. We determine the point of intersection (the point of contact) of the shell's surface with the fluid's free surface, which is approximated by a broken line passing through the nodes located on this surface and through two additional nodes. When making such an approximation of the fluid's surface, a speed of expansion of the wetted surface calculated at each time step corresponds accurately to the real speed of expansion; this procedure leads to almost complete elimination of non-physical oscillations in pressure distributed over the shell's surface (parasitic oscillations of this type appear each time when the shell occupies a new fluid cell).

When determining the hydrodynamic pressure distributed over the shell's wetted surface, it is necessary to take into account an additional force caused by a discrete mode (in the scheme of calculations) of the process of occupation of the new fluid particles by the body. The necessity of introduction of such a force into the calculations is provided by the fact that the law of conservation of momentum will not be obeyed in the cell where the occupation of fluid's free surface takes place. The expansion of the shell's wetted surface during the nth time step causes a variation of the normal velocity of the fluid located under the surface of occupation (i.e., of those cells which got under the shell's wetted surface at the nth time step). Hence, in accounting for the difference approximations accepted, a variation of the momentum of the cell caused by the occupation of the fluid's free surface at the nth time step gives us a required value of additional force. When determining the hydrodynamic pressure applied to the shell, this force should be uniformly distributed over the surface of occupation at the given time step.

When studying fluid motion numerically using Lagrangian coordinates, we can watch an appearance of the non-physical motion of the mesh's vertices that are produced by the short-wave disturbances and cause significant distortions of the cells. In order to smoothen these disturbances, some kind of regularization of the mesh can be used (Butler (1971)).

The reduction of the differential equations for a shell (7.2) to difference equations can be provided by the integro-interpolation method (Samarskii (1971)).

Let us note that when representing the forces and moments via the displacements, the system of equations (7.2) takes the form

$$\frac{\partial^2 U}{\partial \tau^2} = L_1(U) + X_1(W, \chi) + Y_1(U, W, \chi) + Z_1,$$

$$\frac{\partial^2 W}{\partial \tau^2} = L_2(W) + X_2(U, \chi) + Y_2(U, W, \chi) + Z_2,$$

$$\frac{\partial^2 \chi}{\partial \tau^2} = L_3(\chi) + X_3(U, W) + Y_3(U, W, \chi), \tag{7.8}$$

where L_j and X_j, $j = 1, 2, 3$, are the linear terms of (7.2), Y_j are the nonlinear terms of (7.2), and Z_1 and Z_2 are the components of the external load.

Integrating both sides of (7.8) between the limits from $(i - 1/2)\,\Delta\theta$ to $(i + 1/2)\,\Delta\theta$ (here, $\Delta\theta$ is the step of a regular mesh which covers the shell's middle surface) and assuming that the values under the integral sign, except the hydrodynamic force, are constant within the length of the interval, we arrive at the following system of equations:

$$\frac{d^2\varphi}{d\tau^2} + A_1\varphi + A_2\varphi + Y = Z; \tag{7.9}$$

here,

$$\varphi = (U_0, \ldots, U_J; W_0, \ldots, W_J; \chi_0, \ldots, \chi_J)^T$$

is the vector of generalized node displacements of the shell's middle surface, A_1 and A_2 are the matrices obtained by a transformation of the linear part of the differential operators (7.2) (the operators L_j and X_j, respectively) when replacing them by the difference relationships, $\mathbf{Y}$ is the vector of aerodynamic load, and $\mathbf{Z}$ is the vector of hydrodynamic load; the vector $\mathbf{Y}$ is obtained by transformation of the nonlinear terms of equation (7.2). The approximation of forces and moments is made by means of the central differences saving the terms of the second order. Then, for example, for H_1, we obtain

$$H_{1i} = -\left(\frac{\partial \chi}{\partial \theta} + \nu f \chi\right)\bigg|_{i\Delta\theta} = -\left(\frac{\chi_{i+1} - \chi_{i-1}}{2\Delta\theta} + \nu f_i \chi_i\right) + O(\Delta\theta^2),$$

$$H_{1,i+1/2} = -\left(\frac{\partial \chi}{\partial \theta} + \nu f \chi\right)\bigg|_{(i+1/2)\Delta\theta}$$

$$= -\left(\frac{\chi_{i+1} - \chi_i}{\Delta\theta} + \nu \frac{f_{i+1}\chi_{i+1} + f_i \chi_i}{2}\right) + O(\Delta\theta^2), \qquad \Delta\theta \to 0.$$

The finite difference relationships for the other forces and moments can be represented similarly.

Replacing the second derivative with respect to time by the central differences and setting the difference operator A_1 in an implicit form, we obtain the following weighted finite difference scheme:

$$\left(E_1 + \alpha \Delta \tau_1^2 A_1\right) \frac{\varphi^{n+1} - 2\varphi^n + \varphi^{n-1}}{\Delta \tau_1^2} + A\varphi^n + \mathbf{Y}^n = \mathbf{Z}^n; \qquad (7.10)$$

here, α is the weighting parameter, $\Delta\tau_1$ is the time step of integration of the equations of shell, $A = A_1 + A_2$ is the matrix obtained by a transformation of the linear part of the differential operators (7.2), E_1 is the identity matrix, and n is the time step.

The displacements at the points (0) and (J) ($n = 0$ and $n = J$) can be determined from the conditions of symmetry (7.7) and the condition of restriction of the shell's edge (7.6).

In the case of immovable stiff restraint of the shell's edge, the condition (7.6) is followed exactly, that is, we have

$$U_J = W_J = \chi_J = 0. \qquad (7.11)$$

The conditions of symmetry can be reduced to the difference form by means of the expansion of the displacements at the points $\theta = 0$ into the Taylor series via the displacements in the neighboring nodes. At that, the number of terms of the expansion should be chosen from the condition of invariability of the second order of approximation of the initial system of equations (7.2).

The relationships obtained should be accounted for by the replacement of the terms of the matrices A and A_1 corresponding to the displacements at the points (0) and (J) by their expansions in terms of the displacements

at the neighboring points followed by the elimination of the displacements at
these points from the system of finite difference equations. The fact that the
matrix A_1 is a three-diagonal one follows from the method of its derivation.
Hence, the conditions of restriction and symmetry should not upset its three-
diagonal structure.

The matrix A_1 enters into the system of equations (7.10) having the multiplier $\Delta\tau_1^2$. Since the matrix three-diagonal structure remains invariable, a
decrease in the order of approximation of terms of the system of equations
(7.10) related to the points (0) and (J) does not affect the order of approximation of the initial system of equations.

The system of finite difference equations (7.10) is the system of the second order of accuracy with respect to time and with respect to the spatial
coordinate. This system approximates the initial system of differential equations (7.2). In order to study some aspects of the stability, let us consider the
linear part of the system of equations (7.10). The necessary condition of its
stability follows from the Fourier analysis (Marchuk (1977)); thus, we get

$$\Delta\tau_1 \leq \frac{2}{\sqrt{\lambda_0\left[(E_1 + \Delta\tau_1^2\alpha A_1)^{-1}A\right]}}, \tag{7.12}$$

where $\lambda_0(A)$ is the maximum eigenvalue of the matrix A. Assuming that the
matrices $(E_1 + \Delta\tau_1^2\alpha A_1)^{-1}$ and A have $J - 1$ independent eigenvectors and
taking into account that A_1 and A are the positively semi-definite matrices,
it is easy to prove that, for $\alpha > 0$,

$$\lambda_0\left[(E_1 + \Delta\tau_1^2\alpha A_1)^{-1}A\right] \leq \lambda_0(A). \tag{7.13}$$

Let us write the stability condition for the explicit scheme ($\alpha = 0$)

$$\Delta\tau_1 \leq \frac{2}{\sqrt{\lambda_0(A)}}. $$

Hence, from the condition (7.12), it follows that, for $\alpha > 0$, the scheme (7.10)
gives us an opportunity to perform calculations using a larger time step than
that in the explicit form.

It is impossible to estimate the eigenvalues of the matrices analytically;
for this reason, in the case of explicit schemes (recall that the system of
equations (7.2) is of a hyperbolic type (Ogibalov (1963))), it is common to
use the condition of stability determined by the Courant number. Thus, we
get

$$\Delta\tau_1 \leq \frac{\Delta\theta}{\xi}. \tag{7.14}$$

Accounting for the nonlinear terms in (7.10) should not affect significantly
the conditions of stability and, as the calculations have proven, for $\alpha = 1$,

the scheme (7.10) gives us an opportunity to perform calculations using the time step larger than that determined by the Courant number (the explicit scheme).

The solution of the system of linear algebraic equations obtained by the reduction of equations (7.10) at each time step should be obtained by the double sweep method (Samarsky (1974), Drobyshevsky (1980), Drobyshevskii (1984)).

Equation (7.3) used for determination of the speed of submersion of the body can be reduced to a finite difference form similarly.

A general scheme of calculations can be reduced to a series of transitions from the state at the instant of time τ^n to the state at the instant of time $\tau^{n+1} = \tau^n + \Delta\tau$. At that, at each time step of the transition, we solve sequentially the equations of motion of the fluid and of the shell. As a result of the restrictions imposed by the requirement of stability of the schemes, the maximum permissible time step for the difference equations of dynamics of the shell is smaller than that in the case of the equations describing the fluid motion; hence, when making one step in the calculations for the fluid, we should make several steps in the calculations for the shell. At that, we assume that the hydrodynamic load does not vary at these steps of transition from φ^n to φ^{n+1}.

When solving particular problems, we should choose the necessary number of nodes of the finite difference meshes and the values of time steps for integration of the equation of motion based on the numerical experiments. At the initial instant of time, the half-space occupied by the fluid should be covered by a quadrangular nonuniform mesh (square at the coordinate origin and then expanding). The fluid region neighboring to the part of free surface, which, within the time period under consideration, occupies the body's surface, should be covered by a uniform mesh having the constant step Δr (the presence of the nonuniform cells in this region at the nonzero instant of time leads to significant non-physical oscillations of the resultant hydrodynamic load). The appearance of 10...15 mesh's nodes on the shell's surface within the time interval under consideration provides sufficient accuracy for the determination of the distributed hydrodynamic pressure. As the comparisons made have proven, the further decrease in the mesh step (the increase in the number of nodes) does not affect the results significantly.

The shell's middle surface should be divided into 50...70 intervals; the further increase in the number of intervals does not affect the results significantly. The application of the operator A_1 in the finite difference equations of motion of the shell in the implicit form allows us to use one and the same time step for integration of the equations of motion of the fluid and for integration of the equations of motion of the shell in most of the problems considered.

As a test example, let us consider the process of submersion of a spherical shell into the water; at that, the shell is simply supported at a rigid body ($V_0 = 6.67 \cdot 10^{-3}$; $m = 0.075$; $\delta = 0.0255$; $\theta_0 = 0.42$; $\xi = 2.07$; $\zeta = 0.6$).

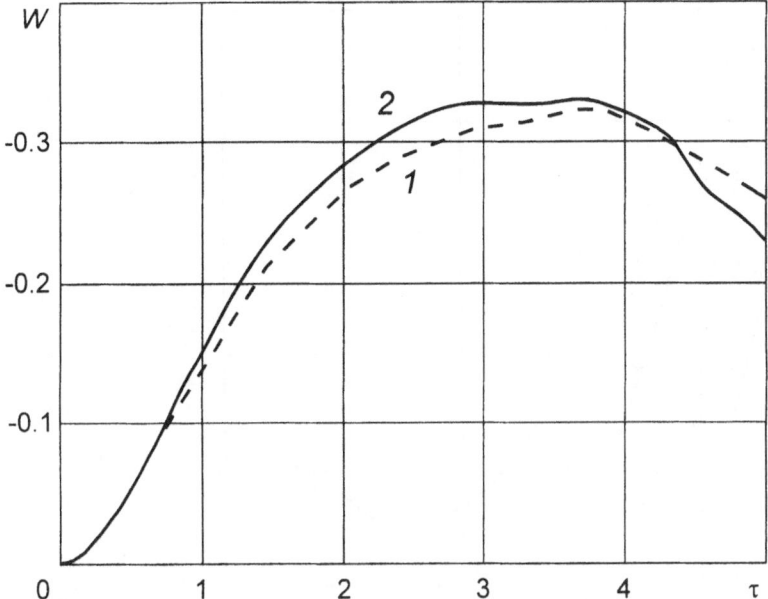

Fig. 7.2. Vertical penetration of a spherical shell into a compressible fluid. The displacement at the pole versus time

In Fig. 7.2, we present the normal displacement at the shell's pole $\theta = 0$ versus time (curve 1); here, $m = M/(\rho_1 R^3)$, where M is the mass of the whole system. Curve 2 corresponds to the approximate solution obtained for the incompressible fluid.

Let us now consider the submersion of a half-spherical shell into the water; at that, the shell is rigidly connected to a rigid body ($V_0 = 6.67 \cdot 10^{-3}$; $m = 3.0$; $\delta = 0.03$; $\theta_0 = \pi/2$; $\xi = 3.77$; $\zeta = 9.9 \cdot 10^{-3}$; $\nu = 0.3$). In Fig. 7.3, we present the distributions of the normal displacements of the shell along the shell's generatrix at different instants of time. At the initial stage of penetration, the normal displacements achieve the higher values at the shell's pole. The value of maximum displacement decreases with an increase in the submersion depth (for $\tau > 0.9$) and the point of its location moves away from the pole to the point of support.

In order to study the shell's stress–strain state, let us investigate the character of the variation in time of the membrane stresses

$$\bar{\sigma}_i^m = \frac{\sigma_i^m (1 - \nu^2)}{E} = N_i , \qquad i = 1, 2 ,$$

and the bending stresses (in the face filaments)

$$\bar{\sigma}_i^n = \frac{\sigma_i^n (1 - \nu^2)}{E} = \frac{H_i}{2} , \qquad i = 1, 2 ,$$

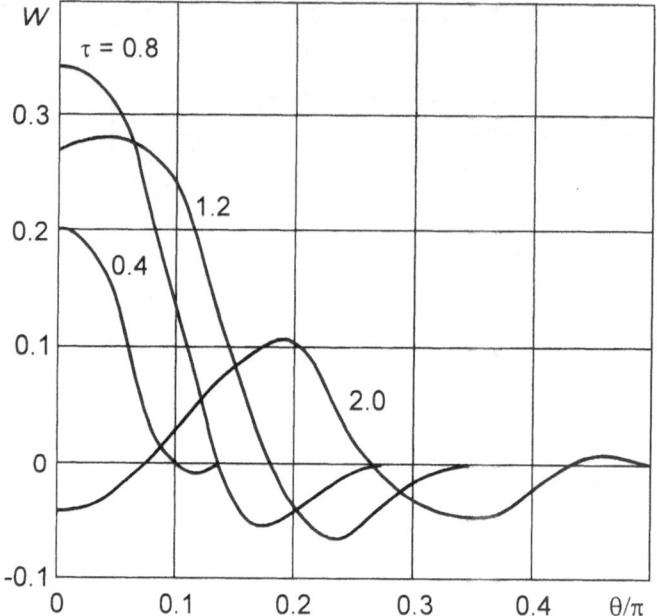

Fig. 7.3. Vertical penetration of a spherical shell into a compressible fluid. Distribution of the normal displacements along the shell's generatrix at different instants of time

at the characteristic points of the shell. In Fig. 7.4, curve *2* corresponds to $\bar{\sigma}_1^m = \bar{\sigma}_2^m$, $\theta = 0$; curve *4* corresponds to $\bar{\sigma}_1^m = \bar{\sigma}_2^m/0.3$, $\theta = \theta_0$; curve *1* corresponds to $\bar{\sigma}_1^n = \bar{\sigma}_2^n$, $\theta = 0$, and curve *3* corresponds to $\bar{\sigma}_1^n = \bar{\sigma}_2^n/0.3$, $\theta = \theta_0$.

The stress–strain state of the shell has the clearly defined moment properties. Both, the membrane and bending stresses reach their maximums at the shell's pole. At the initial stage of penetration, quite significant deformations take place in a small region neighboring to the shell's pole. At that, the bending deformations exceed in magnitude the membrane ones. During the followed submerging of the shell into the fluid ($\tau > 0.5$), the membrane stresses dominate.

When we want to apply the analytical methods for determination of the dynamic loads, one of the main assumptions to be used is the linearization of the boundary condition on the shell's wetted surface, that is, we should replace the condition (7.5) by the following one:

$$u_2|_{S_2} = V + \delta \frac{\partial W}{\partial \tau}.$$ (7.15)

In Figs. 7.4 and 7.5, we used the dashed lines to plot the curves corresponding to the stress at the shell's pole and the acceleration of the body

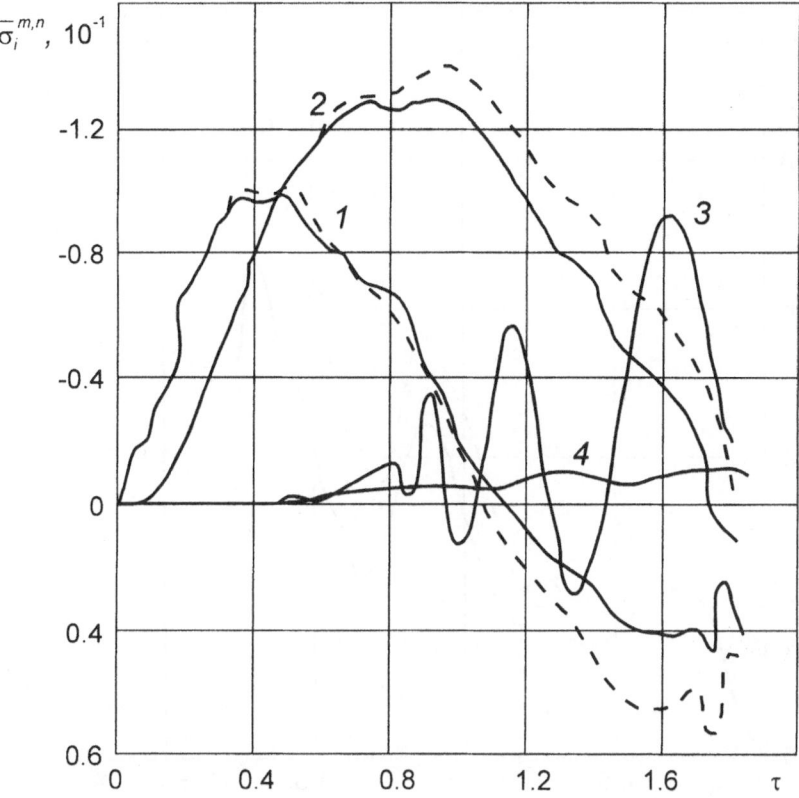

Fig. 7.4. Vertical penetration of a spherical shell into a compressible fluid. The membrane stress $\bar{\sigma}_i^m$ and the bending stress $\bar{\sigma}_i^n$ (at the faces of the shell) at the characteristic points of the spherical shell (the dashed curves correspond to the calculations made using the approximate condition at the shell's wetted surface)

attached to the shell, respectively. The curves were obtained using the boundary condition (7.15) for the wetted surface. Linearization of the boundary condition leads to an overestimation of the parameters, which characterize the shell's stress–strain state. At that, the overestimation is most conspicuous when considering the acceleration of the body attached to the shell.

In the case when a closed spherical shell of mass M submerges into the fluid, the origin of the coordinate system to be used for description of the shell's behavior should be placed into the shell's center of mass. Then, we can obtain the velocity of the shall from the following equation:

$$m \frac{dV}{d\tau} = -\iint_{S_2} p\,dS\,, \qquad m = \frac{M}{\rho_1 R^3}\,. \tag{7.16}$$

For the closed shell, the stresses reach their maximum values at the shell's pole (the point which is the first to get in touch with the fluid). At that, the

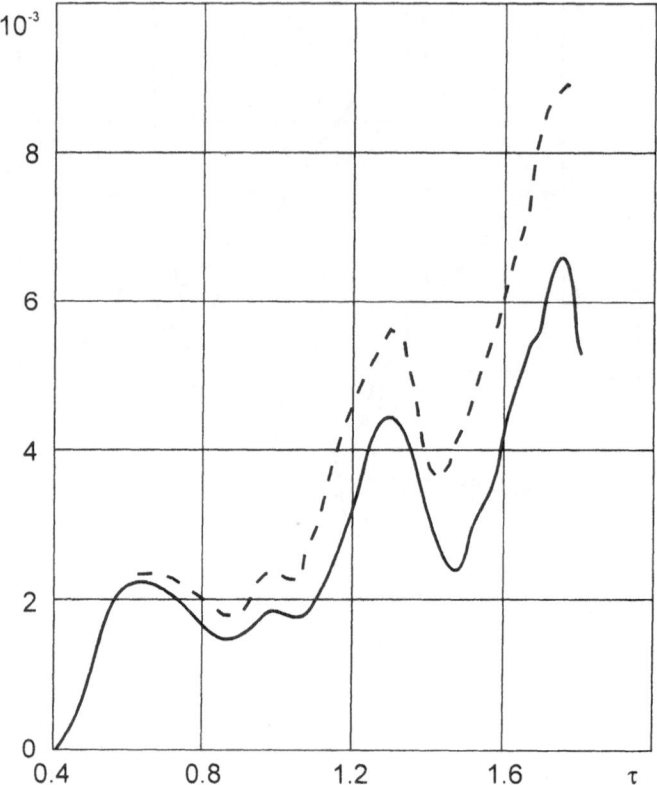

Fig. 7.5. Penetration of the shell attached to the body into the compressible fluid. The acceleration of the body $\dot{V}(\tau)$ versus time (the dashed curve corresponds to the estimations obtained applying the approximate, linearized condition for the shell's wetted surface)

character of the variation of the stresses and the values of stresses practically coincide with those produced at the pole of a shell which have the half-angle $\pi/2$ when the latter is attached to a rigid body whose mass is equal to a half of the mass of the closed shell (see Fig. 7.4).

7.2 Penetration of Spherical Shells into an Incompressible Fluid

When the velocities of submersion of deformable bodies (shells) into a fluid through its free surface are small (i.e., $V_0 \ll c_1$), the influence of the fluid compressibility can be detected only at the initial stage of penetration (when the compression wave has not left the body, yet). For the bodies of revolution, which have no plane boundaries, this period of time is very short. In this case, we can describe the fluid motion by the Laplace equation

$$\Delta \Phi = 0 \,, \tag{7.17}$$

and thus we can obtain the analytical expressions for the hydrodynamic loads applied to the body (the shell). Substituting these expressions into the equations of motion of the shell and integrating them using any method known, we can determine the characteristics of the reaction. In the case of submersion of spherical and cylindrical shells into a fluid, the results derived are in agreement with the experimental data and the numerical solutions obtained for the compressible fluids (Grigolyuk, Gorshkov (1976), Drobyshevsky (1980)).

Similarly to Sect. 7.1, let us confine ourselves to a study of vertical submersion of a spherical shell into an ideal incompressible fluid; we shall also assume that the shell is attached to a rigid body of mass M_0.

In the case of high-speed vertical submersion of elastic shells into fluid, the hydrodynamic loads reach their maximum values at small depths of submersion. For this reason, we can use the same considerations as in the case of rigid bodies (Grigolyuk, Gorshkov (1976)). We shall suppose that under the assumptions made earlier, the fluid flow around the elastic spherical shell is similar to the flow around an elastic disk of radius $b = b(\tau)$, which makes the translational motion at the speed $V + \dot{w}$, where V is the velocity of the system as a solid unit and w is the normal displacement of the shell's surface (Fig. 7.1). The less is the depth of submersion, the more accurate is the approximation of the shell's wetted surface.

Let us introduce the coordinates of an oblate ellipsoid of revolution (an oblate spheroid) ξ_1, ξ_2, and θ defined by the following relationships:

$$x = b \left(1 + \xi_1^2\right),$$

$$y = b \, \xi_1 \xi_2 \,,$$

$$z = b \sqrt{\left(1 + \xi_1^2\right)\left(1 - \xi_2^2\right)} \cos \theta \,,$$

$$0 \le \xi_1 < \infty \,, \qquad -1 \le \xi_2 \le 1 \,, \qquad 0 \le \theta \le 2\pi \,. \tag{7.18}$$

Here, $Oxyz$ is the rectangular coordinate system, the origin of which coincides with the initial point of contact; at that, the y axis is directed downwards, and two others are on a free surface of the fluid.

The surfaces $\xi_1 = $ const make up a family of confocal oblate ellipsoids of revolution, the surfaces $\xi_2 = $ const make up a family of confocal one-sheet hyperboloids, and the surfaces $\theta = $ const make up the family of meridianal planes. The surface $\xi_2 = 0$ corresponds to the fluid's free surface and the surface $\xi_1 = 0$ coincides with the disk.

The Laplace equation in the new coordinates takes the form

$$\left(1 + \xi_1^2\right) \frac{\partial^2 \Phi}{\partial \xi_1^2} + 2\xi_1 \frac{\partial \Phi}{\partial \xi_1} + \left(1 - \xi_2^2\right) \frac{\partial^2 \Phi}{\partial \xi_2^2}$$

$$-2\xi_2 \frac{\partial \Phi}{\partial \xi_2} + \frac{\xi_1^2 + \xi_2^2}{(1 + \xi_2^2)(1 + \xi_1^2)} \frac{\partial^2 \Phi}{\partial \theta^2} = 0.$$ (7.19)

In the future, we shall confine our studies to the axially symmetric problem. In that case, the potential Φ will depend on the coordinates ξ_1 and ξ_2 only. Thus, the problem of determination of the potential Φ will be reduced to solution of the equation

$$(1 + \xi_1^2) \frac{\partial^2 \Phi}{\partial \xi_1^2} + 2\xi_1 \frac{\partial \Phi}{\partial \xi_1} + (1 - \xi_2^2) \frac{\partial^2 \Phi}{\partial \xi_2^2} - 2\xi_2 \frac{\partial \Phi}{\partial \xi_2} = 0$$ (7.20)

under the following boundary conditions:

$$\frac{\partial \Phi}{\partial \xi_1}\bigg|_{\xi_1=0} = -b\xi_2 \left[V(t) + \frac{\partial w(\xi_2, t)}{\partial t} \right],$$ (7.21)

$$\Phi|_{\xi_2=0} = 0.$$ (7.22)

At infinity, we assume the condition of absence of fluid motion. After the separation of variables, we can write the solution of (7.20) as follows:

$$\Phi = \sum_{n=0}^{\infty} A_n P_{2n+1}(\xi_2) Q_{2n+1}(i\xi_1);$$ (7.23)

here, P_{2n+1} and Q_{2n+1} are the Legendre functions of the first and second kind.

The solution in the form (7.23) satisfies the boundary conditions at the fluid's surface (7.22) and at infinity. We can determine the constants A_n from the boundary condition (7.21). In order to do this, let us multiply the right-hand and left-hand sides of the boundary condition (7.21) by $P_{2n+1}(\xi_2)$ and make integration with respect to ξ_2 between the limits from 0 to 1. In accounting for the condition of orthogonality for the Legendre polynomials

$$\int_0^1 P_j(x) P_q(x) \, dx$$

$$= \begin{cases} \dfrac{1}{2j+1}, & q = j, \\[2mm] 0, & j - q \text{ even}, \quad q \neq j, \\[2mm] \dfrac{(-1)^{1/2(q+j-1)} q! \, j!}{2^{q+j-1}(j-q)(j+q+1) \left[\left(\dfrac{j}{2}\right)! \left(\dfrac{q-1}{2}\right)! \right]^2}, & \begin{array}{l} j \text{ even}, \\ q \text{ odd} \end{array} \end{cases},$$ (7.24)

we obtain

$$A_n = -b \, \frac{4n + 3}{\left. \frac{\partial}{\partial \xi_1} Q_{2n+1}(i\xi_1) \right|_{\xi_1 = 0}} \left[\frac{1}{3} V(t) \delta_{n0} + B_n \right] ,$$

$$B_n = \int_0^1 \xi_0 \dot{w}(\xi_2, t) P_{2n+1}(\xi_2) \, d\xi_2 , \qquad (7.25)$$

where δ_{nm} is the Kronecker delta.

Using the known representations for the Legendre functions of the first and second kind

$$Q_1(i\xi_1) = \xi_1 \arctan \xi_1 - 1 , \qquad (7.26)$$

we can write the solution (7.23) in the form of the sum of two potentials:

$$\Phi = \Phi_1 + \Phi_2 ,$$

$$\Phi_1 = \frac{2V}{\pi} b\xi_2(1 - \xi_1 \arctan \xi_1) ,$$

$$\Phi_2 = -b \sum_{n=0}^{\infty} \frac{4n + 3}{\left. \frac{\partial}{\partial \xi_1} Q_{2n+1}(i\xi_1) \right|_{\xi_1 = 0}} B_n P_{2n+1}(\xi_2) Q_{2n+1}(i\xi_1) . \qquad (7.27)$$

The potential Φ_1 corresponds to the translational motion of a rigid disk of radius b in a boundless fluid at the speed V (the Lamb solution) and the potential Φ is related to the presence of the additional field of velocities $\partial w / \partial t$ produced by the elastic vibrations of the shell.

In the future, when determining the pressure p, we shall need to use the following relationships for the Legendre functions of the second kind:

$$\left. \frac{Q_{2n+1}(i\xi_1)}{\frac{\partial}{\partial \xi_1} Q_{2n+1}(i\xi_1)} \right|_{\xi_1 = 0} = -\frac{[(n!)^2 2^{2n+1}]^2}{2\pi[(2n + 1)!]^2} . \qquad (7.28)$$

The hydrodynamic pressure p applied to the shell (disk) is related to the potential Φ by the Cauchy–Lagrange relationships (1.36). As a result, we obtain the dimensionless relationship

$$p = p_1 + p_2 , \qquad (7.29)$$

where

$$p_1 = \frac{2}{\pi} \left[\frac{V^2 \nu_0}{\kappa \sqrt{\nu_0^2 - r_0^2}} + \sqrt{\nu_0^2 - r_0^2} \, \frac{dV}{dt} \right],$$

$$p_2 = \frac{\delta}{2\nu_0^2} \sum_{n=0}^{\infty} f_n \left\{ \frac{V}{\kappa} \left[P_{2n+1}(\xi_2) \right. \right.$$

$$+ \left. \frac{1 - \xi_2^2}{\xi_2} P'_{2n+1}(\xi_2) \right] \int_0^{\nu_0} r_0 \frac{\partial W}{\partial \tau} P_{2n+1}(\xi_2) \, dr_0$$

$$+ \left. \nu_0 P_{2n+1}(\xi_2) \int_0^{\nu_0} r_0 \frac{\partial^2 W}{\partial \tau^2} P_{2n+1}(\xi_2) \, dr_0 \right\} ; \quad (7.30)$$

$$\nu_0 = \frac{b}{R}, \qquad \kappa = \frac{V}{\dfrac{d\nu_0}{d\tau}}, \qquad \xi_2 = \frac{\sqrt{\nu_0^2 - r_0^2}}{\nu_0}, \qquad r_0 = \frac{\sqrt{x^2 + z^2}}{R},$$

$$f_n = \frac{4n + 3}{\pi} \left[\frac{(n!)^2 2^{2n+1}}{(2n + 1)!} \right]^2 . \tag{7.31}$$

The function $\kappa(\tau)$ in (7.30) was introduced formally by analogy to the problem of impact of rigid bodies.

For a free surface $r > b$, we assume that the vertical velocities at each instant of time are the same as in the case of vertical impact of the disk of radius $b(\tau)$. Then, we can write the integral equation for determination of $\kappa(\tau)$ and, hence, $\nu_0(\tau)$, as follows:

$$\lambda = H - \int_0^r \frac{\kappa}{V} \left. \frac{\partial \Phi}{\partial y} \right|_{y=0} db_1 , \qquad r = \sqrt{x^2 + z^2} ; \tag{7.32}$$

here, λ is the equation of the contour of the deformed body and H is the depth of submersion.

To solve the problem (7.32) for the specified values of parameters requires many computations. When deriving the formulas (7.30) and (7.31), we have omitted some terms in the Cauchy–Lagrange integral which are proportional to the square of the fluid velocity. The role of these terms increases with an increase in the depth of submersion.

The formulas (7.29)–(7.32) give us the distribution of pressure over the elastic disk; these formulas are valid everywhere except a small region at the very boundaries of the impact surface, where a *splash jet* of fluid appears. In the regions where such a jet appears, the fluid flow (relative one) will not be similar to the flow around the disk; hence, we cannot use the formulas for pressure obtained earlier.

In order to study the phenomena which take place in the zone of the splash jet, we should analyze the fluid flow in this region and determine the characteristic function of this flow.

The peculiarities at $r = b$ mentioned above can be removed since the resultant force of resistance F takes the finite value, that is,

$$F(t) = 2 \int_0^\pi \int_0^b pr \, \mathrm{d}r \, \mathrm{d}\theta \,. \tag{7.33}$$

Introducing dimensionless variables

$$\tilde{F}_i = \frac{F_i}{\rho_0 R^2 c_0}\,, \qquad i = 1, 2\,,$$

and substituting (7.30) and (7.31) into (7.33), we obtain (a tilde is omitted)

$$F = F_1 + F_2\,,$$

$$F_1 = 4\nu_0^2 \left(\frac{V}{\kappa} + \frac{\nu_0}{3} \frac{\mathrm{d}V}{\mathrm{d}\tau} \right)\,,$$

$$F_2 = 12\delta \left(\frac{V}{\nu_0 \kappa} S_1(\nu, \tau) + \frac{1}{3} S_2(\nu, \tau) \right)\,; \tag{7.34}$$

$$S_1 = \int_0^{\nu_0} r_0 \sqrt{\nu_0^2 - r_0^2} \, \frac{\partial W}{\partial \tau} \, \mathrm{d}r_0\,,$$

$$S_2 = \int_0^{\nu_0} r_0 \sqrt{\nu_0^2 - r_0^2} \, \frac{\partial^2 W}{\partial \tau^2} \, \mathrm{d}r_0\,. \tag{7.35}$$

Thus, when taking into account the elasticity of the impacting body, the hydrodynamic forces F_1 and F_2 turn out to be interrelated via the function κ; this function should be determined as part of the solution process.

Similarly, we can derive the expressions for the hydrodynamic loads in the case of asymmetrical deformations of the shell. Then, the potential Φ should be determined from the solution of equation (7.19) (the boundary conditions remain the same form). Following Grigolyuk, Gorshkov (1976), let us expand the normal displacement of the shell $w(\xi_2, \theta, \tau)$ into the Fourier series with respect to the angular coordinate θ:

$$w = \sum_{m=0}^\infty w_m(\xi_2, t) \cos m\theta\,; \tag{7.36}$$

then, we can write the solution of (7.19) as follows:

$$\Phi = \sum_{m=0}^\infty \sum_{n=0}^\infty A_{nm} P_q^m(\xi_2) Q_q^m(\mathrm{i}\xi_1) \cos m\theta\,, \qquad q = m + 2n + 1\,; \tag{7.37}$$

here, $P_q^m(\xi_2)$ and $Q_q^m(\mathrm{i}\xi_1)$ are the associated Legendre functions of the first and second kind.

For $m = 0$, the solution (7.37) switches to (7.23). For $m \neq 0$, the hydrodynamic pressure applied to the shell can also be represented in the form of a sum of two terms (7.29); at that, p_1 can be determined by (7.30) and p_2 is as follows:

$$p_2 = \frac{\delta}{2\nu_0^2} \sum_{m=0}^{\infty} \sum_{n=0}^{\infty} f_{nm} \left\{ \frac{V}{\kappa} \left[P_q^m(\xi_2) \right. \right.$$
$$\left. + \frac{1 - \xi_2^2}{\xi_2} P_q^{m'}(\xi_2) \right] \int_0^{\nu_0} r_0 \dot{W}_m P_q^m(\xi_2)\, dr_0$$
$$\left. + \nu_0 P_q^m(\xi_2) \int_0^{\nu_0} r_0 \ddot{W}_m P_q^m(\xi_2)\, dr_0 \right\} \cos m\theta . \quad (7.38)$$

Here,

$$f_{nm} = \frac{2m + 4m + 3}{\pi} \left[\frac{(m+n)!\, n!\, 2^{2n+m+1}}{(2m+2n+1)!} \right]^2 , \qquad W_m = \frac{w_m}{h}. \quad (7.39)$$

The coefficient f_{nm} was derived based on the following relationship for the functions $Q_q^m(i\xi_1)$:

$$\left. \frac{Q_q^m(i\xi_1)}{i\, Q_q^{m'}(i\xi_1)} \right|_{\xi_1=0} = -\frac{[n!\,(n+m)!\, 2^{2n+m+1}]^2}{2\pi(2n+1)!\,(2m+2n+1)!} .$$

In Wilkinson et al. (1968), when considering the vertical entry of a spherical shell into the water, the authors did not consider how the elastic deformations and the backflow of the fluid affect the value of b. The relationship between the radius of wetted shell b and the depth of submersion H was assumed in the following form:

$$b = b_0 = \sqrt{2RH - H^2} \approx \sqrt{2RH}. \quad (7.40)$$

In this case, we have

$$\kappa = \frac{b_0}{R} \quad (7.41)$$

and we should replace ν_0 by $\nu_0^* = b_0/R$ in (7.30) and (7.38).

When approximating the shell's wetted surface by a plane disk, we can represent the potential Φ in the forms which differ from those derived above. For example, we can obtain the value of Φ at one side of the boundless plane ($y = 0$) via the values of Φ or $\partial\Phi/\partial n$ at the points of this plane (Lamb (1931)). Let us consider the Laplace equation in the cylindrical coordinate system yOr when the y axis is directed downwards. Then, taking into account the condition

$$\left. \frac{\partial \Phi}{\partial y} \right|_{y=0} = -f(r), \quad (7.42)$$

we arrive at

$$\Phi = \int_0^{\infty} e^{-ky} J_0(kr)\, dk \int_0^{\infty} f(r_1) J_0(kr_1)\, r_1\, dr_1 . \quad (7.43)$$

A similar formula can be derived in the case when there is no symmetry with respect to the y axis.

The formula (7.43) was used in Hirano (1973) for determination of the potential Φ_2 when solving the problem of vertical submersion into the water of a shallow spherical shell; the potential Φ_1 was calculated similarly to (7.27). At that, the boundary conditions for $y = 0$ were assumed in the following form:

$$f(r) = -w(r, t), \qquad r < b,$$
$$f(r) = 0, \qquad r > b; \tag{7.44}$$

here, b is the radius of the disk which approximates the shell's wetted surface. Then,

$$\Phi_2|_{y=0} = -\int_0^\infty J_0(kr)\, dk \int_0^b w(r_1 t)\, J_0(kr_1)\, r_1\, dr_1 . \tag{7.45}$$

We should note that the second condition (7.44) for the potential Φ_2, when $r > b$, corresponds to the model in which a rigid weightless screen covers the fluid's free surface. When determining the potential Φ_1 on the fluid's free surface, we assumed a more realistic boundary condition $\Phi_1 = 0$.

In Hirano (1973), the radius of the shell's wetted surface b was determined using the line of intersection of the fluid's undisturbed free surface with the shell's deformed contour

$$l + w|_{r=b} = \int_0^t V\, dt_1 , \qquad l = R - \sqrt{R^2 - b^2} ; \tag{7.46}$$

here, $w|_{r=b}$ is the displacement of the shell at the point $r = b$.

Thus, when using this approach, we do not take into account the increase in the area of the wetted surface caused by the backflow of the fluid; moreover, we determine the potential Φ_2 using the simplified boundary condition at the fluid's free surface.

The results of calculations based on the application of the expressions (7.29)–(7.31) are presented in Fig. 7.2 (curve 2). In order to describe the shells' motion, we integrated the equations of the theory of shallow shells (1.58) using the Bubnov method (Grigolyuk, Gorshkov (1976)); at that, we used seven terms of the series

$$W = \sum_{i=1}^\infty W_i(\tau)\, J_0(\xi_i \alpha) ,$$

where W is the normal displacements, $J_0(\xi_i \alpha)$ is the Bessel function of a nonzero order, and ξ_i is the root of the equation $J_0(\xi_i) = 0$.

For the given parameters of the system, the type of equations of motion does not play a significant role when estimating the stress–strain state. The comparisons made for the initial dimensionless speed of entry $V_0 = 6.67 \cdot 10^{-3}$ and the small depths of submersion $H/R < 0.055$ show that the influence of the fluid compressibility is insignificant.

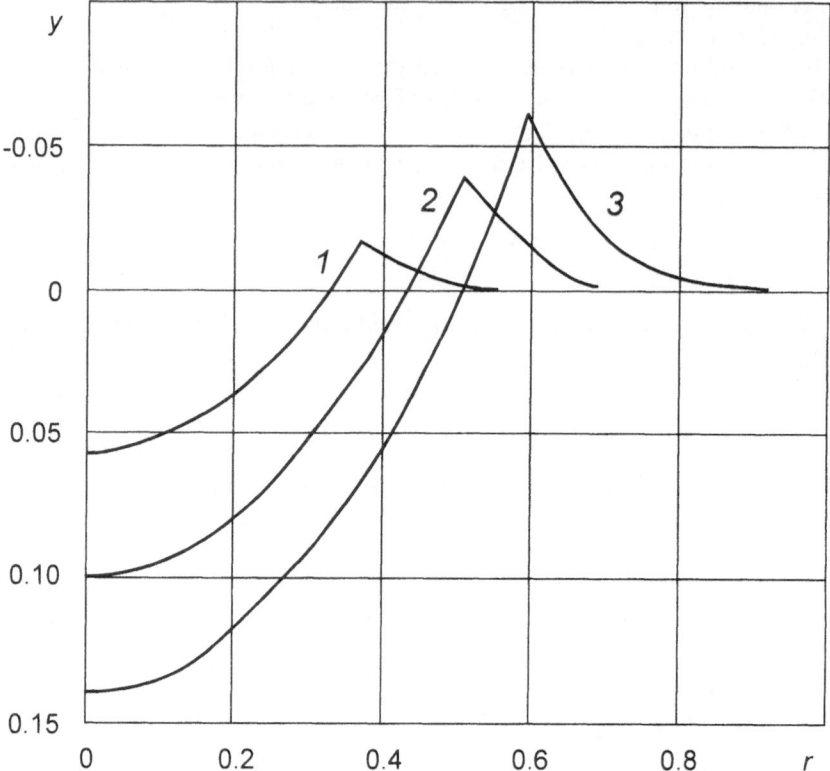

Fig. 7.6. Penetration of a rigid sphere into a compressible fluid. The shape of the fluid's surface at different instants of time

7.3 Penetration of Rigid Spherical Bodies into a Fluid Half-Space

The solution of the problem of submersion of rigid bodies into a fluid is necessary for testing the accuracy of various schemes of analysis applied in the more complicated problems of penetration of deformable structures into a fluid. Besides, the solutions obtained can be of practical interest (Grigolyuk, Gorshkov (1976)). In the case of vertical submersion of a rigid sphere (a ball) into a half-space occupied by a compressible fluid, we can obtain the solution of the problem numerically (see Sect. 7.1). From a mathematical point of view, the problem reduces to simultaneous integration of the system of equations (7.1), which describes the axially symmetric motion of fluid, and the following equation of motion of the rigid sphere:

$$m \frac{dV}{d\tau} = -2\pi \int_0^{r_*} pr\,dr, \qquad r_* = \frac{r^*}{R}, \qquad m = \frac{M}{\rho_1 R^3} \, ; \tag{7.47}$$

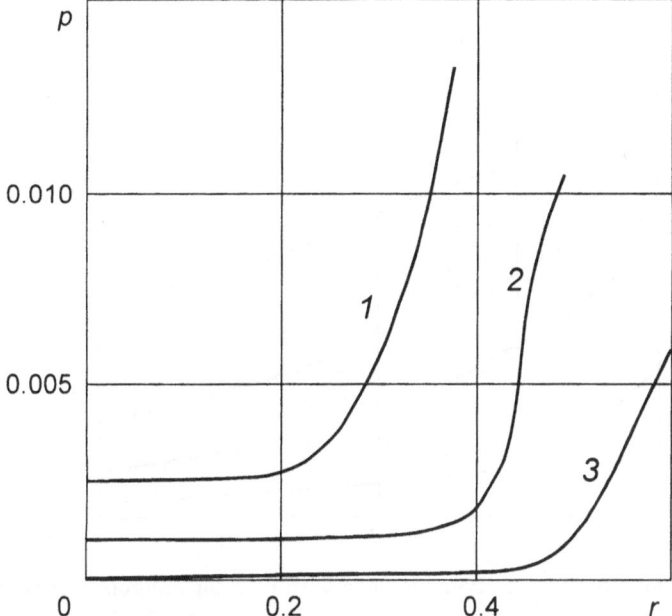

Fig. 7.7. The pressure distribution over the sphere's wetted surface at different instants of time

here, M is the body's mass and r^* is the radius of the body's wetted surface.

This system should be supplemented by the initial and boundary conditions (7.4) and (7.5); at that, the terms describing the elastic behavior of the shell must be omitted.

As an example, let us consider a problem of submersion of a sphere when $V_0 = 6.67 \cdot 10^{-2}$ and $m = 0.3$ (the entry at the speed $100 \, \mathrm{m/s}$). The shape of the fluid's surface at different instants of time is presented in Fig. 7.6: curve *1* corresponds to $\tau = 0.9$, curve *2* corresponds to $\tau = 1.8$, and curve *3* corresponds to $\tau = 2.7$.

During the submersion of the body, the free surface deforms; at that the radius of the sphere's wetted surface increases significantly. The pressure distribution over the sphere's wetted surface at different instants of time is presented in Fig. 7.7: curve *1* corresponds to $\tau = 0.9$, curve *2* corresponds to $\tau = 1.8$, and curve *3* corresponds to $\tau = 2.7$.

We can see that the pressure reaches its minimum values at the point of the first touch of the body and the fluid $r = 0$ and reaches its maximum values at the points where the fluid's free surface intersects with the body's wetted surface. The pressure at all of the points of the body's wetted surface decreases monotonically with an increase in the depth of submersion.

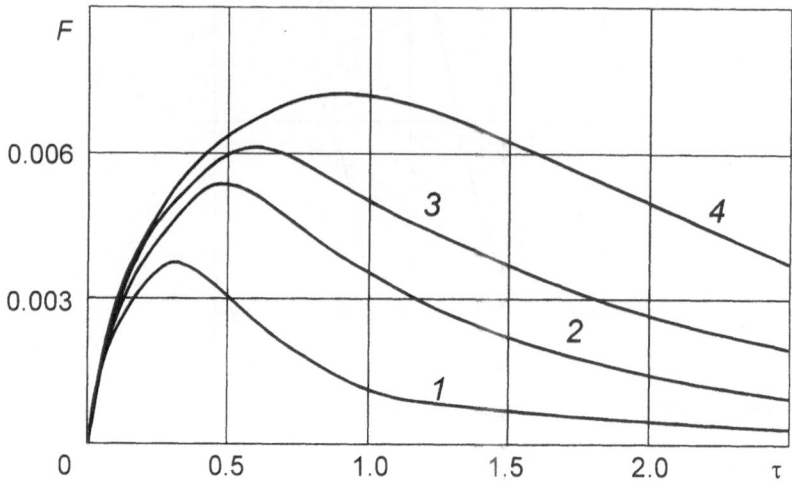

Fig. 7.8. The variation of the hydrodynamic force with time for different values of the parameter m

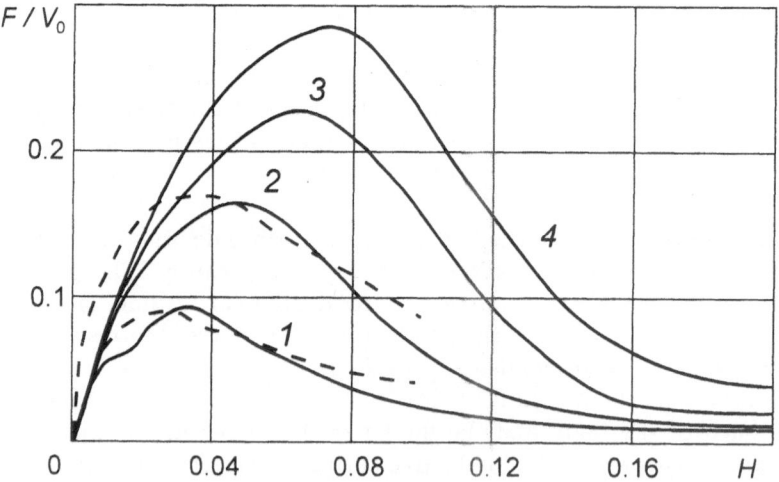

Fig. 7.9. The penetration of a rigid sphere into a compressible fluid. The variation of the hydrodynamic force with the depth of submersion

The variation of the hydrodynamic force

$$\tilde{F} = \frac{F}{\rho_0 c_0^2 R^2}$$

with time for different values of the parameter m is presented in Fig. 7.8: curve 1 corresponds to $m = 0.075$, curve 2 corresponds to $m = 0.3$, curve 3 corresponds to $m = 0.75$, and curve 4 corresponds to $m = \infty$. The maximum value of the hydrodynamic force for all of the values of the parameter m considered are reached at the initial stage of penetration ($\tilde{H} = H/R < 0.07$). The value of maximum force and the depth of submersion corresponding to it increase with an increase in the parameter m; at that, the limiting values of these parameters correspond to the submersion at a constant speed ($m = \infty$).

The hydrodynamic force F/V_0 versus the depth of submersion for different values of V_0 are presented in Fig. 7.9: curve 1 corresponds to $V_0 = 0.0667$, curve 2 corresponds to $V_0 = 0.1333$, curve 3 corresponds to $V_0 = 0.2$, and curve 4 corresponds to $V_0 = 0.2667$. The hydrodynamic force increases significantly with an increase in the initial speed of submersion; at that, the most intensive increase takes pace at the slow speeds of submersion.

The dashed lines in Fig. 7.9 correspond to the solution for a rigid sphere submerging into an incompressible fluid when the sphere's wetted surface is approximated by a plane expanding disk (see Sect. 7.2). In this case, we get $F = F_1$ (7.35). For a rigid sphere, equation (7.32) can be reduced to the Abel integral equation; as a result, we can find that (Grigolyuk, Gorshkov (1976))

$$\kappa = \frac{1 + \nu_0^2}{4\nu_0} \ln \left| \frac{1 + \nu_0}{1 - \nu_0} \right| - \frac{1}{2\nu_0} . \tag{7.48}$$

These considerations show that accounting for the fluid compressibility at the speeds of entry $V_0 < 0.0667$ (for the water, it is less than $100\,\text{m/s}$) does not make any significant effect on the hydrodynamic load. When the depths of submersion reach $H > 0.08$, the replacement of the sphere submersion by the flow about the expanding disk brings us to a significant overestimation of the hydrodynamic load.

The variations of the coefficient of resistance C_F (curve 1) and the depth of submersion $H_{\max}$ (curve 2) with the speed of entry are shown in Fig. 7.10; here, $H_{\max}$ is the depth, up to which the hydrodynamic force increases, and

$$C_F = \frac{F_{\max}}{\rho_0 V_0^2 R^2} ,$$

where $F_{\max}$ is the maximum value of the hydrodynamic force. The dashed lines are used to plot the results obtained in the experiments when the bodies strike against a fluid at a low speed of sound. The light circles are used to plot the results for the case of submersion of a rigid sphere into an incompressible fluid. In the example under consideration, the variation of the speed

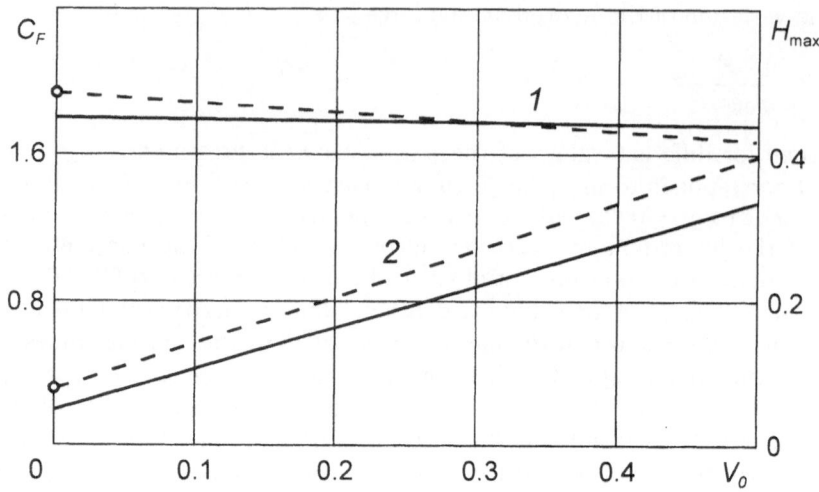

Fig. 7.10. The penetration of a rigid sphere into a compressible fluid. The variation of the coefficient of resistance C_F and the depth of submersion H_{max} with the speed of entry; the dashed curves correspond to the experimental data, and the circles correspond to the case of submersion of the sphere into an incompressible fluid

of submersion of the sphere at the initial stage of penetration is insignificant, that is, $m = \infty$.

We can see that the coefficient of resistance C_F depends only slightly on the dimensionless speed of entry, though the compressibility of the fluid significantly affected this parameter at the initial stage of submersion (Fig. 7.9).

8. Spherical Waves in Media with Complicated Properties

The problem of transient interaction of deformable bodies with surrounding media is versatile since, when making the studies, we deal with the bodies of different geometry and use various models of the media. Considering a separate class of boundary surfaces (for example, the spherical ones), we have an opportunity to make a sufficiently complete study of the particular problem and to reveal the main laws typical for other similar problems. The number of problems whose solution presumes accounting for the influence of the media on the behavior of structures, buildings, and systems, multiplies continuously. These are the problems for bio-mechanics, pipelines, diagnostics of defects, analysis of the nuclear reactors' components, design of the structures at sea shore and offshore, seismic effects, interactions of liquids and gasses with soft and permeable surfaces, etc.

Currently, there is no such general method which gives us an opportunity to derive a solution for the problems of transient mechanics for different kinds of media and body shapes using one and the same (or similar) approach. Even when we apply the numerical methods (the finite element method, the method of finite differences, the method of characteristics, etc.), we should consider accurately the specifics of geometry of the region occupied by the medium. For this reason, it is possible to develop methods for solution of the problems involving obstacles of canonic shape: cylinders, ellipsoids, paraboloids, etc. In the case of real structures, the solutions for these idealized obstacles can be used for qualitative estimations.

In this book, we use the following three models of continuum: an acoustic fluid, a linearly elastic homogeneous medium, and an isotropic medium; we also use the model of a thin-walled elastic shell. The methods of solution developed earlier can be used without serious modification for the analysis of the materials described by the linear equations of hyperbolic type, e.g., anisotropic materials, multicomponent (including the elastically-porous) media, etc. It is reasonable to extend the results obtained to the inhomogeneous elastic media, particularly to those having special types of singularities. The methods presented above require some modification when being applied to the models of linear viscoelastic and thermoelastic media and when accounting for the combined (forced and kinematic) interactions.

To make an example of application of the methods presented in this book, let us consider in brief the corresponding problems for the multicomponent media. When solving the wave problems, complexity of the processes caused by the deformation of media brings us to a large variety of the models and methods which can be used. Let us dwell on the linear theory proposed in Biot (1962) for an elastic-porous medium impregnated by fluid. In the absence of dissipative forces, the equations of motion are as follows:

$$N\Delta \mathbf{u} + (A + N)\, \text{grad div } \mathbf{u} + Q\, \text{grad div } \mathbf{U} = \rho_{11}\frac{\partial^2 \mathbf{u}}{\partial t^2} + \rho_{12}\frac{\partial^2 \mathbf{U}}{\partial t^2}\,,$$

$$Q\, \text{grad div } \mathbf{u} + R\, \text{grad div } \mathbf{U} = \rho_{12}\frac{\partial^2 \mathbf{u}}{\partial t^2} + \rho_{22}\frac{\partial^2 \mathbf{U}}{\partial t^2}\,,$$

$$\rho_{11} = (1 - \beta_0)\,\rho_\mathrm{s} - \rho_{12}\,, \qquad \rho_{22} = \beta_0\rho_\mathrm{f} - \rho_{12}\,; \tag{8.1}$$

here, $\mathbf{u}$ and $\mathbf{U}$ are the displacement vectors of the skeleton and fluid, respectively, A and N are the elastic constants of the skeleton medium, R is the pressure to be applied to the fluid in order to fill the porous volume (at that, the total volume remains invariable), Q is the value of the adhesion force between the solid and fluid components, ρ_{11} and ρ_{22} are the effective masses of the components under their relative motion, ρ_{12} is the coefficient of dynamical adhesion between solid and fluid components, ρ_s and ρ_f are the densities of the solid and fluid components, respectively, and β_0 is the porosity factor of the medium.

Let us introduce the scalar potentials φ_1 and φ_2 and the vector potential ψ of the medium displacements:

$$\mathbf{u} = \text{grad}(\varphi_1 + \varphi_2) + \text{rot } \psi\,,$$

$$\mathbf{U} = \text{grad}(\beta_1\varphi_1 + \beta_2\varphi_2) + \text{rot } \beta_3\psi\,, \qquad \beta_3 = -\frac{\rho_{12}}{\rho_{22}}\,,$$

$$(\rho_{22}Q - \rho_{12}R)\,\beta_k^2 + (\rho_{22}P - \rho_{11}R)\,\beta_k + \rho_{12}P - \rho_{11}Q = 0\,, \qquad k = 1, 2\,,$$

$$P = A + 2N\,; \tag{8.2}$$

then, we can reduce (8.1) to the following system of three independent wave equations:

$$c_k^2\Delta\varphi_k = \frac{\partial^2\varphi_k}{\partial t^2}\,, \qquad c_k^2 = \frac{P + Q\beta_k}{\rho_{11} + \rho_{12}\beta_k}\,, \qquad k = 1, 2\,,$$

$$c_3^2\Delta\psi = \frac{\partial^2\psi}{\partial t^2}\,, \qquad c_3^2 = \frac{N}{\rho_{11} + \rho_{12}\beta_3}\,. \tag{8.3}$$

Hence, in the medium described by the model under consideration, we have two longitudinal waves P_1 and P_2 propagating at the speeds c_1 and c_2, respectively, and the transverse wave S propagating at the speed c_3.

Equations (8.3) give us an opportunity to obtain the models of acoustic and elastic media making the passages to the limits. As $Q \to 0$, $R \to 0$, $\rho_{12} \to 0$, and $\rho_{22} \to 0$ (i.e., as $c_1 \to (\lambda + \mu)/\rho$, $c_2 \to 0$, $c_3 \to \mu/\rho$), we arrive at the equations of motion of an elastic medium, and, as $N \to 0$, $Q \to 0$, $R \to 0$, $A \to \lambda$, $\rho_{11} \to 0$, and $\rho_{12} \to 0$ (i.e., as $c_1 \to 0$, $c_2 \to \lambda/\rho$, $c_3 \to 0$), we obtain a model of an acoustic medium.

The fact that the motion of an elastic-porous medium can be described by the wave equation (8.3) allows us to apply the methods presented in this book without serious modification. We shall confine ourselves to a study of the axially symmetric problem on propagation of the disturbances from a spherical cavity (Gorshkov et al. (1987)) located in an infinite elastic-porous medium (see Sect. 4.2).

Similarly to (1.11), let us introduce the dimensionless values $c_* = c_1$ and $\lambda + 2\mu = H = P + 2Q + R$ and the following parameters:

$$\xi = \frac{A}{H}, \qquad \eta_1 = \frac{P}{H}, \qquad \eta_2 = \frac{Q}{H}, \qquad \zeta = \frac{R}{H},$$

$$\gamma_k = \frac{c_1}{c_k}, \qquad k = 1, 2, 3. \tag{8.4}$$

We shall use the spherical coordinate system (R, θ, ϑ). Taking into account the axial symmetry of the problem, we arrive at the equations of motion of the medium

$$\gamma_k^2 \frac{\partial^2 \varphi_k}{\partial \tau^2} = \Delta \varphi_k, \qquad k = 1, 2,$$

$$\gamma_3^2 \frac{\partial^2 \varphi_3}{\partial \tau^2} = \Delta \varphi_3 - \frac{\varphi_3}{r^2 \sin^2 \theta}, \tag{8.5}$$

where φ_3 is the nonzero component of the vector potential ψ.

At that, the radial displacements u_1 and u_2 and the tangential displacements v_1 and v_2 of the skeleton and fluid, respectively, the stresses in the skeleton $\sigma_{\alpha\beta}$, the pressure in the fluid σ_m, and the potentials φ_k are interrelated by the following differential expressions:

$$u_1 = \frac{\partial}{\partial r}(\varphi_1 + \varphi_2) + \frac{1}{r \sin \theta} \frac{\partial}{\partial \theta}(\varphi_3 \sin \theta),$$

$$v_1 = \frac{1}{r}\left[\frac{\partial}{\partial \theta}(\varphi_1 + \varphi_2) - \frac{\partial}{\partial r}(r\varphi_3)\right],$$

$$u_2 = \frac{\partial}{\partial r}(\beta_1 \varphi_1 + \beta_2 \varphi_2) + \frac{1}{r \sin \theta} \frac{\partial}{\partial \theta}(\beta_3 r \varphi_3),$$

$$v_2 = \frac{1}{r}\left[\frac{\partial}{\partial \theta}(\beta_1 \varphi_1 + \beta_2 \varphi_2) - \frac{\partial}{\partial r}(\beta_3 r \varphi_3)\right],$$

$$\sigma_{rr} = \eta_1 \frac{\partial u_1}{\partial r} + 2\xi \frac{u_1}{r} + \frac{\xi}{r}\left(\frac{\partial v_1}{\partial \theta} + v_1 \cot \theta\right)$$

$$+ \eta_2 \left[\frac{\partial u_2}{\partial r} + \frac{2}{r} u_2 + \frac{1}{r} \left(\frac{\partial v_2}{\partial \theta} + v_2 \cot \theta \right) \right],$$

$$\sigma_{r\theta} = \frac{\eta_1 - \xi}{2} \left(\frac{1}{r} \frac{\partial u_1}{\partial \theta} + \frac{\partial v_1}{\partial r} - \frac{v_1}{r} \right),$$

$$\sigma_m = \eta_2 \left[\frac{\partial u_1}{\partial r} + \frac{2}{r} u_1 + \frac{1}{r} \left(\frac{\partial v_1}{\partial \theta} + v_1 \cot \theta \right) \right]$$

$$+ \zeta \left[\frac{\partial u_2}{\partial r} + \frac{2}{r} u_2 + \frac{1}{r} \left(\frac{\partial v_2}{\partial \theta} + v_2 \cot \theta \right) \right]. \tag{8.6}$$

Similarly to Sect. 4.2, we shall consider two variants of boundary conditions at the cavity's surface.

Problem A. The stresses are predetermined at the cavity's surface:

$$\sigma_{rr}|_{r=1} = -(1 - \beta_0) p(\tau, \theta),$$
$$\sigma_m|_{r=1} = -\beta_0 p(\tau, \theta),$$
$$\sigma_{r\theta}|_{r=1} = q(\tau, \theta). \tag{8.7}$$

Problem B. The displacements are predetermined at the cavity's surface:

$$u_1|_{r=1} = W_1(\tau, \theta),$$
$$v_1|_{r=1} = W_2(\tau, \theta),$$
$$u_2|_{r=1} = W_3(\tau, \theta). \tag{8.8}$$

At infinity (as $r \to \infty$), the disturbances are absent.

We shall consider the zero initial conditions:

$$\varphi_k|_{\tau=0} = \left. \frac{\partial \varphi_k}{\partial \tau} \right|_{\tau=0} = 0, \qquad k = 1, 2, 3. \tag{8.9}$$

Thus, we state the mixed initial and boundary value problem (8.5)–(8.9). Similarly to (1.78), (3.6), and (4.41), let us represent the functions φ_1, φ_2, u_1, u_2, σ_{rr}, σ_m, and p in the form of the series in terms of the Legendre polynomials $P_n(\cos \theta)$ and the functions φ_3, v_1, v_2, $\sigma_{r\theta}$, and q in the form of the series in terms of the Gegenbauer polynomials $C_{n-1}^{3/2}(\cos \theta)$.

Then, we arrive at the following problem with respect to the coefficients of these series ($n = 0, 1, 2, \ldots$):

$$\gamma_k^2 \frac{\partial^2 \varphi_{kn}}{\partial \tau^2} = \Delta_n \varphi_{kn},$$

$$\varphi_{kn}|_{\tau=0} = \left. \frac{\partial \varphi_{kn}}{\partial \tau} \right|_{\tau=0} = 0, \qquad k = 1, 2, 3. \tag{8.10}$$

In the case of Problem A, we get the boundary conditions

$$\sigma_{rrn}|_{r=1} = -(1 - \beta_0)\, p_n(\tau),$$
$$\sigma_{mn}|_{r=1} = -\beta_0 p_n(\tau),$$
$$\sigma_{r\theta n}|_{r=1} = q_n(\tau), \tag{8.11}$$

and in the case of Problem B, we get the boundary conditions

$$u_{1n}|_{r=1} = W_{1n}(\tau),$$
$$v_{1n}|_{r=1} = W_{2n}(\tau),$$
$$u_{2n}|_{r=1} = W_{3n}(\tau). \tag{8.12}$$

The coefficients of the series for the potentials, displacements, and stresses are interrelated as follows:

$$u_{1n} = \frac{\partial}{\partial r}(\varphi_{1n} + \varphi_{2n}) - \frac{n(n+1)}{r}\varphi_{3n},$$

$$v_{1n} = \frac{\varphi_{1n} + \varphi_{2n} - \varphi_{3n}}{r} - \frac{\partial \varphi_{3n}}{\partial r},$$

$$u_{2n} = \frac{\partial}{\partial r}(\beta_1\varphi_{1n} + \beta_2\varphi_{2n}) - \frac{n(n+1)}{r}\beta_3\varphi_{3n},$$

$$v_{2n} = \frac{\beta_1\varphi_{1n} + \beta_2\varphi_{2n} - \beta_3\varphi_{3n}}{r} - \beta_3\frac{\partial \varphi_{3n}}{\partial r},$$

$$\sigma_{rrn} = \eta_1\frac{\partial u_{1n}}{\partial r} - n(n-1)\frac{\xi}{r}v_{1n} + \frac{2\xi}{r}u_{1n}$$
$$+\eta_2\left[\frac{\partial u_{2n}}{\partial r} + \frac{2}{r}u_{2n} - \frac{n(n+1)}{r}v_{2n}\right],$$

$$\sigma_{r\theta n} = -\frac{\eta_1 - \xi}{2}\left(\frac{u_{1n} - v_{1n}}{r} + \frac{\partial u_{1n}}{\partial r}\right),$$

$$\sigma_{mn} = \eta_2\left[\frac{\partial u_{1n}}{\partial r} + \frac{2}{r}u_{1n} - \frac{n(n+1)}{r}v_{1n}\right]$$
$$+\zeta\left[\frac{\partial u_{2n}}{\partial r} + \frac{2}{r}u_{2n} - n(n+1)\frac{v_{2n}}{r}\right]. \tag{8.13}$$

In order to solve the mixed initial and boundary value problem (8.10)–(8.12), we can use two methods: the method of generalized spherical waves (see, for example, Sect. 4.2) or the direct application of the Laplace transform with respect to time (see Sect. 3.2). In the case under consideration, these methods are equivalent. Let us demonstrate the second approach.

Similarly to (3.20), we can write the solution of (8.10) in the space of the Laplace transform as follows:

$$\varphi_{kn}^{L}(r,\,s) = \frac{1}{\sqrt{r}}\,[B_{kn}(s)I_{n+1/2}(\gamma_k rs) + D_{kn}(s)K_{n+1/2}(\gamma_k rs)]. \tag{8.14}$$

Considering the relationship between the Bessel function and the elementary functions (1.126) and taking into account the boundedness of the solution at infinity ($B_{kn} = 0$), we can write φ_{kn}^{L} as follows:

$$\varphi_{kn}^{L}(r,\, s) = \frac{C_{kn}(s)}{(\gamma_k rs)^{n+1}}\, R_{n0}(\gamma_k rs)\, e^{-\gamma_k rs},$$

$$C_{kn} = \frac{D_{kn}}{(\gamma_k s)^{n+1}}\, \sqrt{\frac{\pi}{2\gamma_k s}}\, e^{-\gamma_k s}. \tag{8.15}$$

Then, from (8.13) and (8.15) we can derive the following expressions for the coefficients of the series expansion of the displacements and stresses:

$$u_{1n}^{L}(s) = -\frac{1}{r^{n+2}}\left[\sum_{k=1}^{2} C_{kn}(s)\, R_{n1}(\gamma_k rs)\, e^{-\gamma_k(r-1)s}\right.$$
$$\left. + n(n+1)\, C_{3n}(s)\, R_{n0}(\gamma_3 rs)\, e^{-\gamma_3(r-1)s}\right],$$

$$u_{2n}^{L}(s) = -\frac{1}{r^{n+2}}\left[\sum_{k=1}^{2} \beta_k C_{kn}(s)\, R_{n1}(\gamma_k rs)\, e^{-\gamma_k(r-1)s}\right.$$
$$\left. + n(n+1)\, \beta_3 C_{3n}(s)\, R_{n0}(\gamma_3 rs)\, e^{-\gamma_3(r-1)s}\right],$$

$$v_{1n}^{L}(r,\, s) = \frac{1}{r^{n+2}}\left[\sum_{k=1}^{2} C_{kn}(s)\, R_{n0}(\gamma_k rs)\, e^{-\gamma_k(r-1)s}\right.$$
$$\left. + C_{3n}(s)\, R_{n3}(\gamma_3 rs)\, e^{-\gamma_3(r-1)s}\right],$$

$$v_{2n}^{L}(r,\, s) = \frac{1}{r^{n+2}}\left[\sum_{k=1}^{2} \beta_k C_{kn}(s)\, R_{n0}(\gamma_k rs)\, e^{-\gamma_k(r-1)s}\right.$$
$$\left. + \beta_3 C_{3n}(s)\, R_{n3}(\gamma_3 rs)\, e^{-\gamma_3(r-1)s}\right],$$

$$\sigma_{rrn}^{L}(r,\, s) = \frac{1}{r^{n+3}} \sum_{k=1}^{3} C_{kn}(s)\, Q_{1kn}(\gamma_k rs)\, e^{-\gamma_k(r-1)s},$$

$$\sigma_{r\theta n}^{L}(r,\, s) = \frac{1}{r^{n+3}} \sum_{k=1}^{3} C_{kn}(s)\, Q_{2kn}(\gamma_k rs)\, e^{-\gamma_k(r-1)s},$$

$$\sigma_{mn}^{L}(r,\, s) = \frac{1}{r^{n+3}} \sum_{k=1}^{3} C_{kn}(s)\, Q_{3kn}(\gamma_k rs)\, e^{-\gamma_k(r-1)s}; \tag{8.16}$$

here,

$$Q_{1kn}(s) = (\eta_1 + \beta_k \eta_2) R_{n2}(s) - (\xi + \beta_k \eta_2) [2R_{n1}(s) + n(n+1) R_{n0}(s)],$$
$$Q_{3kn}(s) = (\eta_2 + \beta_k \xi) [R_{n2}(s) - 2R_{n1}(s) - n(n+1) R_{n0}(s)],$$
$$Q_{2kn}(s) = (\eta_1 - \xi) [R_{n1}(s) + R_{n0}(s)], \qquad k = 1, 2,$$
$$Q_{13n}(s) = n(n+1) (\eta_1 - \xi) [R_{n1}(s) + R_{n0}(s)],$$
$$Q_{23n}(s) = \frac{\eta_1 - \xi}{2} [R_{n2}(s) + (n+2) (n-1) R_{n0}(s)],$$
$$Q_{33n}(s) = 0.$$

The polynomials $R_{nj}(s)$, $j = 1, 2, 3$, are defined by (3.22).

Substituting (8.16) into the boundary conditions (8.11) or (8.12), we obtain the system of linear algebraic equations with respect to C_{kn}, $k = 1, 2, 3$:

$$\mathbf{AC} = \mathbf{B},$$
$$\mathbf{A} = (a_{ij})_{3 \times 3}, \qquad a_{ij} = Q_{ijn}(\gamma_i s),$$
$$\mathbf{C} = (C_{1n}, C_{2n}, C_{3n})^{\mathrm{T}}, \qquad \mathbf{B} = (b_1, b_2, b_3)^{\mathrm{T}}. \tag{8.17}$$

Let us write the elements of the matrix $\mathbf{A}$ and of the column of free terms $\mathbf{B}$. In the case of Problem A, we have

$$a_{ik}(s) = Q_{ikn}(s),$$
$$b_1(s) = -(1 - \beta_0) p_n^{\mathrm{L}}(s), \qquad b_2(s) = q_n^{\mathrm{L}}(s),$$
$$b_3(s) = -\beta_0 p_n^{\mathrm{L}}(s). \tag{8.18}$$

Similarly, in the case of Problem B, we have

$$a_{1k}(s) = -R_{n1}(\gamma_k s), \qquad a_{2k}(s) = R_{n0}(\gamma_k s),$$
$$a_{3k}(s) = -\beta_1 R_{n1}(\gamma_k s), \qquad k = 1, 2,$$
$$a_{13}(s) = -n(n+1) R_{n0}(\gamma_3 s), \qquad a_{23}(s) = R_{n3}(\gamma_3 s),$$
$$a_{33}(s) = -n(n+1) \beta_3 R_{n0}(\gamma_3 s),$$
$$b_j(s) = W_{jn}^{\mathrm{L}}(s), \qquad j = 1, 2, 3. \tag{8.19}$$

Let us write the solution of the system of equations (8.17) as follows:

$$C_{kn}(s) = \frac{\Delta_k(s)}{\Delta(s)}, \qquad \Delta(s) = \det \mathbf{A}. \tag{8.20}$$

Here, Δ_k is the determinant derived from the determinant Δ by a replacement of the kth column by the column $\mathbf{B}$. Taking into account the polynomial structure of the matrix $\mathbf{A}$, we obtain that $\Delta_k(s)$ and $\Delta(s)$ are the polynomials of the degree $3n + 6$.

Substituting (8.20) into (8.16), we obtain a solution of the problem (8.10)–(8.12) in the space of the Laplace transform. Since the functions $C_{kn}(s)$ are rational, we can determine their originals similarly to Sect. 4.2 without any difficulties. Hence, we can also obtain the originals of the coefficients of the series expansion of the displacements and stresses.

Using the solution derived, we can obtain the results of Sect. 4.2 for elastic and acoustic media by means of the passages to the limits. Thus, in the case of elastic medium, it is sufficient to set

$$\gamma_2 \to \infty, \qquad \eta_1 = 1, \qquad \zeta = \eta_2 = \beta_1 = \beta_2 = \beta_3 = 0$$

in the formulas (8.16)–(8.20). At that, it is necessary to eliminate the second columns and the third rows in the determinants $\Delta(s)$ and $\Delta_k(s)$. A detailed analysis of the expressions (8.16)–(8.20) brings us to the relationships (4.45), (4.47)–(4.49).

In order to pass to the model of acoustic medium, it is sufficient to set

$$\gamma_1 \to \infty, \qquad \gamma_3 \to \infty, \qquad \xi = \eta_1 = \zeta = 1,$$
$$\eta_2 = \beta_1 = \beta_2 = \beta_3 = 0.$$

Then, the formulas (8.16)–(8.20) for the elastic-porous medium transform into the solution (4.45), (4.52), and (4.53) of the corresponding problem for the acoustic medium.

Let us note that not only the problem presented above but all of the problems presented in Chaps. 1–6 can be generalized to elastic porous media. Application of the methods presented will not make any difficulties. A difference in the solutions will be caused mainly by an increase in the degree of the corresponding polynomials and the dimensions of the matrices.

Though, currently, the area of mechanics under study has advanced greatly, there are still many problems not yet solved. Especially virgin is the area of transient interaction of arbitrary deformable bodies (systems) with elastic media and soils (multicomponent media). Another interesting area is the combined interaction of the fields of various physical nature with the inhomogeneous structures.

In the future, it will be necessary to pay more attention to the following items. Construction of the experimental equipment and measuring complexes for study of the transient behavior of the deformable bodies and structures contacting different media; development of the more accurate schemes (models) for simulation of the nonlinear interaction of waves of various intensity, physical-mechanical fields, and bodies with the deformable obstacles; development of the powerful software for solution of the applied problems of dynamics of structures and buildings contacting the surrounding media.

References

Abramowitz, M., Stegun, I. (1965): Handbook of mathematical functions with formulas, graphs and mathematical tables. Dover, New York

Achenbach, J.D. (1973): Wave propagation in elastic solids. Appl. Math. and Mech. **16**. North-Holland, Amsterdam London, Elsievier, New York

Achenbach, J.D., Sun, C.T. (1966): Propagation of waves from a spherical surface of time-dependent radius. J. Acoust. Soc. Amer. **40**, 4, 877–882

Adischev, V.V., Kornev, V.M. (1979): On the design of walls of blasting chambers. Fizika Goreniya i Vzryva **15**, 6, 108–114 (Russian)

Akkas, N. (1977): Transient response of a moving spherical shell in an acoustic medium. Int. J. Solids Struct. **13**, 3, 211–220

Akkas, N. (1979): Residual potential method in spherical coordinates and related approximations. Mech. Res. Comm. **6**, 5, 257–262

Akkas, N., Engin, A.E. (1980): Transient response of spherical shell in an acoustic medium. Comparison of exact and approximate solutions. J. Sound and Vibr. **73**, 3, 447–460

Aleksandrova, N.A. (1982): On the effect of weak shock waves on an elastic spherical shell. Izvestiya AN SSSR. Mekhanika Tverdogo Tela [Mechanics of Solids], 1, 176–182 (Russian)

Alexits, G. (1961): Convergence problems of orthogonal series. Akademiai Kiado, Budapest

Allen, D.E., Robinson, A.R. (1966): International of plane stress waves with spherical cavity in elastic and viscoelastic media. Depart. Civil. Eng. Struct. Res. Ser. 302. Ill. Univ, Urbana

Allen, W.A., Goldsmith, W. (1955): Elastic description of a high-amplitude spherical pulse in steel. J. Appl. Phys. **26**, 1, 69–74

Alterman, Z., Abramovici, F. (1965): Propagation of a P-pulse in a solid sphere. Bull. Seismol. Soc. Amer. **55**, 5, 821–861

Alterman, Z., Abramovici, F. (1966): Effect of depth of a point source on the motion of the surface of an elastic solid sphere. Geophys. J. Roy. Astron. Soc. **11**, 1–2, 189–224

Alterman, Z., Abramovici, F. (1967): The motion of sphere caused by an impulsive force and by explosive point-source. Geophys. J. Roy. Astron. Soc. **13**, 1–3, 117–148

Alterman, Z., Kornfeld, P. (1963): Propagation of a pulse within a sphere. J. Acoust. Soc. Amer. **35**, 10, 1649–1662

Alterman, Z., Kornfeld, P. (1964): Propagation of a pulse in a fluid sphere. Geophysics **29**, 2, 259–287

Alterman, Z., Kornfeld, P. (1965a): Shallow focus explosion in a liquid sphere. Geophys. J. Roy. Astron. Soc. **9**, 2–3, 121–151

Alterman, Z., Kornfeld, P. (1965b): Propagation of an SH-torque pulse in a sphere. Revs. Geophys. **3**, 1, 55–82

Amenadze, Yu.A. (1976): Theory of elasticity. Vysshaya Shkola, Moscow (Russian)

Ansell, J.H., Tupholme, G.E. (1972): Use of Clemmon functions in the study of an acoustic pulse generated by a deformable sphere. J. Sound and Vibr. **25**, 2, 185–195

Arsenin, V.Ya. (1974): Methods of mathematical physics and special functions. Nauka, Moscow (Russian)

Auphan, M., Matthys, J. (1978): Reflection of a plane impulsive acoustic pressure wave by a rigid sphere. J. Sound and Vibr. **66**, 2, 227–237

Azizov, K.O. (1973): On propagation of spherical waves in an isotropic elastic medium. Doklady AN UzSSR, 8, 2–9 (Russian)

Babaev, A.E. (1974a): Transient interaction of an internal spherical shock wave with a spherical surface. Prikladnaya Mekhanika [Soviet Applied Mechanics] **10**, 6, 114–118 (Russian)

Babaev, A.E. (1974b): The effect of internal spherical pressure wave on a spherical cavity. Prikladnaya Mekhanika [Soviet Applied Mechanics] **10**, 8, 50–55 (Russian)

Babaev, A.E. (1980): Deformation of a spherical shell containing a wave source of varying size. Doklady AN USSR, Ser. A, 8, 29–33 (Russian)

Babaev, A.E. (1981): Rigid cavities under action of internal acoustic shock waves irradiated by a surface having variable boundary. Prikladnaya Mekhanika [Soviet Applied Mechanics] **17**, 6, 36–44 (Russian)

Babaev, A.E. (1983): The effect of an internal transient acoustic pressure wave on a three-layer sphere. Matematicheskie Issledovaniya, 75, 3–7 (Russian)

Babaev, A.E. (1984): The internal transient wave problem for a multilayer sphere in an acoustic medium. Prikladnaya Mekhanika [Soviet Applied Mechanics] **20**, 5, 32–40 (Russian)

Babaev, A.E. (1985): The transient problem for an acoustic pressure wave produced by an eccentrically located point source interacting with a multilayer sphere. Matematicheskaya Fizika i Nelineinaya Mekhanika, 4(38), 59–65 (Russian)

Babaev, A.E., Kubenko, V.D. (1976): Deformation of a spherical shell having a rigid core produced by a weak shock wave. Doklady AN USSR, Ser. A, 11, 984–990 (Russian)

Babaev, A.E., Kubenko, V.D. (1977a): Interaction of a transient pressure wave with a system of two spherical shells. Izvestiya AN SSSR. Mekhanika Tverdogo Tela [Mechanics of Solids], 2, 135–141 (Russian)

Babaev, A.E., Kubenko, V.D. (1977b): The effect of internal shock wave on an elastic spherical shell. Prikladnaya Mekhanika [Soviet Applied Mechanics] **13**, 5, 73–78 (Russian)

Babaev, A.E., Guz, A.N., Kubenko, V.D. (1974): Determination of the transient loads produced by internal wave sources. In: Selected problems of applied mechanics. VINITI, Moscow, 53–62 (Russian)

Babaev, A.E., Kubenko, V.D., Kurbakov, V.G. (1979): Deformation of a multilayer spherical shell produced by a weak shock wave. Prikladnaya Mekhanika [Soviet Applied Mechanics] **15**, 12, 28–35 (Russian)

Babaev, A.E., Gordienko, V.I., Saprykin, Yu.V. (1980): The internal transient problem for a spherical shell having a core. Prikladnaya Mekhanika [Soviet Applied Mechanics] **16**, 3, 32–37 (Russian)

Babaev, A.E., Kubenko, V.D., Saprykin, Yu.V. (1983): Internal transient problem of hydroelasticity for a system of two spherical shells having a core. Prikladnye Problemy Prochnosti i Plastichnosti, 23, 124–129 (Russian)

Babich, V.M., Molotkov, I.A. (1977): Mathematical methods in the theory of elastic waves. In: Advances in science and technology of VINITI. Mechanics of deformable solids **10**. VINITI, Moscow, 5–62 (Russian)

Babich, V.M., Kapilevich, M.B., Mikhlin, S.G., Natansom, D.I., Riz, P.M., Slobodetsky, L.N., Smirnov, M.M. (1964): Linear equations of mathematical physics. Nauka, Moscow (Russian)

Babichev, A.I. (1966a): Translational motion of a sphere in an elastic medium. Izvestiya AN UzSSR. Seriya Tekhnicheskikh Nauk, 4, 35–40 (Russian)

Babichev, A.I. (1966b): Analysis of interaction of elastic waves with cylindrical or spherical obstacles by means of the method of characteristics. In: Propagation of elastic and elastic–plastic waves. Proc. All-Union Symp., Baku, 1964. Baku, 443–456 (Russian)

Babichev, A.I. (1968): Interaction of an elastic medium and a sphere in the case when the adhesion takes place. In: Problems of cybernetics and computational mathematics. Tashkent, 106–114 (Russian)

Babichev, A.I. (1969a): Plane-parallel motion of a rigid sphere in an elastic medium. In: Propagation of elastic and elastic–plastic waves. Proc. 3rd All-Union Symp. Tashkent, 29–41 (Russian)

Babichev, A.I. (1969b): Analysis of the dynamics of circular cylindrical and spherical elastic layers by means of the method of characteristics. In: Propagation of elastic and elastic–plastic waves. Proc. 3rd All-Union Symp. Tashkent, 19–29 (Russian)

Babichev, A.I. (1973): On the variety of solutions of the equations of elasticity corresponding to the action of concentrated force. In: Propagation of elastic and elastic–plastic waves. Proc. 5th All-Union Symp. Alma-Ata, 55–57 (Russian)

Babichev, A.I., Sarimsakov, U. (1986): Nonstationary problems of propagation of waves and their interactions with solid bodies and deformable media. Fan, Tashkent (Russian)

Babichev, A.I., Khidoyatov, K., Saidov, T.Kh., Shibanova, V.P. (1966): Analysis of transient motion of a cylinder and a sphere in an elastic medium by means of the method of characteristics. In: Problems of cybernetics and computational mathematics. No. 3. Tashkent, 66–81 (Russian)

Babichev, A.I., Saidov, T.Kh., Khidoyatov, K. (1967): Response of an elastic medium to translational motion of a sphere. Doklady AN UzSSR, 2, 8–11 (Russian)

Babichev, A.I., Saidov, T.Kh., Sarimsakov, U. (1976): Determination of the reaction forces for an elastic medium and an immovable sphere, when a stepwise compression wave passes by. In: Problems of computational and applied mathematics. Tashkent, 167–175 (Russian)

Baidak, D.A., Torsky, A.R., (1983): On interaction of a plane acoustic wave with an elastic spherical shell. Fiziko–Khimicheskaya Mekhanika Materialov 19, 4, 114–117 (Russian)

Bak, G. (1978): Dynamika gruboscziennego rbiornika kulistego. In Konstr. powl. teor. i zastosow. Proc. 2nd Conf., S. 1, s.a. Golun, 13–18 (Polish)

Bakarat, R.G. (1960): Transient diffraction of scalar waves by a fixed sphere. J. Acoust. Soc. Amer. 32, 1, 61–66

Baker, W.E., Allen, F.J. (1958): The response of elastic spherical shells to spherically symmetric internal blast loading. In: Proc. Third US National Congress of Appl. Mech., Providence, Rhode Island. New York, 79–87

Bakhrakh, S.Kh., Grigorev, F.V., Kodola, B.E., Kormer, S.B., Koznetsov, O.N., Mokhov, V.N., Pevnitsky, A.V., Sinitsyn, M.V., Tolochko, A.P., Urlin, V.D., Khelemendik, A.M. (1979): The study of dynamic loading of thin-walled cylindrical and spherical shells. In: Proceedings of the All-Union scientific-research institute of physical engineering and measurements in radio engineering, 44/74, 112–114 (Russian)

Baranov, V.M., Gryazev, A.P. (1979): Radiation of sound under expansion of a spherical cavity in an isotropic elastic medium. Defektoskopiya, 11, 28–34 (Russian)

Baron, M.L. (1957): Response of nonlinearly supported spherical boundaries to shock waves. J. Appl. Mech. **24**, 4, 501–505

Batalov, B.A., Svidinsky, V.A. (1971): The study of influence of medium parameters on cavity's dimension under explosion. Izvestiya AN SSSR. Fizika Zemli, 12, 24–31 (Russian)

Bazhenov, V.G., Mikhailov, G.S. (1979): Nonlinear dynamic interaction of thin-walled structures with ideal compressible media. Prikladnye Problemy Prochnosti i Plastichnosti, 10, Gorky, 41–55 (Russian)

Bazhenov, V.G., Kochetkov, A.V., Mikhailov, G.S. (1977): Numerical solution of the plane and axially symmetric problems of interaction of elastic–plastic shells with shock waves. Prikladnye Problemy Prochnosti i Plastichnosti, 7, Gorky, 55–63 (Russian)

Bazhenov, V.G., Kochetkov, A.V., Mikhailov, G.S., Ugodchikov, A.G. (1979): Interaction of elastic–plastic thin-walled structural elements with shock waves in ideal compressible media. Izvestiya AN SSSR. Mekhanika Tverdogo Tela [Mechanics of Solids], 2, 141–149 (Russian)

Bazhenov, V.G., Kochetkov, A.V., Krylov, S.V., Lomunov, V.K., Mikhailov, G.S. (1981): The nonlinear wave problems of interaction of thick-walled structures filled with ideal compressible media. In: Theoretical and applied mechanics. Proc. 5th All-Union Congr. Alma-Ata, 42 (Russian)

Bazhina, I.A. (1957): Transient irradiation of an oscillator, when a traveling wave runs at the azimuth direction. In: Proceedings of the All-Union scientific-research institute of broadcasting radio reception and acoustics, 8, 91–112 (Russian)

Bedrosian, B., Dimaggio, F.L. (1972): Acoustic approximations in fluid–shell interactions. J. Eng. Mech. Div. Pros. ASCE. **98**, EM3, 731–742

Belov, A.I., Klapovsky, V.E., Kornilov, V.A., Mnyeev, V.N., Shiyan, V.S. (1984): Dynamics of a spherical shell under nonsymmetrical internal impulsive loading. Fizika Goreniya i Vzryva **20**, 3, 71–74 (Russian)

Belytschko, T. (1980): Fluid–structure interaction. Comput. and Struct. **12**, 4, 459–469

Belytschko, T., Kennedy, J.M., Schoeberle, D.F. (1980): Quasi-Eulerian finite element formulation for fluid–structure interaction. Trans. ASME. J. Rpessure Vessel Technol. **102**, 1, 62–69

Berezhnoi, A.I. (1966): Glass-ceramic and photo-glass-ceramic materials. Mashinostroenie, Moscow (Russian)

Berger, B.S. (1969): Vibration of the hollow sphere in an acoustic medium. J. Appl. Mech. **36**, 2, 330–333

Berger, B.S., Klein, D. (1972): Application of the Cesaro mean to the transient interaction of a spherical acoustic wave and a spherical elastic shell. J. Appl. Mech. **39**, 2, 623–625

Bhaduri, S., Kanoria, M. (1982): Force vibration of an isotropic sphere having a rigid spherical inclusions. Indian J. Theor. Phys. **30**, 1, 93–98

Bhattacharya, J. (1969): On the propagation of waves in an elastic medium due to various types of pressures and velocities prescribed on the inner surface of a spherical cavity. Gerlands Beitr. Geophys. **78**, 3, 223–231

Biot, M.A. (1962) Mechanics of deformation and acoustic propagation in porous media. J. Appl. Phys. **33**, 4, 1482–1498

Biryulya, I.I. (1983): The stress–strain state of a three-layer spherical shell under impulsive loading. Voprosy Mekhaniki Deformiruemogo Tverdogo Tela, 4, 3–5 (Russian)

Blake, F.G. (1952): Spherical wave propagation in solid media. J. Acoust. Soc. Amer. **24**, 2, 211–215

Broberg, K.B. (1956): Shock waves in elastic and elastic–plastic media. Kgl. Fortifikationsförvalt. Befästningsbyrán, Stokholm

Brodacki, J. (1965): Naprężenia sprezyste w grubošciennym zbiorniku kulistym przy krǫtkotrwalym cišnieniu wewnętrznym. Prace Inst. Mech. Precyzyjnej. **13**, 47, 1–10

Brune, J.N. (1970): Tectonic stress and the spectra of seismic shear waves from earthquakes. J. Geophys. Res. **75**, 26, 4997–5009

Buldyrev, V.S., Molotkov, I.A. (1958): On transient propagation of waves in homogeneous and isotropic media separated by cylindrical or spherical interfaces. Uchenye Zapiski LGU, 246. Seriya Matematicheskikh Nauk, 32, 261–321 (Russian)

Buldyrev, V.S., Yanson, Z.A. (1965): Interference phenomena in mediums having spherical interfaces. Revs. Geophys. **3**, 1, 115–122

Burdun, E.T., Sazonov, I.A. (1985): On transient concentration of stresses at a spherical cavity in elastic massif. Sudostroenie, 34, 11–15 (Russian)

Burridge, R. (1963): The reflection of a pulse in a solid sphere. Proc. Roy. Soc. **A276**, 1368, 367–400

Butler, T.D. (1971): LINC method extension. In: Proceedings of the second international conference on numerical methods in fluid dynamics. (Lecture Notes in Physics **8**) Springer, New York

Chadwick, P., Trowbridge, E.A. (1967a): Oscillations of a rigid sphere embedded in an infinite elastic solid. I. Torsional oscillations. II. Rectilinear oscillations. Proc. Cambridge Phil. Soc. **63**, 4, 1189–1227

Chadwick, P., Trowbridge, E.A. (1967b): Elastic wave fields generated by scalar wave functions. Proc. Cambridge Phil. Soc. **63**, 4, 1177–1187

Chakraborty, S.K. (1958): A disturbance generated by a pulse of pressure on the surface of a spherical cavity in an elastic medium. Indian J. Theor. Phys. **6**, 3, 85–89

Cheban, V.G., Sabodash, P.F. (1972): Elastic and thermoelastic waves in deformable media. Shtiintsa, Kishinev (Russian)

Cherednichenko, R.A. (1975): Propagation of elastic waves in a spherical shell when colliding with an obstacle. Izvestiya AN SSSR. Mekhanika Tverdogo Tela [Mechanics of Solids], 3, 136–143 (Russian)

Chou, P.C., Koenig, H.A. (1966): A united approach to cylindrical and spherical elastic waves by method of characteristics. J. Appl. Mech. **33**, 1, 159–167

Chou, S.-C., Greif, R. (1968): Numerical solution of stress waves in layered media. AIAA Journal **6**, 6, 1067–1074

Cinelli, G. (1966): Dynamic vibrations and stresses in elastic cylinders and spheres. J. Appl. Mech. **33**, 4, 825–830

Clebsch, A. (1862): Theorie der Elastizität fester Körper. Druck und Verlag von B.G.Teubner, Leipzig

Cohen, D.S., Handelman, G.H. (1965): Scattering of a plane acoustical wave by a spherical obstacle. J. Acoust. Soc. Amer. **38**, 5, 827–834

Courant, R., Hilbert, D. (1966): Methods of mathematical physics. Interscience Publ., New York

Courbon, J. (1971): Vibration radiales d'une enveloppe sphérique. Anncl. Ponts et Chaussées **141**, 11, 59–79

Darboux, G. (1915): Lecons sur la théorie générale des surfaces **2**. Paris

Das Gupta, S.C. (1954): Waves and stresses produced in an elastic medium due to impulsive radial forces and twist on the surface of a spherical cavity. Geofis. Pura e Appl. **27**, 1–6

Datta, B.K., Sengupta, P.R. (1984): Note on wave propagation in an infinite elastic space due to explosion at the cavity center. Rev. Roum. Sci. Techn. Sér. Méc. Appl. **29**, 5, 487–492

Demchuk, A.F. (1968): One method of analysis of blasting chambers. Prikladnaya Mekhanika i Tekhnicheskaya Fizika [Journal of Applied Mechanics and Technical Physics], 5, 47–50 (Russian)

Dikasov, V.M. (1979): Solution of some external boundary value problems for the wave equation. Differentsialnye Uravneniya [Differential Equations], 10, 1863–1872 (Russian)

Dikasov, V.M. (1982): Solution of some problems on propagation of waves in a sphere by means of the method of reflections. Differentsialnye Uravneniya [Differential Equations], 4, 631–638 (Russian)

Diomidov, M.N., Dmitriev, A.N. (1974): Conquest of the depth. Sudostroenie, Leningrad (Russian)

Donea, J., Fasoli–Stella, P., Guiliani, S. (1977): Lagrangian and Eulerian finite element techniques for transient fluid–structure interaction problems. In: Trans. 4th Int. Conf. Struct. Mech. React. Technol., San Francisco, Calif., 1977. **B**. Amsterdam. B1/2

Dötsch, G. (1974): Introduction to the theory and application of the Laplace transformation. Springer, Berlin

Drobyshevsky, N.I. (1980): Penetration of spherical and cylindrical bodies into an ideal compressible fluid. In: Dynamics of elastic and solid bodies interacting with fluid. Tomsk State University Publishing House, Tomsk, 74–84 (Russian)

Drobyshevsky, N.I. (1984): Oblique entry of cylindrical bodies into a fluid. In: Interaction of plates and shells with liquid and gas. Moscow University Publishing House, Moscow, 92–98 (Russian)

Duffey, T.A., Johnson, J.N. (1981): Transient response of a pulsed spherical shell surrounded by a infinite elastic medium. Int. J. Mech. Sci. **23**, 10, 589–593

Efimova, I.V., Stepanenko, M.V. (1984): Dynamics of a spherical hydroelastic system under action of an internal transient source. In: Vibrations of elastic structures filled with fluid. Proc. 5th All-Union Symp., Novosibirsk, 1982. Moscow, 109–113 (Russian)

Eringen, A.C. (1957): Elasto-dynamic problem concerning the spherical cavity. Quart. J. Mech. and Appl. Math. **10**, 3, 257–270

Ershov, N.F., Shakhverdi, G.G. (1984): Finite element method in the problems of hydrodynamics and hydroelasticity. Sudostroenie, Leningrad (Russian)

Fadeev, A.D. (1968): Formation and propagation of stress waves due to explosion in rocks. Fizika Goreniya i Vzryva **4**, 2, 254–259 (Russian)

Filippov, I.G., Bakhramov, B.M. (1978): Waves in elastic homogeneous and inhomogeneous media. Fan, Tashkent (Russian)

Filippov, I.G., Egorychev, O.A. (1977): Transient vibrations and diffraction of waves in acoustic and elastic media. Mashinostroenie, Moscow (Russian)

Forrestal, M.J., Sagartz, M.J. (1971): Radiated pressure in an acoustic medium produced by pulsed cylindrical and spherical shells. J. Appl. Mech. **38**, 4, 1057–1060

Friedlender, F. (1958): Sound pulses. Cambridge University Press, Cambridge

Fridman (1976a): Solution of the dynamic problem of the theory of elasticity in the curvilinear coordinates. Problemy Prochnosti, 5, 56–61 (Russian)

Fridman, L.I. (1976b): Application of the method of travelling wave to solution of the dynamic problem on loading of elastic sphere. Prikladnaya Mekhanika [Soviet Applied Mechanics] **12**, 9, 30–35 (Russian)

Gaek, Yu.V., Shumilo, V.A. (1965): Stress waves having central and axial symmetry in elastic boundless medium, and analysis of modern approaches to their nature. In: Blasting operations, 57/14. Moscow, 90–105 (Russian)

Galazyuk, V.A., Gorechko, A.N. (1980a): On one method of solution of problems of dynamics in the theory of elasticity in the spherical and cylindrical coordinates. Doklady AN USSR, Ser. A, 6, 41–44 (Russian)

Galazyuk, V.A., Gorechko, A.N. (1980b): Transient interaction of a spherical inclusion with an acoustic medium. Prikladnaya Mekhanika [Soviet Applied Mechanics] 16, 2, 119–122 (Russian)

Galazyuk, V.A., Gorechko, A.N. (1983): Application of the Chebyshev–Laguerre polynomials to the studies of problems of interaction of a thick-walled elastic sphere with an acoustic medium. Prikladnaya Mekhanika [Soviet Applied Mechanics] 19, 2, 21–27 (Russian)

Galiev, Sh.U. (1977): Dynamics of interaction of structural elements with a pressure wave in a fluid. Naukova Dumka, Kiev (Russian)

Galiev, Sh.U. (1981): Dynamics of hydroelastic–plastic systems. Naukova Dumka, Kiev (Russian)

Geers, T.L. (1975): Shock-wave attenuation by resilient scatterers. J. Appl. Mech. 42, 2, 390–394

Gelchinsky, B.Ya. (1958): Some aspects of propagation of waves in a homogeneous and isotropic elastic sphere. Part 1. Uchenye Zapiski LGU, 246. Seriya Matematicheskikh Nauk, 32, 322–345 (Russian)

Gladkov, A.A. (1977): Solution of the mixed initial and boundary value problem for the wave equation by means of the wave potential of an ordinary layer. Uchenye Zapiski TsAGI 8, 1, 105–107 (Russian)

Glenn, L.A., Kidder, R.E. (1983): Blast loading of a spherical container surrounded by an infinite elastic medium. J. Appl. Mech. 50, 4a, 723–726

Goldsmith, W., Allen, W.A. (1955): Graphic representation of the spherical propagation of explosive pulses in elastic media. J. Acoust. Soc. Amer. 27, 1, 47–55

Gordienko, V.I., Kubenko, V.D. (1983): On computation of transition functions in the internal problems of transient hydroelasticity of cylindrical and spherical shells. Prikladnaya Mekhanika [Soviet Applied Mechanics] 19, 6, 58–62 (Russian)

Gorshkov, A.G. (1974): Interaction of weak transient pressure waves with elastic shells. Izvestiya AN SSSR. Mekhanika Tverdogo Tela [Mechanics of Solids], 3, 155–164 (Russian)

Gorshkov, A.G. (1976): Dynamic interaction of shells and plates with surrounding media. Izvestiya AN SSSR. Mekhanika Tverdogo Tela [Mechanics of Solids], 2, 177–189 (Russian)

Gorshkov, A.G. (1979): Interaction of shock waves with deformable obstacles. In: Advances in science and technology of VINITI. Mechanics of deformable solids 13. VINITI, Moscow, 105–186 (Russian)

Gorshkov, A.G. (1980): Diffraction of weak shock waves by deformable bodies submersed into a fluid. Prikladnaya Mekhanika [Soviet Applied Mechanics] 16, 5, 3–11 (Russian)

Gorshkov, A.G. (1981): Transient interaction of plates and shells with continuum. Izvestiya AN SSSR. Mekhanika Tverdogo Tela [Mechanics of Solids], 4, 177–189 (Russian)

Gorshkov, A.G., Tarlakovsky, D.V. (1981): Time-dependent radial vibrations of an elastic piecewise-homogeneous space having spherical symmetry. Izvestiya AN SSSR. Mekhanika Tverdogo Tela [Mechanics of Solids], 1, 96–101 (Russian)

Gorshkov, A.G., Tarlakovsky, D.V. (1983a): Spherically symmetric waves in an elastic piecewise-homogeneous space. Izvestiya AN SSSR. Mekhanika Tverdogo Tela [Mechanics of Solids], 4, 178–183 (Russian)

Gorshkov, A.G., Tarlakovsky, D.V. (1983b): Propagation of elastic waves in a piecewise-homogeneous space having spherical interfaces. In: Mechanics of inhomogeneous structures. Proc. 1st All-Union Conf., Lvov. Kiev, 55–57 (Russian)

Gorshkov, A.G., Grigolyuk, E.I., Tarlakovsky, D.V. (1978): The internal problems of dynamics of a thick-walled sphere contacting elastic or acoustic media. Prikladnaya Mekhanika [Soviet Applied Mechanics] **14**, 12, 12–22 (Russian)

Gorshkov, A.G., Grigolyuk, E.I., Tarlakovsky, D.V. (1979a): Application of the generalized spherical waves in the transient problems of diffraction. In: Theory of elasticity. Proc. All-Union Conf. Erevan, 123–125 (Russian)

Gorshkov, A.G., Kuranov, B.A., Konovalova, A.I. (1979b): Analysis of the spherical reservoirs subjected to a weak shock wave by means of the finite element method. In: Modern methods and algorithms for analysis and design of structures using computers. Proc. All-Union Conf. Tallinn, 125–126 (Russian)

Gorshkov, A.G., Konovalova, A.I., Kuranov, B.A. (1980): Analysis of a stress-strain state of spherical reservoirs subjected to weak shock waves. In: Statics and dynamics of thin-walled structures. Moscow, 152–171 (Russian)

Gorshkov, A.G., Poruchikov, V.B, Tarlakovsky, D.V. (1983): On one technique of inversion of the Laplace transform in the problems of interaction of transient waves with spherical inclusions. Izvestiya AN SSSR. Mekhanika Tverdogo Tela [Mechanics of Solids], 1, 82–90 (Russian)

Gorshkov, A.G., Saliev, A.A, Tarlakovsky, D.V. (1987) Propagation of transient disturbances from a spherical cavity in an elastic–porous medium. Doklady AN UzSSR, 7, 15–16 (Russian)

Gradshtein, I.S., Ryzhik, I.M. (1971): The tables of integrals, sums, series, and products. Nauka, Moscow (Russian)

Gray, R.M. (1955): The elastic sphere under dynamic and shock loads. Dissert. Abstracts **15**, 12, 2506–2507

Gross, M.B., Hofmann, R. (1977): Fluid–structure interaction calculation. Summary Trans. 4th Int. Conf. Struct. Mech. React. Technol., San Francisco, Calif., 1977. **B**. Amsterdam. B1.5/1

Grigolyuk, E.I., Gorshkov, A.G. (1967): Displacement of a rigid sphere due to an acoustic pressure wave. Doklady AN SSSR **177**, 3, 539–541 (Russian)

Grigolyuk, E.I., Gorshkov, A.G. (1968): Action of an acoustic pressure wave on a shallow spherical shell. Doklady AN SSSR **182**, 4, 787–789 (Russian)

Grigolyuk, E.I., Gorshkov, A.G. (1974): Transient hydroelasticity of shells. Sudostroenie, Leningrad (Russian)

Grigolyuk, E.I., Gorshkov, A.G. (1976): Interaction of elastic structures with fluid. Shock and submersion. Sudostroenie, Leningrad (Russian)

Grigolyuk, E.I., Kuznetsov, E.B. (1975): Response of a three-layer spherical shell joint to the rigid masses to an acoustic pressure wave. In: Dynamics of elastic and solid bodies interacting with fluid. Tomsk State University Publishing House, Tomsk, 53–59 (Russian)

Grigolyuk, E.I., Gorshkov, A.G., Khromushkin, A.V. (1968): Response of spherical and cylindrical shells to an acoustic pressure wave. In: Propagation of elastic and elastic–plastic waves. Proc. 4th All-Union Symp. Kishinev, 33 (Russian)

Grigolyuk, E.I., Gorshkov, A.G., Khromushkin, A.V. (1974): Response of spherical and cylindrical shells to an acoustic pressure wave. In: Selected problems of applied mechanics. VINITI, Moscow, 259–269 (Russian)

Grigolyuk, E.I., Gorshkov, A.G., Tarlakovsky, D.V. (1976): To determination of hydrodynamic forces of interaction of weak shock waves with an elastic sphere. Doklady AN SSSR **230**, 1, 60–63 (Russian)

Grigolyuk, E.I., Gorshkov, A.G., Tarlakovsky, D.V. (1977a): Transient hydroelastic vibrations of a thick-walled sphere. Doklady AN SSSR **233**, 5, 812–815 (Russian)

Grigolyuk, E.I., Gorshkov, A.G., Tarlakovsky, D.V. (1977b): Response of an elastic sphere to the action of an acoustic pressure wave. In: Theory of shells and plates. Proc. 11th All-Union Conf., Kharkov. Moscow, 27 (Russian)

Grigolyuk, E.I., Gorshkov, A.G., Tarlakovsky, D.V. (1979): Interaction of weak shock waves with an elastic sphere submerged into an acoustic medium. In: Design of spatial structures. No. 18. Moscow, 21–55 (Russian)

Grilitsky, D.V., Onischuk, V.Ya. (1980a): Transient interaction of a spherical shell with an acoustic wave. Visnik Lvivskogo Universitetu. Mekhanika, Matematika, 16, 43–49 (Ukranian)

Grilitsky, D.V., Onischuk, V.Ya. (1980b): Radiation of sound waves by a hollow elastic sphere with a filler. Izvestiya AN SSSR. Mekhanika Tverdogo Tela [Mechanics of Solids], 4, 179–186 (Russian)

Grilitsky, D.V., Onischuk, V.Ya. (1980c): Transient sound field produced by a hollow elastic sphere. Matemematicheskie Metody i Fiziko–Mekhanicheskie Polya, 11, 51–56 (Russian)

Grilitsky, D.V., Poddubnyak, A.P. (1980): Scattering of a transient torsional wave by a rigid immovable sphere in an elastic medium. Izvestiya AN SSSR. Mekhanika Tverdogo Tela [Mechanics of Solids], 5, 86–92 (Russian)

Gross, M.B., Hofmann, R. (1977): Fluid–structure interaction calculation. Summary Trans. 4th Int. Conf. Struct. Mech. React. Technol., San Francisco, Calif., 1977. B. Amsterdam. B1.5/1

Guz, A.N., Kubenko, V.D. (1982): Methods of design of shells. In: Theory of transient aerohydroelasticity of shells 5. Naukova Dumka, Kiev (Russian)

Guz, A.N., Kubenko, V.D., Cherevko, M.A. (1978a): Diffraction of elastic waves. Naukova Dumka, Kiev (Russian)

Guz, A.N., Kubenko, V.D., Cherevko, M.A. (1978b): Diffraction of elastic waves. Prikladnaya Mekhanika [Soviet Applied Mechanics] 14, 8, 3–15 (Russian)

Guz, A.N., Kubenko, V.D., Panasyuk, N.N. (1980): Transient diffraction of a spherical pressure wave by a spherical cavity in an elastic medium. Izvestiya AN SSSR. Mekhanika Tverdogo Tela [Mechanics of Solids], 3, 88–92 (Russian)

Guz, A.N., Kubenko, V.D., Babaev, A.E. (1984): Hydroelasticity of the system of shells. Vischa Shkola, Kiev (Russian)

Heale, D.G., Raddy, D.V. (1975): Response of a spherical cavity to a plane wave. CANCAM 75 Pros. 5th Can. Congr. Appl. Mech., Frederiction, N.B. 1975. Frederiction, 307–308

Henneberg, L. (1878–1879): Ueber die elastischen Schwirgungen einer isotropen Kugel ohne Einwirkung von äussern Kräften. Annuli Matem. Pura ed Appl., Serie 2, 9, 3, 193–209

Hickling, R., Means, R.W. (1968): Scattering of frequency-modulated pulses by spherical elastic shell in water. J. Acoust. Soc. Amer. 44, 5, 1246–1252

Hirano, Y. (1973): Hydrodynamic shock of flexible spherical shells. J. Japan. Soc. Aeronaut. and Space Sci. 21, 228, 14–31

Hirasawa, H. (1964): Elastic waves from a spherical source: Aperiodic solutions for Scholtes model. Bull. Seismol. Soc. Amer. 54, 3, 897–908

Hirt, C.W. (1971): An arbitrary Lagrangian–Euler computational technique. In: Proceedings of the second international conference on numerical methods in fluid dynamics. (Lecture Notes in Physics 8) Springer, New York

Hobson, E.W. (1955): The theory of spherical and ellipsoidal harmonics. Chelsea, New York

Holzer, F. (1966): Calculation of seismic source mechanisms. Proc. Roy. Soc. A290, 1422, 408–429

Honda, H. (1960a): The generation of seismic waves. Publ. Dominion Observatory, Ottawa 24, 10, 327–334

Honda, H. (1960b): The elastic waves generated from a spherical source. Sci. Rep. Tohoku Univ., Ser. 5: Geophysics 11, 3, 178–183

Hopkins, H.G. (1960): Dynamic expansion of spherical cavities in metals. Progr. Solid. Mech. **1**. North-Holland Publ., Amsterdam; Interscience, New York, 83–164

Huang, H. (1969): Transient interaction of plane acoustic waves with a spherical elastic shell. J. Acoust. Soc. Amer. **45**, 3, 661–670

Huang, H. (1979): Transient response of two fluid-coupled spherical elastic shells to an incident pressure pulse. J. Acoust. Soc. Amer. **65**, 4, 881–887

Huang, H., Wang, Y.F. (1972): Transient stress concentration by a spherical cavity in an elastic medium. J. Appl. Mech. **39**, 4, 1002–1004

Huang, H., Lu, Y.P., Wang, Y.F. (1971): Transient interaction of spherical acoustic waves and a spherical elastic shells. J. Appl. Mech. **38**, 1, 71–74

Huth, J.H., Cole, J.D. (1955): Elastic-stress waves produced by pressure loads on a spherical shell. J. Appl. Mech. **2**, 4, 473–478

Ilyushin, A.A. (1978): Mechanics of continuum. Moscow University Publishing House, Moscow (Russian)

Inouye, W. (1937): Notes on the origin of earthquakes (Second paper). Bull. Earthq. Res. Univ. Tokyo **15**, 1, 90–101

Inouye, W. (1938): Notes on the origin of earthquakes (Sixth paper). Bull. Earthq. Res. Univ. Tokyo **16**, 3, 595–631

Israilov, M.Sh. (1977): To the problem of motion of rigid bodies under action of an acoustic pressure wave. In.: The studies in mechanics of fluid and solid body. Moscow, 85–88 (Russian)

Israilov, M.Sh. (1981): Diffraction of elastic waves by heterogeneities. In.: Dynamics of constructions, foundations, and underground structures. Proc. 5th All-Union Conf. **2**, Tashkent. Moscow, 100–103 (Russian)

Israilov, M.Sh. (1982): The existence theorems for transient problems of diffraction of elastic waves. Doklady AN UzSSR, 6, 16–18 (Russian)

Israilov, M.Sh. (1983): The general method of solution of the problems of diffraction of elastic waves by deformable obstacles. Doklady AN SSSR **268**, 5, 1082–1086 (Russian)

Jaerisch, P. (1880): Ueber die elastischen Schwingungen einer isotropen kugel. J. Reine und Angew. Math. **88**, 4, 131–145

Jeffreys, H. (1931): On the cause of oscillatory movement in seismograms. Monthly Notices Roy. Astron. Soc.: Geophys. Sypl. **2**, 407–417

Jeffreys, H. (1959): The Earth, its origin, history and physical constitution. Cambridge University Press, Cambridge

Jeffreys, H., Lapwood, E.R. (1957): The reflection of a pulse within a sphere. Proc. Roy. Soc. **A241**, 1227, 455–479

Jeffreys, H., Swirles, B. (1966): Methods of mathematical physics **1–3**. Cambridge University Press, Cambridge

Jingu, T., Nezu, K. (1985a): Transient stress in an elastic sphere under diametrical concentrated shock loads. Bull. JSME **28**, 245, 2553–2561

Jingu, T., Nezu, K. (1985b): Stress waves in an infinite medium under the diametral concentrated shock loads on the spherical cavity. Bull. JSME **28**, 245, 2592–2598

Kamen, V.B. (1984): Propagation of axially symmetric and relatively undistorted elastic waves through a spherical interface. In: Differential equations and their applications. Dnepropetrovsk, 21–27 (Russian)

Kamen, V.B., Ostapenko, V.A. (1982a): Reflection and refraction of spherically symmetric displacement waves. In: Differential equations and their applications. Dnepropetrovsk, 38–41 (Russian)

Kamen, V.B., Ostapenko, V.A. (1982b): The generalized solutions of spherically symmetric problems of the dynamic theory of elasticity. Dinamika i Prochnost Tyazh. Mashin, 6, 81–89 (Russian)

Karmishin, A.V., Startsev, V.G., Skurlatov, E.D., Feldshtein, V.A. (1979): Experimental and theoretical methods in the dynamics of thin-walled structures. Prikladnye Problemy Prochnosti i Plastichnosti, 11, Gorky, 62–72 (Russian)

Karmishin, A.V., Skurlatov, E.D., Startsev, V.G., Feldshtein, V.A. (1982): Transient aeroelasticity of thin-walled structures. Mashinostroenie, Moscow (Russian)

Kasiak, M., Włodarczyk, E. (1975): Odbicie koncentrycznej kulistej fali naprezenia w osrodku sprezystym. Biul. WAT. J. Dabrowskiego 24, 4, 29–42

Kasiak, M., Włodarczyk, E. (1976): O powiększeniu prędkiści napędzania cial w warstwowym koncentryernym układzie miotającym. Cz. I, II. Biul. WAT. J. Dabrowskiego 25, 2, 19–38; 9, 15–22

Kasiak, M., Włodarczyk, E. (1980): Reflection of concentric spherical stress wave in elastic medium. Proc. Vibr. Probl. 21, 1, 97–110

Kasiak, M., Włodarczyk, E., Zoń, S. (1977): Analiza stanu naprężeń w pręzystym górotworze wygenerowanych natychmiastowym wybuchem łoadunków symetrycznych. Biul. WAT. J. Dabrowskiego 26, 8, 19–40

Kawasumi, H., Yosiyama, R. (1935): On the elastic wave animated by the potential energy of initial strain. Bull. Earthq. Res. Inst., Univ. Tokyo 13, 3, 496–503

Kenner, V.H., Goldsmith, W. (1972): Dynamic loading of a fluid–fluid shell. Int. J. Mech. Sci. 14, 9, 557–568

Kharkevich, A.A. (1950): Transient wave phenomena. Gostekhizdat, Moscow, Leningrad (Russian)

Kheinloo (1979): Designing of multilayer spherical vessels, cylindrical pipes, and round disks having a purpose to decrease the amplitude of jumps in stress waves. Uchenye Zapiski Tartusskogo Universiteta, 487/23, 72–74 (Russian)

Kheisin, D.E. (1967): Radial vibrations of an elastic sphere in a compressible fluid. Izvestiya AN SSSR. Mekanika Zhidkosti i Gaza, 2, 79–82 (Russian)

Kirchhoff, G. (1876): Vorlesungen über Mathematiche Physik. Mechanik. Verlag von B.G. Teubner

Kirillov, F.A. (1947): Seismic effect of explosion. In: Proceedings of the Institute of seismology of the Academy of Science of the USSR. No. 121 (Russian)

Kochin, N.E., Kibel, I.A., Rose, N.V. (1963): Theoretical hydromechanics. Gostekhizdat, Moscow (Russian)

Kokhmanyuk, S.S., Yanyutin, E.G., Romanenko, L.G. (1980): Vibrations of deformable systems produced by shock and movable loads. Naukova Dumka, Kiev (Russian)

Konovalov, Yu.P. (1970): On the shape of seismic pulse produced by explosion in an elastic medium. Izvestiya AN SSSR. Fizika Zemli, 9, 84–87 (Russian)

Kornilov, G.L., Lepikhin, P.P. (1982): Dynamics of interaction of the multilayer elastic cylinders and spheres with a pressure wave in a fluid. In: Problems of design of the airplane structures. No. 3. Kharkov, 113–119 (Russian)

Koshlyakov, N.S., Gliner, E.B., Smirnov, M.N. (1962): Differential equations of mathematical physics. Gostekhizdat, Moscow (Russian)

Kostyuchenko, V.N., Rodionov, V.N. (1974): On radiation of seismic waves produced by powerful underground explosions in a rigid rock. Izvestiya AN SSSR. Fizika Zemli, 10, 65–73 (Russian)

Kovshov, A.N. (1979): Diffraction of an elastic wave by a spherical cavity. Numerical solution. Izvestiya AN SSSR. Mekhanika Tverdogo Tela [Mechanics of Solids], 2, 62–70 (Russian)

Kovshov, A.N., Simonov, I.V. (1967): On the motions of a rigid sphere soldered in a boundless elastic medium. Inzhenernyi Zhurnal. Mekhanika Tverdogo Tela, 5, 155–163 (Russian)

Kreiser, N.D. (1968): Diffraction of transient elastic waves by rigid cylindrical and spherical inclusions. In: Scientific reports of the Institute of mining. No. 56, 110–115 (Russian)

Krylov, V.I., Skoblya, N.S. (1974): Methods of approximate Fourier transforms and inversion of the Laplace transform. Nauka, Moscow (Russian)

Kubenko, V.D. (1972a): Deformation of a spherical shell produced by a transient spherical hydroacoustic wave. Prikladnaya Mekhanika [Soviet Applied Mechanics] 8, 10, 106–110 (Russian)

Kubenko, V.D. (1972b): On translational motions of a spherical shell produced by a transient wave in the compressible fluid. Doklady AN URSR, Ser. A, 1, 73–76 (Ukranian)

Kubenko, V.D. (1975a): Transient deformation of a shell filled with a fluid produced by a weak shock wave. Prikladnaya Mekhanika [Soviet Applied Mechanics] 11, 6, 64–71 (Russian)

Kubenko, V.D. (1975b): On solution of problems of diffraction of transient elastic waves by obstacles of cylindrical and spherical shape. Doklady AN USSR, Ser. A, 10, 901–906 (Russian)

Kubenko, V.D. (1975c): On numerical solution of one type of singular equations used in transient problems of hydroelasticity. In: Mathematical physics. No. 18. Kiev, 95–103 (Russian)

Kubenko, V.D. (1979): Transient interaction of elements of structures with media. Naukova Dumka, Kiev (Russian)

Kubenko, V.D. (1981): Penetration of elastic shells into a compressible fluid. Naukova Dumka, Kiev (Russian)

Kubenko, V.D. (1983): The transient wave problems of propagation of acoustic and elastic waves in the media having cylindrical and spherical interfaces. Matematicheskie Issledovaniya, 75, 27–31 (Russian)

Kubenko, V.D., Babaev, A.E. (1983): Interaction of transient waves with shells. In: Mechanics of composite materials and structural elements. 2: Mechanics of structural elements. Naukova Dumka, Kiev, 431–450 (Russian)

Kubenko, V.D., Panasyuk, N.N. (1977): Transient diffraction of acoustic spherical and cylindrical waves by rigid sphere and cylinder. Prikladnaya Mekhanika [Soviet Applied Mechanics] 13, 9, 14–20 (Russian)

Kubenko, V.D., Stepanenko, M.V. (1980): Numerical solution of the transient problem of dynamics of a shell filled with a fluid, when an internal source acts. In: Vibrations of elastic structures filled with fluid. Proc. 4th Symp., Novosibirsk, 1979. Moscow, 149–157 (Russian)

Kubenko, V.D., Babaev, A.E., Kruglenko, V.P. (1986): The effect of shock loading on a spherical shell filled with a fluid. Prikladnaya Mekhanika [Soviet Applied Mechanics] 22, 9, 33–41 (Russian)

Kulak, R.F. (1981): A finite element formulation for fluid structure interaction in three-dimensional space. Trans. ASME. J. Pressure Vessel Technol. 103, 2, 183–190

Kuranov, B.A., Konovalova, A.I., Samarin, A.V. (1974): The action of an air explosion wave on spherical reservoirs for storage of liquefied gases. Stroitelnaya Mekhanika i Raschet Sooruzhenii, 4, 30–33 (Russian)

Kurochkin, V.A. (1981): Interaction of a shallow spherical shell with an acoustic shock wave. In: Interaction of shells with a fluid. No. XIV. Kazan, 42–50 (Russian)

Kuznetsov, D.S. (1965): Special functions. Vysshaya Shkola, Moscow (Russian)

Lamb, H. (1924): Hydrodynamics. Cambridge University Press, Cambridge

Lamb, H. (1931): The dynamical theory of sound. Edward Arnold and Co., London

Lauvstad, V.R. (1965): Transient scattering of a monochromatic acoustical wave by a scatterer fixed in space. J. Acoust. Soc. Amer. 38, 1, 35–46

Lavrentev, M.A., Shabat, B.V. (1987): Theory of functions of complex variables. Nauka, Moscow (Russian)

Lazarenko, M.A. (1966): Diffraction and scattering of acoustic waves by a sphere placed into a half-space. In: Geophysics and astronomy. No. 9, 24–28 (Russian)

Lepikhin, P.P. (1981): Transient hydroelasticity of the systems of co-axial cylindrical and centrally symmetric spherical shells. In: Problems of mechanics of deformable solid body. No. 2, 17–23 (Russian)

Levitan, B.M., Sargsyan, I.S. (1970): Introduction into the spectrum theory. Self-adjoint differential operators. Nauka, Moscow (Russian)

Lou, Y.K., Klosner, J.M. (1971): Dynamics of a submerged ring-stiffened spherical shell. J. Appl. Mech. **38**, 2, 408–417

Lou, Y.K., Klosner, J.M. (1973): Transient response of a point-excited submerged ring-stiffened spherical shell. J. Appl. Mech. **40**, 4, 1078–1084

Love, A.E.H. (1905): Some illustrations of modes of decay of vibratory motions. Proc. London Math. Soc., Second Ser. **2**, 88–113

Love, A.E.H. (1959): A treatise on the mathematical theory of elasticity. Cambridge University Press, Cambridge

Lurie, A.I. (1950): Operational calculus and its application to problems of mechanics. Gostekhizdat, Moscow, Leningrad (Russian)

Maaz, R. (1964): Theoretische Untersuchung eines im elastischen Medium eingebetteten mechanicshen Empfängers eines Seismographen in einer longitudinalen Planwelle. Pure and Appl. Geophys. **58**, 2, 23–40

McKinney, J.M. (1971): Spherically symmetric vibration of an elastic spherical shell subject to a radial and time-dependent body-force field. J. Appl. Mech. **38**, 3, 702–705

McLeary, R. (1969): The interaction of a plane wave with a spherical cavity. J. Appl. Mech. **36**, 3, 644–646

Maiti, M. (1969): Pulse shapes of once-reflected phases within a sphere. Proc. Vibr. Probl. Polish Acad. Sci. **10**, 2, 149–167

Mann–Nachbar, P. (1957): The interaction of an acoustic wave and on elastic spherical shell. Quaterly J. Appl. Math. **15**, 1, 83–89

Makarov, G.I., Petrashen, G.I. (1953): Transient diffraction of acoustic and electromagnetic waves by a sphere. Uchenye Zapiski LGU, 170. Seriya Matematicheskikh Nauk, 27, 266–301 (Russian)

Maltsev, V.A., Stepanov, G.V., Konon, Yu.A., Pervukhin, L.B. (1985): Stress state of thin-walled structures subjected to short pressure pulse. Problemy Prochnosti, 12, 100–103 (Russian)

Marchuk, G.I. (1977): Methods of computational mathematics. Nauka, Moscow (Russian)

Matsumoto, H., Nakahara, I. (1971): Hollow sphere subjected to dynamic pressure of its inner or outer surface. Bull. Tokyo Inst. Technol., 104, 27–55

Mehta, P.K., Davids, N. (1966): A direct numerical analysis method for cylindrical and spherical elastic waves. AIAA Journal **4**, 1, 112–117

Metsaveer, Kh.A., Veksler, N.D., Stulov, A.S. (1979): Diffraction of acoustic pulses by elastic bodies. Nauka, Moscow (Russian)

Mindlin, Ya.A. (1940a): On three-dimensional propagation of waves. Doklady AN SSSR **26**, 6, 572–576 (Russian)

Mindlin, Ya.A. (1940b): Solution of the external Cauchy–Dirichlet problem for the wave equation in the case of a sphere. Doklady AN SSSR **26**, 6, 577–580 (Russian)

Mindlin, Ya.A. (1947): General representation of the solution of the wave equation. Doklady AN SSSR **58**, 1, 17–20 (Russian)

Mises, R. (1966): Mathematical theory of compressible fluid flow. Acad. Press, New York

Mnev, E.N., Pertsev, A.K. (1970): Hydroelasticity of shells. Sudostroenie, Leningrad (Russian)

Molodtsov, I.N. (1981): To the studies of dynamics of a multilayer inhomogeneous hollow sphere. Vestnik MGU. Matematika, Mekhanika, 1, 82–86 (Russian)

Moodie, T.B., Mioduchowski, A., Haddow, J.B., Tait, R.J. (1983): Elastic waves generated by loading applied to surface of spherical cavity. Int. J. Eng. Sci. **21**, 11, 1369–1378

Morris, G. (1950): Some considerations of the mechanism of the generation of seismic waves by explosives. Geophysics **15**, 1, 61–69

Moskalenko, V.N. (1965): On the shock action on a shell located in elastic medium. In: Results of the scientific-research works made in 1964–1965. Moscow institute of power-engineering, Moscow, 189–196 (Russian)

Mow, C.C. (1965): Transient response of a rigid spherical inclusions in an elastic medium. J. Appl. Mech. **32**, 3, 637–642

Mow, C.C. (1966): On the transient motion of a rigid spherical inclusions in an elastic medium and its inverse problem. J. Appl. Mech. **33**, 4, 807–813

Naimark, M.A. (1969): Linear differential operators. Nauka, Moscow (Russian)

Natrashvili, D.G. (1980): Solution of dynamic problems for a sphere. In: Problems of the theory of elasticity. Tbilisi, 72–88 (Russian)

Natrashvili, D.G., Dzhagmaidze, A.Ya. (1977): Some problems of statics and dynamics in the theory of elasticity for piecewise homogeneous bodies. In: Theoretical and applied mechanics. Proc. 3rd National Congr. Sofia, 551–555 (Russian)

Nayfeh, A.H. (1979): Stress wave propagation in bilayered media. J. Acoust. Soc. Amer. **66**, 1, 291–295

Neilson, H.C., Everstine, G.C., Wang, Y.F. (1981): Transient response of a submerged fluid-coupled double-walled shell structure to a pressure pulse. J. Acoust. Soc. Amer. **70**, 6, 1776–1782

Neishlos, H., Israeli, M., Kivity, Y. (1983): The stability of explicit difference schemes for solving the problem of interaction between a compressible fluid and elastic shell. Comput. Meth. Appl. Mech. and Eng. **41**, 2, 129–143

Nigul, U.K. (1976): Echo-signals from elastic objects. Valgus, Tallinn (Russian)

Nigul, U.K., Metsaveer, Ya.A., Veksler, N.D., Kutser, M.E. (1974): Echo-signals from elastic objects. Valgus, Tallinn (Russian)

Nishimura, G. (1937): On the elastic waves due to pressure variation on the inner surface of a spherical cavity in an elastic solid. Bull. Earthq. Res. Inst., Univ. Tokyo **15**, 3, 614–635

Nishimura, G., Takayama, T. (1938): Seismic waves due to traction applied to the inner surface of a spherical cavity in an elastic earth. Bull. Earthq. Res. Inst., Univ. Tokyo **16**, 2, 317–354

Noh, W.F. (1964): CEL: a time-dependent, two-space-dimensional, coupled Eulerian–Lagrange code. Methods of Computational Physics **3**, 117–180

Norwood, F.R., Miklowitz, J. (1967): Diffraction of transient elastic waves by a spherical cavity. J. Appl. Mech. **34**, 3, 735–744

Novozhilov, V.V. (1959): On the displacement of a rigid body produced by an acoustic pressure wave. Prikladnaya Matematika i Mekhanika [Journal of Applied Mathematics and Mechanics] **23**, 4, 794–796 (Russian)

Nowacki, W. (1970): Theory of elasticity. PWN, Warsaw (Polish)

Ogibalov, P.M. (1963): Problems of dynamics and stability of shells. Moscow University Publishing House, Moscow (Russian)

Ogurtsov, K.I., Uspensky, I.N., Ermilova, V.I. (1957): The qualitative studies of propagation of waves in the simplest elastic media. In: Problems of dynamic theory of propagation of seismic waves. Moscow, 296–365 (Russian)

Onishchuk, V.Ya. (1980): Acoustic field produced by an elastic solid sphere. Visnik Lvivskogo Universitetu. Mekhanika, Matematika, 16, 54–60 (Ukranian)

Panasyuk, N.N. (1978): Transient diffraction of stress waves by a spherical cavity in an elastic medium. In: Waves in continuum. Kiev, 79–85 (Russian)

Panichkin, V.I. (1984): Transient deformation of a system of spherical shells filled with a fluid. In: Vibrations of elastic structures filled with fluids. Proc. 5th All-Union Symp., Novosibirsk, 1982. Moscow, 207–212 (Russian)

Pao, Y.-H. (1983): Elastic waves in solids. J. Appl. Mech. 50, 4b, 1152–1164

Pao, Y.-H., Mow, C.-C. (1973): Diffraction of elastic waves and dynamic stress concentrations. Crane, Russak, New York

Pao, Y.-H., Varatharajulu, Y. (1976): Hygens principle, radiation conditions, and integral formulas for the scattering of elastic waves. J. Acoust. Soc. Amer. 59, 6, 1361–1371

Pec, K. (1957): The mixed external boundary value problem for a sphere. Geofys. Sbor., Praha, 45, 367–382

Pečinka, L., Klàtil, J., Dràbek, P. (1983): Dynamic fluid effects in liquid-filled horizontally accelerated rigid spherical shell. Dynamics of Machines. Proc. 14th Conf. INTERDYNAMICS 83, Liblice, 12–16 Sept., 1983. Praha, 147–149

Pekurovsky, L.E., Poruchikov, V.B., Sozonenko, Yu.A. (1983): Interaction of acoustic waves with bodies coated by a thin compressible layer. Prikladnaya Matematika i Mekhanika [Journal of Applied Mathematics and Mechanics] 47, 5, 823–831 (Russian)

Peralta, L.A., Carrier, G.F., Mow, C.C. (1966): An approximate procedure for the solution of a class transient-wave diffraction problems. J. Appl. Mech. 33, 1, 168–172

Petrashen, G.I. (1945): Solution of the vector limiting problems of mathematical physics in the case of a sphere. Doklady AN SSSR 46, 7, 291–294 (Russian)

Petrashen, G.I. (1949): Problems of dynamics in the theory of elasticity for the case of isotropic sphere. Part 1. Uchenye Zapiski LGU, 114. Seriya Matematicheskikh Nauk, 17, 3–27 (Russian)

Petrashen, G.I. (1950): Problems of dynamics in the theory of elasticity for the case of isotropic sphere. Part 2. Uchenye Zapiski LGU, 135. Seriya Matematicheskikh Nauk, 21, 24–70 (Russian)

Petrashen, G.I. (1953): Methods of study of wave processes in the media having spherical and cylindrical interfaces. Uchenye Zapiski LGU, 170. Seriya Matematicheskikh Nauk, 27, 96–220 (Russian)

Petrashen, G.I., Smirnova, N.S., Galchinsky, B.Ya. (1953a): Problems of the dynamic theory of elasticity for the media having cylindrical or spherical interfaces. Uchenye Zapiski LGU, 170. Seriya Matematicheskikh Nauk, 27, 221–265 (Russian)

Petrashen, G.I., Smirnova, N.S., Makarov, G.I. (1953b): On asymptotic representations of the cylindrical functions. Uchenye Zapiski LGU, 170. Seriya Matematicheskikh Nauk, 27, 7–95 (Russian)

Petrova, S.S. (1974): To the history of the Laplace method of cascades. In: Historical–mathematical studies. No. 19. Moscow, 125–131 (Russian)

Plakhotnyi, P.I. (1971): To the problem of propagation of spherical stress waves in an elastic medium. Prikladnaya Mekhanika [Soviet Applied Mechanics] 7, 5, 112–116 (Russian)

Poddubnyak, A.P. (1979): The echo-signal produced by an elastic sphere caused by a pencil-beam sound pulse. Matemema150ticheskie Metody i Fiziko–Mekhanicheskie Polya, 9, 92–95 (Russian)

Poddubnyak, A.P. (1980): Reflection and refraction of a sound pulse by a two-layer acoustic sphere. Matemematicheskie Metody i Fiziko–Mekhanicheskie Polya, 12, 46–50 (Russian)

Poddubnyak, A.P. (1984): Transient reflection and refraction of a torsional wave by a two-layer elastic sphere. Matemematicheskie Metody i Fiziko–Mekhanicheskie Polya, 19, 82–86 (Russian)

Poddubnyak, A.P., Podstrigach, Ya.S., Grilitsky, D.V. (1978): The problem of hydroacoustics for an elastic body of revolution. Matemematicheskie Metody i Fiziko–Mekhanicheskie Polya, 7, 3–6 (Russian)

Poddubnyak, A.P., Porokhovsky, V.V., Pozdeev. V.A., Dykhta, V.V. (1985): Radiation of an acoustic signal, which was generated by an internal source, by an elastic spherical shell. Prikladnaya Mekhanika [Soviet Applied Mechanics] 21, 5, 97–102 (Russian)

Podilchuk, Yu.N., Rubtsov, Yu.K. (1979): On propagation of transient elastic waves from axially symmetric cavities. Doklady AN USSR, Ser. A, 12, 1006–1010 (Russian)

Podilchuk, Yu.N., Rubtsov, Yu.K. (1981): Propagation of transient elastic waves from a spherical cavity under axially symmetric disturbance. Prikladnaya Mekhanika [Soviet Applied Mechanics] 17, 7, 3–9 (Russian)

Podilchuk, Yu.N., Rubtsov, Yu.K. (1986): Application of the beam-series method for study of the axially symmetric transient problems of the dynamic theory of elasticity. Prikladnaya Mekhanika [Soviet Applied Mechanics] 22, 3, 3–9 (Russian)

Podstrigach, Ya.S., Poddubnyak, A.P. (1986): Scattering of sound beams by elastic bodies of spherical and cylindrical shape. Naukova Dumka, Kiev (Russian)

Podstrigach, Ya.S., Poddubnyak, A.P., Pyrev, Yu.A. (1979): Transient interaction of sound pulses with elastic spherical objects. In: Theory of elasticity Proc. All-Union Conf. Erevan, 280–281 (Russian)

Podstrigach, Ya.S., Porokhovsky, V.V., Poddubnyak, A.P. (1980): Analysis of the re-radiated transient signal from an elastic sphere subjected to action of a sound beam. Matemematicheskie Metody i Fiziko–Mekhanicheskie Polya, 12, 28–31 (Russian)

Podstrugach, Ya.S., Galazyuk, V.A., Gorechko, A.N. (1981): Transient interaction of an elastic spherical body with an acoustic medium. Prikladnaya Mekhanika [Soviet Applied Mechanics] 17, 2, 22–28 (Russian)

Ponomarev, P.V. (1971): On propagation of spherical waves in solid bodies. Uchenye Zapiski Kurskogo Gosudarstvennogo Pedagogicheskogo Instituta 91, 57–77 (Russian)

Porokhovsky, V.V. (1982): Scattering of a sound beam by an elastic spherical shell in the water. Prikladnaya Mekhanika [Soviet Applied Mechanics] 18, 8, 111–113 (Russian)

Poruchikov, V.B. (1981): Expansions of the Legendre differential equation in terms of eigenfunctions and their application to solution of the diffraction problem. In: Interaction of acoustic and shock waves with elastic structures. Moscow, 45–62 (Russian)

Poruchikov, V.B. (1986): Methods of the dynamic theory of elasticity. Nauka, Moscow (Russian)

Poruchikov, V.B., Sazonenko, Yu.A., Pekurovsky, L.E. (1981): Diffraction of an acoustical wave by a sphere of finite mass with soft coating. In: Interaction of acoustic and shock waves with elastic structures. Moscow, 12–34 (Russian)

Poverus, L., Myannil, A. (1975): The study of propagation of elastic waves in a spherical shell. In: Theory of shells and plates. Proc. 9th All-Union Conf., 1973. Leningrad, 212–215 (Russian)

Rakhmatulin, Kh.A., Babichev, A.I., Saidov, T.Kh., Khidoyatov, K. (1967a): The study of dynamics of the multilayer spherical and cylindrical elastic shells using direct integration of the dynamic equations of elasticity. In: Transient processes of deformation of shells and plates. Proc. All-Union Symp., Tartu. Tallinn, 113–133 (Russian)

Rakhmatulin, Kh.A., Babichev, A.I., Saidov, T.Kh., Khidoyatov, K. (1967b): Propagation of elastic waves from a spherical cavity, the boundary of which is subjected to nonuniform stress loading. In: Problems of cybernetics and computational mathematics. No. 10. Tashkent, 98–104 (Russian)

Reismann, H., Gideon, J. (1971): Forced wave motion in an unbounded space surrounding a lined, spherical cavity. Pure and Appl. Geophys. **85**, 2, 189–213

Rodean, H.C. (1971): Nuclear-explosion seismology. University of California, US Atomic Energy Commission

Rose, J.L., Chou, S.C., Chou, P.C. (1973): Vibration analysis of thick-walled spheres and cylinders. J. Acoust. Soc. Amer. **53**, 3, 771–776

Ryayamet, R.K. (1975): Propagation of elastic waves in a thick-walled spherical shell. Trudy Tallinskogo Politekhnicheskogo Instituta, 375, 33–45 (Russian)

Saakyan, S.G. (1973): On propagation of elastic waves in media in the case of axial symmetry. Doklady AN ArmSSR **57**, 4, 225–331 (Russian)

Sabodash, P.F., Tsurpal, O.A. (1969): On propagation of waves in a physically nonlinear medium weakened by a cylindrical or a spherical cavity. In: Propagation of elastic and elastic–plastic waves. Proc. 3rd All-Union Symp. Tashkent, 149–162 (Russian)

Sadykov, R. (1972): Analysis of transient motion of a sphere in an elastic medium using the method of particles. In: Problems of computational and applied mathematics. No. 11. Tashkent, 170–177 (Russian)

Sagomonyan, A.Ya. (1986): Shock and penetration of bodies into a fluid. Moscow University Publishing House, Moscow (Russian)

Saidov, T.Kh., Khidoyatov, K. (1968): Response of an elastic medium to a translationally moving smooth sphere. In: Problems of cybernetics and computational mathematics. No. 14. Tashkent, 141–157 (Russian)

Saidov, T.Kh., Khidoyatov, K. (1969): Estimation of accuracy of computations in the problems of motion of cylindrical and spherical bodies in elastic media using the method of characteristics. In: Propagation of elastic and elastic–plastic waves. Proc. 3rd All-Union Symp. Tashkent, 385–393 (Russian)

Samarsky, A.A. (1974): Introduction into the theory of finite difference schemes. Nauka, Moscow (Russian)

Saprykin, Yu.V. (1984): Transient interaction of an internal pressure wave with a system of two spherical shells. Prikladnaya Mekhanika [Soviet Applied Mechanics] **20**, 6, 40–46 (Russian)

Sarimsakov, U. (1976): Motion of a rigid sphere produced by the action of an elastic compression wave. Doklady AN UzSSR, 6, 13–15 (Russian)

Sato, Y. (1963): Propagation of torsional disturbances in a homogeneous elastic sphere. IBM J. Res. and Developm. **7**, 2, 117–121

Sato, Y., Usami, T. (1964): Propagation of spheroidal disturbances on an elastic sphere with a homogeneous mantle and a core. Bull. Earthq. Res. Inst., Univ. Tokyo **42**, 3, 407–425

Sato, Y., Usami, T., Ewing, M. (1962): Basic study on the oscillation of a homogeneous elastic sphere. Part IV: Propagation of disturbances on the sphere. Geophys. Magazine **31**, 2, 237–242

Sato, Y., Usami, T., Landisman, M., Ewing, M. (1963): Basic study on the oscillation of a sphere. Part V: Propagation of torsional disturbances on a radially heterogeneous sphere. Geophys. J. Roy. Astron. Soc. **8**, 1, 44–63

Sedov, L.I. (1984): Mechanics of continua. Nauka, Moscow (Russian)

Selberg, H. (1952): Transient compression waves from spherical and cylindrical cavities. Aktiv. för Fysik **5**, 1–2, 7, 97–108

Sengupta, P,R., Roy, R. (1971): Generation of waves in an infinite space due to an explosion of the centre of cavity. Pure and Appl. Geophys. **86**, 3, 18–27

Senitsky, Yu.E. (1971): Dynamic problems for a cylinder and a sphere. In: Analysis of spatial building structures. No. 2. Kuibyshev, 138–170 (Russian)

Senitsky, Yu.E., Syromyatnikova, G.V. (1976): The study of stress–strain state of a thick-walled spherical shell subjected to symmetric dynamic loading. In: Analysis of spatial building structures. No. 6. Kuibyshev, 48–57 (Russian)

Sezawa, K. (1935): Elastic wave produced by applying statical force to a body or by releasing it from a body. Bull. Earthq. Res. Inst., Univ. Tokyo **13**, 4, 740–746

Sezawa, K., Kanai, K. (1936): Elastic waves formed by local stress changes of different rapidities. Bull. Earthq. Res. Inst., Univ. Tokyo **14**, 1, 10–17

Sezawa, K., Kanai, K. (1941–1942): Transmission of arbitrary elastic waves from a spherical source, solved with operational calculus I–III. Bull. Earthq. Res. Inst., Univ. Tokyo **19**, 2, 151–161; **19**, 3, 417–442; **20**, 1, 1–19

Sharpe, J.H. (1942): The production of elastic waves by explosion pressures I–II. Geophysics **7**, 2, 144–154; **7**, 3, 311–321

Shaw, R.P. (1973): Integral equation formulation of dynamic acoustic fluid–elastic solid interaction problems. J. Acoust. Soc. Amer. **53**, 2, 514–520

Shaw, R.P., English, J.A. (1972): Transient acoustic scattering by a free (pressure release) sphere. J. Sound and Vibr. **20**, 3, 321–331

Shchipitsina, E.M. (1972): The study of deformations of a hollow sphere subjected to impulsive loading using the three-dimensional theory of elasticity and the theory of shells. Prikladnaya Mekhanika [Soviet Applied Mechanics] **8**, 11, 28–32 (Russian)

Singh, S.J. (1973): Generation of SH-type motion by torsion-free sources. Bull. Seismol. Soc. Amer. **63**, 4, 1189–1200

Singh, S.J., Rosenmann, M. (1973): On the disturbance due to a spherical distortional pulse in an elastic medium. Pure and Appl. Geophys. **110**, 9, 1946–1954

Skurlatov, E.D. (1979): The study of dynamics of shells. Nonlinear problems of aerohydroelasticity. In: Theory of shells. No. II. Kazan, 135–146 (Russian)

Slepyan, L.I. (1963): On the displacement of a deformable body in an acoustic medium. Prikladnaya Matematika i Mekhanika [Journal of Applied Mathematics and Mechanics] **27**, 5, 918–923 (Russian)

Slepyan, L.I. (1972): Transient elastic waves. Sudostroenie, Leningrad (Russian)

Slepyan, L.I., Yakovlev, Yu.S. (1980): Integral transforms in transient problems of mechanics. Sudostroenie, Leningrad (Russian)

Smirnov, V.I. (1937a): Solution of the limiting problem for the wave equation in the case of a circle and a sphere. Doklady AN SSSR **14**, 1, 13–16 (Russian)

Smirnov, V.I. (1937b): Solution of the limiting problems of the theory of elasticity in the case of a circle and a sphere. Doklady AN SSSR **14**, 2, 69–72 (Russian)

Stepanischen, P.K. (1973): The impulse response and time-dependent force on a baffled circular piston and a sphere. J. Sound and Vibr. **26**, 3, 287–298

Suzuki, S.-I. (1967): On the stress concentration of a hollow sphere under uniformly distributed shock loads along inner and outer surfaces. Bull. JSME **10**, 37, 23–27

Sysoev, N.N., Shugaev, F.V. (1979): Transient reflection of a shock wave by a sphere or a cylinder. Vestnik MGU. Fizika, Astronomiya **20**, 3, 90–91 (Russian)

Tang, S.-C. (1968): Transient response of a rigid sphere to a plane acoustic step wave. J. Acoust. Soc. Amer. **44**, 1, 289–291

Tanyi, G.E. (1966): Pressure and shear waves generated by a explosion in an elastic sphere. Geophys. J. Roy. Astron. Soc. **10**, 5, 465–483

Tarlakovsky, D.V. (1975): Transient radial vibrations of a hollow elastic sphere submerged into a fluid. Izvestiya AN SSSR. Mekhanika Tverdogo Tela [Mechanics of Solids], 5, 202–203 (Russian)

Tarlakovsky, D.V. (1977): Transient behavior of a thick-walled elastic sphere in a fluid. Izvestiya AN SSSR. Mekhanika Tverdogo Tela [Mechanics of Solids], 2, 209 (Russian)

Tarlakovsky, D.V. (1981): Transient problems of dynamics of a thick-walled sphere contacting elastic or acoustic media. Izvestiya AN SSSR. Mekhanika Tverdogo Tela [Mechanics of Solids], 1, 205 (Russian)

Taylor, R.E. (1981): A review of hydrodynamic load analysis for submerged structures excited by earthquakes. Eng. Struct. 3, 3, 131–139

Temkin, S. (1972): On the response of a sphere to an acoustic pulse. J. Fluid Mech. 54, 2, 339–349

Ting, T.C.T., Lee, E.H. (1969): Response of nonlinearly supported spherical boundaries to shock waves. J. Appl. Mech. 36, 3, 497–504

Titarev, V.G. (1984): Impulsive loading of a hollow multilayer elastic sphere. Problemy Mashinostroeniya, 21, 32–36 (Russian)

Tranter, C.J. (1942): The application of the Laplace transformation to a problem on elastic vibrations. Phil. Mag. and J. Science 33, 223, 614–622

Tricomi, F. (1954): Lezioni sulle equazioni a derivate partiali. Gheroni, Torino

Tsai, I.P. (1973): On the dynamical problems of propagation of elastic waves in the spherical coordinate system. In: Theoretical mechanics and higher mathematics. No. 4. Jambul, 8–15 (Russian)

Tupholme, G.E. (1967): Generation of an axisymmetric acoustical pulse by a deformable sphere. Proc. Cambridge Phil. Soc., Mathem. and Physical Ser. 63, 4, 1285–1308

Tupholme, G.E. (1983): Elastic pulse generation traction applied to a spherical cavity. Appl. Sci. Res. 40, 4, 299–325

Usami, T. (1962): Some remarks on the solution of the equation of motion for a homogeneous elastic medium. Geophys. Magazine 31, 1, 1–13

Usami, T., Sato, Y. (1964): Propagation of spheroidal disturbances on a homogeneous elastic sphere. Bull. Earthq. Res. Inst., Univ. Tokyo 42, 2, 273–287

Vaněk, J. (1953): K teorii elastických vin vitvořených nárazem. Ceskosl. Časop. Pěstov. Fys. 3, 2, 93–112

Vaněk, J. (1955): On the magnitude of the transitional zone for elastic waves produced by different shock-exciting functions. Geofys. Sbor., Praha, 25, 79–89

Vaněk, J. (1956): The theory of elastic waves produced by a spherical source for generalized boundary conditions. Ceskosl. Fys. Zurn. 6, 4, 303–309

Vasudevan, R., DiMaggio, F. (1981): Transient response of submerged shells using improved acoustic approximations. Comput. and Struct. 14, 3–4, 187–194

Veksler, N.D. (1973): On the method of computation and the structure of echo-signals propagating from an acoustically rigid immovable sphere in a boundless ideal compressible fluid. Izvestiya AN SSSR. Mekhanika Tverdogo Tela [Mechanics of Solids], 1, 83–96 (Russian)

Veksler, N.D. (1974): Diffraction of a plane sound wave by a thin-walled elastic spherical shell. Izvestiya AN SSSR. Mekhanika Tverdogo Tela [Mechanics of Solids], 3, 130–138 (Russian)

Veksler, N.D. (1975): Diffraction of a plane sound wave by a hollow elastic sphere. Akust. Zhurnal 21, 5, 694–700 (Russian)

Veksler, N.D., Metsaveer, Ya.A. (1981): Echo-signals propagating from elastic shells. In: Theoretical and applied mechanics. Proc. 5th All-Union Congr. Alma-Ata, 92 (Russian)

Veksler, N.D., Nigul, U.K., Pukk, R.A. (1970): On the algorithm of computation of the Fourier series for the echo-signals propagating from elastic spherical bodies in an ideal fluid. Izvestiya AN SSSR. Mekhanika Tverdogo Tela [Mechanics of Solids], 6, 71–83 (Russian)

Verma (1957): On the stresses produced by impulsive displacements applied to the inner surface of a spherical cavity. Geofis. Pura e Appl. 37, 2, 16–20

Vestyak, A.V., Gorshkov, A.G., Tarlakovsky, D.V. (1983): Transient interaction of deformable bodies and surrounding media. In: Advances in science and technology of VINITI. Mechanics of deformable solids 15. VINITI, Moscow, 69–148 (Russian)

Vestyak, A.V., Gorshkov, A.G., Tarlakovsky, D.V. (1984): Transient vibrations of elastic (acoustic) media having spherical interfaces. In: Interaction of plates and shells with liquid and gas. Moscow University Publishing House, Moscow, 3–26 (Russian)

Viswanathan, K., Biswas, R.N. (1970): On stress waves in a spherical shell. Indian J. Pure and Appl. Math. 1, 4, 568–578

Vodicka, V. (1963): Radial vibrations of an infinite medium with a spherical cavity. Z. angew. Math. und Physik 14, 6, 745–748

Volmir, A.S. (1976): Shells in a flow of liquid and gas. Problems of aeroelasticity. Nauka, Moscow (Russian)

Volmir, A.S. (1979): Shells in a flow of liquid and gas. Problems of hydroelasticity. Nauka, Moscow (Russian)

Wankhede, P.C., Bhonsle, B.R. (1980): Elastic vibrations in composite cylinders or spheres. Proc. Nat. Acad. Sci., India, Sec. A 50, 1, 37–46

Watson, G.N. (1945): A treatise on the theory of Bessel functions. Cambridge University Press, Cambridge

Wheeler, L. (1973): Focusing in stress waves in an elastic sphere. J. Acoust. Soc. Amer. 53, 2, 521–524

Wilkinson, J.P.D., Cappelli, A.P., Selzman, R.N. (1968): Hydroelastic interaction of shells of revolution during water shock. AIAA Journal 6, 5, 792–797

Wolf, A. (1945): Motion of a rigid sphere in an acoustic wave field. Geophysics 10, 1, 91–109

Yang, J.C.S., Achenbach, J.D. (1970): Stresses in multilayered structures under high-rate pressure loads. Pap. ASME, 1070, WA/Unt–14

Yanyutin, E.G. (1980): Transient spherically symmetric deformation of elastic bodies. Problemy Mashinostroeniya, 10, 7–11 (Russian)

Yanyutin, E.G. (1983): Transient deformation of an elastic half-space having an expanding spherical cavity. Izvestiya AN SSSR. Mekhanika Tverdogo Tela [Mechanics of Solids], 6, 86–89 (Russian)

Yanyutin, E.G. (1984): Transient deformation of an elastic space supported by a closed spherical shell. Prikladnaya Mekhanika [Soviet Applied Mechanics] 20, 4, 23–27 (Russian)

Yanyutin, E.G., Titarev, V.G. (1979): Impulsive loading of a spherical layer. In: Dynamics and strength of machines. No. 30. Kharkov, 76–81 (Russian)

Yanyutin, E.G., Titarev, V.G. (1982): Collision of elastic spherical layers. Prikladnaya Mekhanika [Soviet Applied Mechanics] 18, 6, 28–33 (Russian)

Yen, G. (1964): The diffraction of plane compressional wave by a spherical cavity in an elastic medium. Z. angew. Math. und Physik 15, 3, 237–252

Zamyshlyaev, B.V., Yakovlev, Yu.S. (1967): Dynamic loads due to underwater explosion. Sudostroenie, Leningrad (Russian)

Zaripov, R.G., Ilgamov, M.A. (1979): The experimental study of fracture of shells under the action of shock waves. Problemy Prochnosti, 1, 28–33 (Russian)

Index

Foundations of Engineering Mechanics

Series Editors: Vladimir I. Babitsky, Loughborough University
Jens Wittenburg, Karlsruhe University

Foundations of Engineering Mechanics

Series Editors: Vladimir I. Babitsky, Loughborough University
Jens Wittenburg, Karlsruhe University

Svetlitsky	Statics of Rods (2000, ISBN 3-540-67452-7)
Landa	Regular and Chaotic Oscillations (2001, ISBN 3-540-41001-5)
Muravskii	Mechanics of Non-Homogeneous and Anisotropic Foundations (2001, ISBN 3-540-41631-5)
Gorshkov/ Tarlakovsky	Transient Aerohydroelasticity of Spherical Bodies (2001, ISBN 3-540-42151-3)